QA 279.5 .K96
Kwon, Ik-Whan, 1937-
Statistical decision theory with
business and economic applications : a
Bayesian approach

DATE	ISSUED TO

QA 279.5 .K96
Kwon, Ik-Whan, 1937-
  Statistical decision theory with
business and economic applications : a
Bayesian approach

Statistical Decision Theory with Business and Economic Applications

A BAYESIAN APPROACH

**The PETROCELLI/CHARTER
Modern Decision Analysis Series**

Sang M. Lee, editor

Introduction to Decision Science
Sang M. Lee and Laurence J. Moore

Linear Optimization for Management
Sang M. Lee

Gert Modeling and Simulation:
Fundamentals and Applications
Laurence J. Moore and Edward R. Clayton

The Corporate Role and Ethical Behavior:
Concepts and Cases
Robert J. Litschert and Edward A. Nicholson

Strategic Planning and Policy
William R. King and David I. Cleland

IK-WHAN KWON

Saint Louis University
with cooperation from Maurice N. Shriber

Statistical Decision Theory with Business and Economic Applications

A BAYESIAN APPROACH

PETROCELLI /CHARTER NEW YORK 1978

First Printing

Printed in the United States of America

Library of Congress Cataloging in Publication Data

Kwon, Ik-Whan, 1937–
    Statistical decision theory with business and economic applications.

    Includes bibliographies and index.
    1. Bayesian statistical decision theory.   2. Commercial statistics.
    3. Economics, Mathematical.
I. Title.
QA279.5.K96   519.5'4   77-9948
ISBN 0-88405-502-7

To my parents, Jae-Sang and Myung-Soon
my wife, Chung-Soon
my daughters, Nancy, Christy and Gracy

CONTENTS

PREFACE

This book is intended for students of applied business and economic statistics. It is also designed as an introductory text in decision theory for students in an MBA program whose undergraduate backgrounds are in fields other than business or economics. While this book contains sufficient material for an entire academic year, its contents may easily be adapted for use in a one-semester or one-quarter course by deleting various sections or entire chapters to fit the needs and objectives of the course.

The use of statistical decision models is no longer confined to theoretical use in the classroom. The growing size and complexity of modern governmental and private business spheres have resulted in the movement of many of the more sophisticated decision techniques out of the classroom and into practical everyday use in nonacademic situations. Because of this movement, those students planning to enter private industry or government should be familiar with at least the basic techniques and terminology associated with modern decision theory.

The purpose of this book is to present the total structure of modern decision theory. Although the history of modern decision theory (also referred to as Bayesian decision theory) is relatively young, as far as its practical use in both the academic and business environments is concerned, its applicability and usefulness have been successfully proved in many areas of both the private and public sectors. The objective here is to give students and business practitioners, through the use of modern statistical concepts and procedures, a systematic process for determining an optimum decision under conditions of uncertainty.

Because of the basically nonmathematical presentation of the material contained in this text, the student is not required to have a working knowledge of either calculus or advanced algebra. For this reason, mathematical formulations and the derivation of the statistical models contained in the text were deliberately avoided in the body of each chapter. For those students who have a mathematical background or who desire to review the mathematical processes involved in the various statistical models contained in the text, the mathematical proofs for many of these models are contained either in footnotes or included in the notes located at the ends of selected chapters.

The first section of each chapter contains an introduction that explains the basic concepts and techniques to be discussed. This preliminary is designed to help those students who have had only a minimum amount of statistics in their educational background. Many sample cases, examples, and illustrations are included in the body of each chapter. These are augmented by additional discussions, case-oriented questions, and exercises at the end of each chapter. The exercises may be used by readers to develop their ability to apply the basic statistical techniques discussed. The cases and examples used in the text were selected purposely to provide a broad variety of real-life problems in business, government, and other sectors of the economy. Most of the examples and exercises were tested in the classroom at both graduate and undergraduate levels over an extended period of time. Finally, a summary is included at the end of each chapter as an aid in the review of the data presented in each text unit.

ACKNOWLEDGMENTS

I am greatly indebted to many people who assisted me in preparing this book. Special thanks should go to my teachers at the University of Georgia: Professor David M. Wright, James L. Green, Eugene Holshouser and Donald Escarraz. My colleagues at Saint Louis University provided many valuable suggestions and help at early stages of this manuscript. I am deeply grateful to many undergraduate and graduate students who worked with me in every problem of this book; I benefited from their corrections and suggestions to make it readable at the student level.

Among the many students who helped me, I would like to give a special credit to Mr. Maurice N. Shriber at the Pentagon, who stayed with me from the beginning to the end, correcting grammar and checking mistakes. Without his assistance, I could not have completed this book. Finally, Pat Feldmann handled the most difficult job, typing the manuscript many times over and over again without even a single complaint. Her skillful handling of the manuscript made my work considerably easier and less painful. I am grateful to both of them for their unselfish assistance to make this book publishable.

If there is one person in this world who should receive sole credit for this book, it is undoubtedly my wife, Chung-Soon. When I was frustrated with problems in writing and when I needed someone just to talk to, she was always with me and was a good partner in conversation, mostly listening to my complaints and frustrations. It is painful for me to think that she had to sacrifice so many things, including her own profession as a medical doctor, to take care of a home and three children. There are no appropriate words for me to express my sincere thanks to my wife, to whom this book is dedicated.

Ik-Whan Kwon
Ballwin, Missouri

1 / An Introduction to Modern Decision Theory

1.1 / Purpose

The decision-making process described in this book refers to an analytical procedure that uses a systematic approach to the selection of an optimum decision under conditions of uncertainty. Inherent in all subsequent discussions is an analysis concerning the choice of courses of action among two or more alternatives.

In many instances, the problem of choice is not a serious issue, either because the selection of an appropriate course of action is obvious, such as routine desk work in an office, or because the selection of an alternative choice is relatively unimportant, such as the selection of a necktie or the purchase of groceries. Under these circumstances, the decision is not important enough to justify the time and effort that would be required to achieve an actual optimum decision.

In many cases, however, the selection of a proper course of action is a serious issue, especially when the choice in question or the problem being analyzed has a significant monetary or social impact and is accompanied by some degree of uncertainty. When this type of situation arises, decision theory is an important analytical tool that can be used to provide a reasonable approach to the solution of a problem and the achievement of an optimum decision.

The purpose of this chapter is to outline the concepts and to present an overview of the decision-making processes contained in modern decision theory.

1.2 / Examples of Decision Problems

Problems concerning alternative choices that result in some type of decision arise in every facet or phase of both private and public sectors. The following cases and examples from various areas of business and government will be used to illustrate some of the more common types of problems that require a decision and which, as such, involve the use of the decision-making process.

Marketing. The manager of the marketing department in a leading bicycle manufacturing company is analyzing whether or not it would be worthwhile to expand the company's sales operation from only the domestic into the world market. The success of an increased sales base should result in additional profits to the company and increased personal reward for the manager; on the other hand, the failure of the plan could result in great financial loss to the company and, in all likelihood, the subsequent loss of the manager's job. The decision criterion in this problem is the anticipated increase in market share for the company's product as a result of entry into the world market.

Production. A company is experiencing an increase in the number of defective parts being produced by its metalwork machines. The production manager has two mutually exclusive options available to correct the problem: (1) close down the entire manufacturing operation and thoroughly inspect all machines, or (2) continue production as it is currently being performed and allow for replacement of the defective part in those products that are subsequently returned by the customers. The decision criterion in this problem is the proportion of defective parts being produced in relation to the total parts production. In this case, the economic penalty of an incorrect decision is either an unnecessary economic loss as a result of an unwarranted shutdown when the proportion of defectives is actually relatively small, or the loss of customers and the associated costs of replacing parts as a result of the continuous production of defective products.

Finance. A corporation is considering whether or not it should finance a new business venture. Two options are available to the company for acquiring the additional capital required to finance the project: (1) borrow funds from a bank, or (2) issue common stock. The first option involves a fixed payment of interest regardless of the income that is actually received from the new venture, whereas the second choice involves an increase of corporate equity with the possibility of decreased returns per share of stock. The decision criterion in this example is a correct estimate of the future outlook in terms of the potential success of the new venture. From the existing stockholder's viewpoint, if the project is successful, debt financing would probably be preferred, and vice versa. The economic penalty of a wrong decision is an unacceptably high level of fixed-debt financing costs if the new venture fails, or an unnecessarily high rate of additional stockholder's equity if the new venture is successful.

Personnel. A personnel manager must decide whether to send a memorandum to the union representative informing him of the company's intention to terminate the employment of several workers; the alternative is to keep the employees on the payroll for an additional 12 months. According to the agreement between the union and management of the company, the manager must notify the union of the company's decision in less than 30 days. The final decision actually depends upon the outcome of pending negotiations for an additional contract with the company's largest customer.

If these negotiations are successful, the number of people that the company can continue to employ will be affected. The economic penalty of a wrong decision is either to incur unnecessary recruiting expenses (if the employees are dismissed and the negotiations subsequently end successfully) or to continue to expend funds unnecessarily for wages and salaries (if the negotiations result in failure to obtain the contract).

Government sector. The current administration must decide whether it should reduce government spending in an effort to control inflation. The alternative choice (if spending is not reduced) is to maintain the current level of spending (and thereby tacitly accept some rate of inflation) in an effort to keep the unemployment rate down. The decision criterion in this case is the sensitivity of inflation and unemployment levels to variations in fiscal policy. The economic penalty of a wrong decision is either a continuation of inflation or an unacceptable rate of unemployment, or both.

1.3 / The Decision-Making Environment

The selection of a terminal decision is influenced mainly by the states of nature that surround a problem. In this regard, the decision-making process differs in that both concept and operational procedures depend upon the degree of knowledge that the decision maker has concerning the problem environment. In general, based upon this degree of knowledge, there are three basic types of decision-making process: (1) certainty, (2) risk, and (3) uncertainty, each of which is examined below.

The decision-making process under conditions of certainty.

When the decision maker knows the exact states of nature in advance, it is relatively easy for him to maximize either monetary return or some other type of utility. For example, consider the following case.

A company produces two office machines, a calculator and a typewriter. The company has reliable estimates of unit cost, production capacity, estimated demand, price per unit, and optimum inventory level that should be maintained for each item. Since the company has knowledge of all relevant information (states of nature) required to estimate the likelihood that each event will occur, the only decision to be made in this problem is what optimum mix of existing production levels of calculators and typewriters will maximize profit.

The decision-making process under conditions of risk.

In a situation where the decision-making process is conducted in an environment of risk, the degree of knowledge associated with each state of nature is not known. Appropriate estimates of likelihood can be determined,

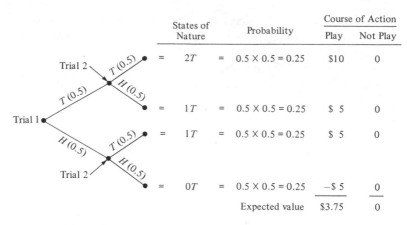

	States of Nature	Probability	Course of Action	
			Play	Not Play
= 2T		= 0.5 × 0.5 = 0.25	$10	0
= 1T		= 0.5 × 0.5 = 0.25	$ 5	0
= 1T		= 0.5 × 0.5 = 0.25	$ 5	0
= 0T		= 0.5 × 0.5 = 0.25	−$ 5	0
		Expected value	$3.75	0

Figure 1.1 / Decision tree: T = tail; H = head.

however, on one of two bases: (1) subjective judgment, or (2) use of certain mathematical functions or relationships. This type of decision environment is illustrated in the case below.

Suppose that you will receive $10.00 if you obtain two consecutive "tails" (T) on two flips of a balanced coin; on the other hand, only $5.00 will be given if only one "tail" appears on the two tosses (1T + 1H). Finally, you must pay $5.00 if you receive no "tails" (2H). Considering only the monetary return, should you play the game?

This problem is depicted graphically in the decision-tree diagram of Figure 1.1, where the results of the first trial can be either a T or an H; the results of the second trial are also either a T or an H. Since a balanced coin was used in the problem,[1] the probability of obtaining a tail in any given trial is assumed to be 0.5, or 50%. Simple intuition and algebra tell us that the probability of obtaining two consecutive tails when flipping a balanced coin twice is 0.5 × 0.5 = 0.25. The other probabilities in Figure 1.1 were similarly computed.[2]

In this case, there are only two alternatives; i.e., either you do or do not play the game. The expected monetary value for the decision "to play" is determined by $10.00 × 0.25 + $5.00 × 0.25 + (−$5.00) × 0.25 − $3.75. The expected monetary value for the alternative of "not to play" is, of course, zero. Accordingly, since the alternative "to play" yields a larger expected monetary value than the alternative of "not to play," the optimum decision is "to play."[3]

---

[1] The significance of the term "balanced coin" will be explained in subsequent chapters.

[2] A complete explanation of the basic rules and computations concerning probability will be presented in Chapter 2.

[3] A complete explanation of the computation and interpretation of the expected value concept will be presented in Chapter 3.

In summary, in the decision-making process under conditions of risk, the probability of occurrence for each event can be generated or determined by various probability rules, and therefore there is a tangible objective basis upon which to make a decision.

The decision-making process under conditions of uncertainty.

Uncertainty is defined as a state of knowledge in which the probability that each outcome or event in a problem will actually occur is either not known or is not based on a meaningful statistical base. Under these conditions, the decision maker must make a terminal decision in an environment of incomplete knowledge. Since there is no empirically tested prior knowledge, the decision maker may be forced to use personal (subjective) judgments as to the composition of the states of nature.

The following case is used to illustrate the decision-making process under conditions of uncertainty. Suppose that a government economist is asked for his opinion about the state of the economy. In response to this question, the economist answers that there is a 70% chance that the economy will continue to expand during the next five years and there is a 30% chance that it will experience a recession during the same time period. Unlike the coin-flipping case, however, the probabilities of 0.7 and 0.3 are not objective probabilities. In other words, these probabilities cannot be empirically tested and proved by any reasonable method prior to the actual passing of time. This situation exists because the predictions and expectations in this case are basically of a subjective nature. Accordingly, even though the estimate is based on past experience and is performed by an expert, the decision maker's forecast of future events may, in reality, be quite different from the outcome that actually occurs.

Since the expectations of an individual who is involved in the decision-making process under conditions of uncertainty are subjective in nature, it follows that there will be a different degree of uncertainty on the part of the decision maker, depending upon the conditions or environment inherent in the problem being analyzed. In this regard, there are three classes of uncertainty: complete knowledge, complete ignorance, and partial ignorance.

Complete (perfect) knowledge. When there is a prior probability distribution over the states of nature that are being considered in a problem, it is said that the decision maker has complete or perfect knowledge. The concept of perfect knowledge should not be confused with that of certainty, discussed above. Under conditions of certainty, we assume that the outcome of each future event can be predicted with certainty; i.e., it is a sure bet, whereas under conditions of perfect knowledge, we simply assume that each future event will occur with some degree of chance. If the problem being considered operates under conditions of perfect knowledge, the decision-making process can be transferred to the category discussed previously: "The Decision-Making Process under Conditions of Risk."

Complete ignorance. This type of problem is exactly the opposite of that considered in the decision-making environment under conditions of perfect knowledge. When the decision maker is in complete ignorance, he has neither subjective nor objective knowledge of the probabilities concerning the various states of nature under consideration. Accordingly, the decision maker may use whatever rational decision criteria desired to arrive at an optimum decision. One of these criteria could be the individual's attitude toward risk; e.g., a decision maker who is an optimist by nature will act quite differently from one who is a pessimist when developing a final decision.[4]

Partial ignorance. Most of the actual problems that require a decision fall somewhere between the two extremes of perfect knowledge and complete ignorance. Under conditions of partial ignorance, several probabilities concerning the states of nature in a problem are either missing or are only partially provided. The decision-making process under these circumstances requires the decision maker to generate subjective probabilities and then to assign them to the corresponding or appropriate states of nature. By doing this, the decision environment can be translated from one of uncertainty to one of risk.[5]

1.4 / The Decision-Making Process

As previously mentioned, the purpose of modern decision theory is to provide a systematic approach to the decision-making process under conditions of imperfect knowledge or uncertainty. The following steps are normally followed in order to arrive at an optimum decision under these conditions:

(1) Identify the objectives in the problem toward which the decision-making process should be directed.
(2) Identify the alternative courses of action to be considered.
(3) List a set of possible events (states of nature) that would influence the eventual payoff of each course of action.
(4) Assign a degree of knowledge, in numerical terms (i.e., probabilities), to the occurrence of each possible state of nature. This degree of knowledge can be based on either subjective judgment or objective observation.

---

[4] A complete discussion of this decision criterion will be presented in Chapter 5.
[5] This subject will be extensively discussed in Chapters 5 through 9.

(5) Given each possible event, assign a numerical value to the anticipated payoff of each course of action. This numerical value is often computed in such a way as to form an expected monetary value.

(6) Compute the weighted average of the payoffs (expected monetary value) assigned to each course of action. This step may be performed by using either the computed probabilities of occurrence for each event, or by using the subjective tolerance level of gain and loss associated with each course of action.

(7) Select the optimum decision among the available alternative courses of action. This is performed on the basis of the combination of the expected value and the maximum gain (or minimum loss) that is most consistent with the decision maker's objectives and attitude toward risk.

The individual terms and decision criteria that are described above will be discussed in detail in subsequent chapters. For the time being, however, let us briefly illustrate the steps that are involved in the decision-making process.

EXAMPLE 1.1 / Based upon increased production requirements, a manu-facturing company has determined that it needs to expand its production capacity. The company is considering whether it should purchase the same type of machine currently used or should buy a larger machine that is much more efficient to operate but which initially costs a great deal more. The major factor influencing the decision is the projected state of the economy for the next five years.

If the economy continues to expand, or at least maintains the status quo, the use of the larger machine could be expected to generate $150,000 in annual profits. On the other hand, if the economy experiences a down-ward trend for the next five years, the company could face a forecast annual loss of $100,000 by using the larger, more expensive machine. This loss would be due to the increased overhead cost and financing charges associated with the larger machine. Finally, regardless of the future state of the economy, the continued use of the current machine would be expected to generate $20,000 in annual profits. The latest government forecast indicates that there is a 40% chance that the economy will show a decline and a 60% chance that the economy will show an expansion during the next five-year period. What is the optimum decision in this problem?

SOLUTION: In order to facilitate the solution to this problem, a payoff matrix is constructed, as shown in Table 1.1. Based on the information contained in this matrix, the decision-making process proceeds in the following manner:

Objective. The objective of the company is to buy the machine that will generate the largest monetary return.

Table 1.1 / Decision Matrix for the Machine Problem

States of nature	Probability	Course of action	
		Buy the larger machine (B_1)	Buy the same machine (B_2)
Economy expansion (A_1)	0.6	$ 150,000	$20,000
Economy decline (A_2)	0.4	− 100,000	20,000

Alternative course of action. The company has two alternative courses open to achieve the objective; i.e., either purchase a larger machine (B_1) or purchase the same machine (B_2) being used currently.

State of nature. The selection of an optimum decision in this case depends upon the state of the economy for the next five years; i.e., whether the economy will continue to expand (A_1) or begin to decline (A_2). If A_1 occurs, the optimum decision is B_1; on the other hand, if A_2 occurs, the optimum decision is B_2.

Probability. As given in the problem, based on government reports, the company assigned a 60% probability that event A_1 would occur and a 40% chance of event A_2. This estimate of probabilities could have been just as easily developed on the basis of the management's subjective judgment or that of an outside consultant.

Payoffs. The monetary reward if choice B_1 is selected, given that state of nature A_1 actually occurs (i.e., a correct decision), is a $150,000 annual profit. On the other hand, if choice B_1 is selected, given that A_2 actually occurs, the manager will have made an incorrect decision. As a result, the company will experience a $100,000 loss. The monetary reward that arises from selecting choice B_2, given that either A_1 or A_2 occurs, is $20,000.

Expected return. How would you value a course of action that has a 60% chance of yielding a $150,000 annual profit and a 40% chance of resulting in a $100,000 annual loss? In order to determine the relative value of each course of action, we weigh each of the possible outcomes by its corresponding probability of occurrence and then sum the resulting values. In the current example, the expected values of these two outcomes are

$$E(B_1) = \$150,000 \times 0.6 + (-\$100,000) \times 0.4$$
$$= \$50,000.$$

$$E(B_2) = \$20,000 \times 0.6 + \$20,000 \times 0.4$$
$$= \$20,000.$$

The optimum decision is to select the alternative that yields the largest expected value; in our example, this is choice B_1.

Table 1.2 / Two Alternative Choices

Alternative choices	Possible outcomes, $	Probability	Expected value, $
A	S = 300	0.6	180
	F = 100	0.4	40
			220
B	S = 1,000	0.7	700
	F = −1,600	0.3	−480
			220

In some instances, however, the expected monetary value alone will not automatically produce an optimum decision. For example, consider the two alternatives given in Table 1.2. Since the expected monetary value for both choices is identical ($220), the decision maker may choose either alternative A or B. The choice concerning an optimum decision would, however, be different, depending upon the individual's attitude toward risk and the external constraints surrounding a problem. For example, alternative A would probably be selected if the decision maker is either financially conservative or cannot afford the potential $1,600 loss that could result from selecting alternative B. On the other hand, alternative B, which yields a larger potential return (i.e., $1,000 as opposed to $300) but has a higher probability of success (0.7 as opposed to 0.6), will normally be selected if the individual is either an optimistic decision maker or has a higher loss tolerance (i.e., he can absorb a $1,600 loss without going bankrupt). In other words, the selection of the optimum decision ultimately rests upon the attitude of the decision maker toward risk and the external financial constraints that exist in the problem environment; i.e., the tolerance level.

Optimum decision. As shown in the preceding paragraphs, the optimum decision in Table 1.1 is to select B_1 (buy the larger machine) as the course of action, since the expected value to be derived from that choice is higher than that of B_2. On the other hand, the optimum decision in Table 1.2 could be either alternative A or alternative B, depending upon the internal and external constraints imposed upon the decision maker.

1.5 / A Look Ahead

In Part I of this book, some of the basic concepts of modern decision theory are examined. These include a survey of probability theory and a discussion of probability distributions with both discrete and continuous random variables. It is essential that students understand the contents contained in Part I, since the remaining sections of this text are built upon the concepts contained in that section of the book.

Part II contains the discussion of modern decision theory itself. It includes an analysis of the decision-making process under the following conditions: (1) a discrete random variable with first prior and then posterior information; (2) a continuous random variable also using prior and posterior information; (3) utility theory; and (4) classical decision-making process. In order to place modern decision theory in proper perspective, Chapter 10 will present a comparative study of classical and modern decision theories.

1.6 / Summary

This chapter contains a general outline of the contents of this book. The following summary may help students to understand the concepts presented.

(1) The purpose of modern decision theory is to assist decision makers in deriving an optimum decision under conditions of uncertainty.

(2) Depending upon the problem environment or states of nature that exist in a problem, there are three possible decision climates that may confront a decision maker: (a) conditions of certainty (i.e., there is no uncertainty concerning the states of nature); (b) conditions of risk, where the probabilities corresponding to the states of nature can be generated by certain statistical rules; and (c) conditions of uncertainty, where the required data (probabilities) are either partially missing or are not in a format that is readily accessible for use by a decision maker.

(3) The decision-making process under conditions of uncertainty is further divided into three branches, depending upon the degree of uncertainty that exists in a problem: (a) uncertainty with perfect knowledge (this condition exists when a prior knowledge concerning the states of nature is provided; (b) uncertainty with complete ignorance (i.e., there is no prior knowledge of the states of nature); and (c) uncertainty with partial ignorance. In this case, only partial information is available to the decision maker.

(4) An established sequence of steps is desirable in the decision-making process in order to derive an optimum decision. These steps include (a) identification of the objectives of the problem; (b) identification of all possible alternative courses of action; (c) listing of all possible events that correspond to each choice or course of action; (d) assignment of likelihood to each of the states of nature; (e) assignment of an estimated payoff for each state of nature, given an associated course of action; (f) computation of the expected value or tolerance level for each alternative; and (g) selection of an optimum decision that is based upon the expected monetary value or appropriate tolerance level.

EXERCISES

1. Develop three decision problems that require the type of systematic reasoning discussed in this chapter.

2. What are the differences in decision criteria between the expected monetary value and the tolerance level?

3. Discuss in detail the three types of decision climates. List several examples for each climate.

4. Explain the three degrees of uncertainty as they relate to the decision-making process under conditions of uncertainty.

5. Discuss the difference in the implications of the decision-making process under conditions of certainty and of perfect information.

6. The ABC Soft Drink Company in San Diego is considering whether it should expand its advertising budget in an effort to increase company sales. If the additional advertising campaign is successful, the revenue generated by sales is estimated to increase by $100,000 per year. If the advertising campaign fails, however, it is estimated that annual revenue will increase by only an additional $10,000. The additional cost of the increased advertising is estimated to be $50,000 per year. The management of the company has assigned an even chance of success to each of the states of nature; i.e., there is a 50% chance of either success or failure of the advertising campaign.

   (a) Construct a decision matrix for this problem that is similar to the one contained in Table 1.1 of this chapter.

   (b) Compute the expected monetary return.

   (c) What is the optimum decision?

   (d) Suppose that the odds are revised to show a 60% chance that the advertising campaign will be successful. Recompute the expected value.

SUGGESTED READING

Bross, Irwin D. *Design for Decision.* New York: Free Press, 1953, chapters 1 through 3. (Paperback)

Costis, Harry G. *Statistics for Business.* Columbus, Ohio: Charles E. Merrill, 1972, chapter 23.

Fellner, William. *Probability and Profit.* Homewood, Ill.: Richard D. Irwin, 1965, chapter 3.

Halter, Albert N., and Dean, Gerald W. *Decisions under Uncertainty with Research Applications.* Cincinnati: South-Western Publishing, 1971, chapters 1, 4, and 12.

Newman, Joseph W. *Management Applications of Decision Theory.* New York: Harper & Row, 1973, chapters 1 and 11.

Raiffa, Howard. *Decision Analysis: Introductory Lectures on Choices under Uncertainty.* Reading, Mass.: Addison-Wesley, 1968, chapters 1 and 10.

Schlaifer, Robert. *Analysis of Decisions under Uncertainty.* New York: McGraw-Hill, 1969, chapters 1 through 4.

Spencer, Milton H. *Managerial Economics.* Homewood, Ill.: Richard D. Irwin, 1968, chapter 1.

Thompson, Gerald E. *Statistics for Decisions.* Boston: Little, Brown, 1972, chapters 1 and 2.

PART I

Probability Theory
and
Probability Distributions

2 / Survey of Probability Theory

2.1 / Purpose

The purposes of this chapter are to introduce the basic concepts of probability theory and to show how probability theory can be used to solve management problems. The chapter should serve as a stepping-stone to Chapters 3 and 4, which comprise a survey of probability distributions.

Why do we need probability theory? The answer is simply that we live under conditions of constant uncertainty, and uncertainty is the most significant problem facing decision makers. A minor error by a decision maker that is caused by uncertainty or imperfect information can result not only in personal disappointment and frustration but also in unanticipated monetary losses.

Probability as a measurement of the degree of uncertainty is derived from any one of the following sources: (1) long-run relative frequency (objective probability) such as the probability of producing a defective part in a factory production line; (2) theoretical analysis such as the probability of getting a "head" on the flip of a balanced coin; and (3) an individual's personal feeling (subjective probability) such as the judgment that there is a better than 50% chance that a new client will accept an offer.

The concepts of both objective and subjective probability form the basic foundation of Bayesian decision theory. According to proponents of the objective concept, probability is the relative frequency in the repeated process and therefore it can be applicable only to an event that can be repeated over and over under the *same conditions*, such as drawing a die or flipping a coin. For example, when a balanced die is drawn repeatedly, the probability that a "1" will appear is 1/6. The objectivist interprets this to mean that if a die is tossed many times, a "1" will appear about once in every six tosses.

On the other hand, proponents of the subjective concept interpret probability as a measurement of personal belief in a *particular event*. Therefore, it is quite reasonable for a subjectivist to state that "the chances are 6/4 that the National League will win the World Series." The subjectivists argue that in order to prove the validity of an objective probability, the identical conditions/environment of a previous trial(s) should be maintained. However, according to a subjectivist, since the surrounding conditions or

15

environment changes constantly in the real world, it is practically impossible to maintain the assumption that all trial conditions are identical.

Those who accept the concept of subjective probability in the decision-making process are known as "Bayesians," whereas those who reject the subjectivity of probability are called "non-Bayesians." Roughly speaking, the Bayesians wish to introduce intuitive judgments and feelings directly into the formal analysis of a decision problem. On the other hand, the non-Bayesians feel that these subjective aspects are best left out of the formal probability analysis.[1]

Probability theory enables decision makers to express in quantitative terms the degree of uncertainty that may affect the success of decisions. By using numbers that can be ranked, they make it easier to visualize and cope with the implied degree of uncertainty. For example, a small ice cream producer in a beach resort would regard a change in weather as a very important decision-making factor. If the weather forecast merely stated that "the probability of rain tomorrow is *high*," the ice cream producer would not have sufficient information upon which to base a decision concerning the status of his sales for the next day. If, however, the forecast stated that "the probability of rain tomorrow is *90%*," then the producer would have sufficient information and could reduce ice cream production in anticipation of a lower demand for his product.

2.2 / Set Theory

In order for a student to properly understand probability theory and its potential for practical application, it is helpful to understand the basic principles of set theory.

Set. A set is defined as a collection of elements that have some describable similarities; these similar elements may be either real or imaginary. Some examples of sets include: all students in a particular college classroom, all cards in a deck of cards, and all integer numbers between 1 and 100.

Generally, there are two ways of specifying a set: (1) listing method, which is primarily used to describe a finite set and merely requires listing all the elements bounded by the set; and (2) defining the property method normally used to describe an infinite set. The elements of the set are usually described by enclosing the set elements within braces. For example, a set. S consisting of all even integers between 1 and 10 inclusive is noted as

$$S = \{2,4,6,8,10\}.$$

---

[1] For a good discussion on the subjectivity of probability, see Howard Raiffa, *Decision Analysis: Introductory Lectures on Choices under Uncertainty* (Reading, Mass: Addison-Wesley, 1968), chapter 10.

A set consisting of the possible outcomes resulting from two tosses of a balanced coin may be expressed as

$$S = \{(H,H),\ (H,T),\ (T,T),\ (T,H)\},$$

where T = tail and H = head. Or a set consisting of the two United States senators from the State of Missouri in 1977 may be expressed as

$$S = \{\text{J. Danford, T. Eagleton}\}.$$

In general,

$$S = \{a_1, a_2, a_3\}$$

is a finite set of three elements, and $a_1 \in S$ means that a_1 is an element of set S.[2]

The listing method of specifying a set is used when the sizes or numbers of elements contained in the set have definable limits. When the set does not distinguish the sizes or numbers of elements, the use of the defining property method is appropriate. For example, a set consisting of all odd integers may be expressed as

$$S = \{1,3,5,7 \cdots\}$$

which may be further reduced to

$$S = \{v | v \text{ is odd integers}\},$$

where v is defined as all the odd integers in the set and the symbol "$|$" is interpreted as "such that." Another illustration is a set consisting of the names of all male students in a college. This may be expressed as

$$S = \{m | m \text{ is the name of all male students in a college}\}.$$

In general, $S = \{a_1, a_2, a_3, \ldots, a_n\}$

Subset. Suppose that set A consists of the senators from the State of New York, and set B consists of all United States senators. In this case, set A is called a subset of set B and is expressed as $A \in B$; conversely, if set A is not a subset of B, the expression becomes $A \notin B$. In general, set A is said to be a subset of set B if and only if every element of set A is also an element of set B. In this respect, every set can be considered as a subset of itself.

Empty set. If no element in set A is contained in set B, set A is said to be an empty set of original set B; this relationship is denoted as $A = \emptyset$. For example, if set A consists of names of all female presidents of the United States and set B consists of names of all presidents of the United States, then set A is a subset of the original set B, but the subset is an empty set (at least at the present time), or $A \ \varepsilon \ B$, but $A = \emptyset$.

---

[2] $\in$ is a form of the Greek letter epsilon, and in set notation means "element of" (a subset); $\notin$ means "is not an element of."

Sample space. In statistics, the process of observation is called an "experiment," and a set that includes all possible outcomes of an experiment, whether the experiment is real or conceptual, is called the "total sample space." The total sample space is also called a *universal set, possibility space,* or *event space.* For example, the sample spaces consisting of all possible outcomes when a balanced coin is tossed twice (S_2) and three times (S_3) are, respectively,

$$S_2 = \{(H,H), (H,T), (T,T), (T,H)\},$$
$$S_3 = \{(H,H,H), (H,H,T), (H,T,H), (T,H,H),$$
$$(T,T,T), (T,T,H), (T,H,T), (H,T,T)\}.$$

Operations with sets

New sets can be formed by combining two or more sets in several ways, as illustrated below.

Union. Assume that the sample space has been defined and that A_1 and A_2 are two subsets included in this sample space. The union of these two subsets results in a set whose elements belong to either subsets A_1, A_2, or both. Many different terms and symbols arc used to describe a union; e.g., A_1 OR A_2, A_1 UNION A_2, $A_1 \cup A_2$, or A_1 CUP A_2.

Let us consider the following problem. Suppose that $W = \{a,b,c,d,g\}$ is the set that consists of students who subscribe to the *Wall Street Journal,* and $N = \{b,d,e,f,h\}$ is the set that consists of students who subscribe to the *New York Times.* The union of these two sets is $W \cup N = \{a,b,c,d,e,f,g,h\}$. This new set implies that the students subscribe either to the *Wall Street Journal* or the *New York Times,* or both. The union of the two subsets W and N is shown graphically in Fig. 2.1, called a "Venn diagram." In general, when there are n subsets in S sample space, the union of these subsets is described as

$$A_1 \cup A_2 \cup A_3 \cup \cdots \cup A_n. \tag{2.1}$$

Figure 2.1 / Venn diagram for subsets W and N.

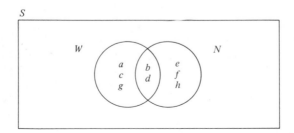

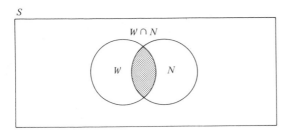

Figure 2.2 / Intersection of subsets W and N.

Intersection. Two sets, A_1 and A_2, will have an intersection if a new set is formed consisting of elements that belong to both sets A_1 and A_2. The intersection is denoted as A_1 AND A_2, $A_1 \cap A_2$, or A_1 CAP A_2. In the example given in Figure 2.1, there are two students (b and d) who subscribe to both the *Wall Street Journal* and the *New York Times*; in this case, the intersection of the new set is $W \cap N = \{b,d\}$. In other words, only students b and d belong to both sets; this is shown in Figure 2.2 as a shaded area.

In general, when there are n subsets in sample space S, the intersection of these subsets is

$$A_1 \cap A_2 \cap A_3 \cap A_4 \cap \cdots \cap A_n. \tag{2.2}$$

Mutually exclusive events. Two subsets, A_1 and A_2, are said to be mutually exclusive, or disjointed, if the occurrence of one event precludes the occurrence of the other event(s) in the same experiment. Consider the Venn diagram in Figure 2.3.

In this example, subset T represents all United States senators and subset R represents all United States congressmen. The sample space S represents the entire body of the U.S. Congress. It follows that subsets T and R are mutually exclusive because they do not contain any common elements; i.e., one individual cannot be both a senator and a congressman at the same time.

Figure 2.3 / Mutually exclusive event.

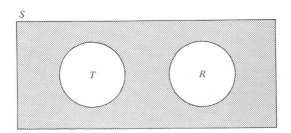

Complementary events. Two sets, A_1 and A_2, are said to be "complementary" if A_2 is a subset that contains all the elements in a sample space that are not included in set A_1. It is clear that the complementary events are mutually exclusive and their union is equal to the sample space S. There are several ways to describe the complementary event, such as $\sim A$, non-A, $\bar{A}$, A^*, or A°. In general,

$$A \cup \sim A = S. \tag{2.3}$$

In Figure 2.3, the shaded area of the sample space S is the complement of both T and R subsets.

2.3 / Basic Rules of Probability Theory

From our discussion of set theory in the preceding section, we can draw a few simple rules concerning probability theory. These basic rules should enhance an understanding of the behavior of probability.

> *Rule 1. A given probability value cannot be greater than 1 or less than zero.*

$$0 \le P(X) \le 1, \tag{2.4}$$

where $P(X)$ is the probability value for a random variable X. According to this rule, the probability value always varies from 0 to 1.0.[3]

EXAMPLE 2.1 / Suppose that there are three departments in a small manufacturing plant. Department A, a delivery area, comprises all male employees; department B, the telephone order-taking section, has all female employees; and department C, the personnel section, consists of four male and six female employees. The probability of selecting randomly a male employee from each department is, respectively,

$P(M)_A = 1.0;$

$P(M)_B = 0.0;$

$P(M)_C = 0.4.$

> *Rule 2. If events A_1 and A_2 are mutually exclusive, the probability that one or the other will occur is the sum of their simple probabilities, or*

$$P(A_1 \cup A_2) = P(A_1) + P(A_2). \tag{2.5}$$

EXAMPLE 2.2 / Assume that a \$100 reward is offered to anyone who obtains either a "1" or a "3" on one roll of a balanced die. What is the chance of receiving the \$100 reward?

---

[3] As far as the implication is concerned, $P(X) = 0$ is identical with $P(X) = 1.0$; both indicate the certainty.

SOLUTION: If you roll a "1," this automatically excludes the occurrence of a "3"; on the other hand, if you roll a "3," the occurrence of a "1" cannot take place. Of course the probability of a "1" is 1/6, since any side of the die has an equal chance of appearing. It is also true that the probability of rolling a "3" is 1/6. Since you are eligible to receive \$100 if you roll either a "1" or a "3," the probability of rolling a "1" *or* "3" is the sum of the two probabilities, or

$$P(1 \cup 3) = P(1) + P(3)$$

$$= \frac{1}{6} + \frac{1}{6} = \frac{2}{6}.$$

EXAMPLE 2.3 / What is the probability of selecting either a diamond or a heart on one draw from a deck of 52 cards?

SOLUTION: Once a diamond is selected, the occurrence of the other event (heart) is automatically eliminated. Since there are 13 diamonds and 13 hearts in a deck of cards, the probability of selecting either a diamond or a heart is

$$P(\text{diamond} \cup \text{heart}) = P(\text{diamond}) + P(\text{heart})$$

$$= \frac{13}{52} + \frac{13}{52} = \frac{26}{52}.$$

Rule 3.   If events $A_1, A_2, A_3 \cdots A_n$ are mutually exclusive and exhaustive in the sample space S, the sum of their simple probabilities is equal to 1, or[4]

$$P(S) = P(A_1 \cup A_2 \cup \cdots \cup A_n)$$
$$= P(A_1) + P(A_2) + \cdots + P(A_n) = 1.0,$$

or

$$\sum_i^n P(A_i) = 1.0, \tag{2.6}$$

where $i = 1, 2, 3 \cdots n$.

EXAMPLE 2.4 / Suppose that you toss a balanced coin one time. The possible outcome of this trial will be either a head or a tail. Therefore, the probability that a head or a tail will appear is $0.5 + 0.5 = 1.0$; i.e., there is no possible event that can occur other than a head or a tail.

---

[4] The summation notation "Σ" (a form of the capital Greek letter sigma) means to add all possible numbers from the beginning number to the last number specified in the equation. For example, if $\sum_1^3 P(A_i)$, where $P(A_1) = 0.1$, $P(A_2) = 0.7$, and $P(A_3) = 0.2$, then $\sum_1^3 P(A_i) = P(A_1) + P(A_2) + P(A_3) = 0.1 + 0.7 + 0.2 = 1.0$.

2.4 / The Calculus of Probability

From our previous discussion concerning the behavior of probabilities, we can explain some of the algebra of probability theory.

Conditional probability

Consider the following case: Suppose that a company planned to hire several college graduates for a new management intern program. In a total of 100 applicants, there were 60 males and 40 females. The information on applicants further disclosed that 70 applicants were majoring in business and 30 applicants had a major in psychology. Table 2.1 illustrates these academic characteristics in summary form.

Table 2.1 / Applicants by Sex and
Academic Major

Sex	Business major	Psychology major	Total
Male	55	5	60
Female	15	25	40
Total	70	30	100

The simple probabilities in the problem can be easily determined from Table 2.1. For example, the probability of selecting a male applicant at random is 60/100; the probability of selecting an applicant with a business major at random is 70/100, and so on. In the preceding illustration, we assumed implicitly that the sample space was already defined (100 applicants). However, since the selection of a sample space varies, depending upon the scope of the problem in question, it is always helpful to define in explicit terms the appropriate sample space being considered. Therefore, the probability of selecting a male applicant from the total sample space can be denoted explicitly by

$$P(\text{male}|\text{sample space}) = 60/100 = 0.6$$

The term $P(\text{male}|\text{sample space})$ is called a "conditional probability" and is stated as "the probability of selecting a male, given a total sample space." The use of the conditional probability concept helps us to express explicitly the sample space from which given sets are selected.

Consider the following conditional probability problem.
Let

$P(M)$ = probability that the company will hire a male

$P(F)$ = probability that the company will hire a female

$P(B)$ = probability that the company will hire an individual with a business major

$P(P_s)$ = probability that the company will hire an individual with a psychology major

What is the probability of hiring an individual with a psychology major, given that the prospective employee is a male?

In the preceding problem, we want to determine the probability of hiring a person with a psychology major, but not from just any group. The problem explicitly states that the selection of a psychology major should be from the male group only. We are, in fact, saying that all female applicants will be excluded from consideration; this has the effect of reducing the sample space from 100 to 60 applicants.

The answer to the preceding question can be expressed in statistical form as:

$$P(P_s|M) = \frac{5}{60}.$$

There are five candidates with a psychology major of a total of 60 male applicants. The probability of hiring an applicant with a psychology major from the list of male applicants is therefore 5/60.

EXAMPLE 2.5 / What is the probability that the company will hire a business major, given that he is male? The answer is

$$P(B|M) = \frac{55}{60}.$$

Likewise, the probability that the company will hire a psychology major, given that the applicant is a female, is

$$P(P_s|F) = \frac{25}{40},$$

and the probability that the company will hire a business major, given that the applicant is a female, is

$$P(B|F) = \frac{15}{40}.$$

In narrative form, the conditional probability $P(P_s|M) = 5/60$ may be restated in the following way: "There are 5 *male psychology* majors in a total sample of 100 applicants, and 60 of them are *males*." Therefore,

$$P(P_s|M) = \frac{5/100}{60/100} = \frac{5}{60}.$$

Conditional probability: General form. Let us discuss a bit further the conditional probability of $P(P_s|M) = 5/60$. If we divide the numerator and the denominator of $P(P_s|M) = 5/60$ by the total sample size of 100 applicants,

23

the result becomes

$$P(P_s|M) = \frac{5/100}{60/100}.$$

An examination of the denominator tells us that this result represents the probability of selecting a set that consists of a male applicant with a psychology major, or $P(M \cap P_s)$. On the other hand, the denominator represents the probability of selecting a male candidate from the *total* sample space. Accordingly, it follows that

$$P(P_s|M) = \frac{P(M \cap P_s)}{P(M)}.$$

In general, the conditional probability of two sets, A and B, is given by

$$P(A|B) = \frac{P(A \cap B)}{P(B)}. \tag{2.7}$$

Equation 2.7 states that the probability of selecting set A, given set B, is equal to the probability of selecting a joint event, $P(A \cap B)$, and then dividing this by the simple probability of selecting set B; i.e., $P(B)$.

In order to assure an understanding of conditional probabilities, the student should verify the meaning and solution of the following conditional probabilities:

(a) $\quad P(F|P_s) = \dfrac{P(F \cap P_s)}{P(P_s)} = \dfrac{25/100}{30/100} = \dfrac{25}{30}$

$\qquad P(M|P_s) = \dfrac{P(M \cap P_s)}{P(P_s)} = \dfrac{5/100}{30/100} = \dfrac{5}{30}$

$\qquad P(F|B) = \dfrac{P(F \cap B)}{P(B)} = \dfrac{15/100}{70/100} = \dfrac{15}{70}$

$\qquad P(M|B) = \dfrac{P(M \cap B)}{P(B)} = \dfrac{55/100}{70/100} = \dfrac{55}{70}$

(b) $P(P_s|M) + P(B|M) = \dfrac{5}{60} + \dfrac{55}{60} = 1.0$

$\qquad P(P_s|F) + P(B|F) = \dfrac{25}{40} + \dfrac{15}{40} = 1.0$

$\qquad P(F|P_s) + P(M|P_s) = \dfrac{25}{30} + \dfrac{5}{30} = 1.0$

$\qquad P(F|B) + P(M|B) = \dfrac{15}{70} + \dfrac{55}{70} = 1.0$

Equation 2.7 provides a convenient tool to test whether any given set is statistically dependent on or independent of the other set(s). Two events are said to be dependent when the result of one experiment is influenced by the result of another experiment. On the other hand, two events are said to be independent if the result of one experiment is not in any way influenced by the result of any other experiment. For example, when tossing a balanced coin twice, the probability of getting a "tail" on the second trial, given that a "tail" already was observed in the first trial, is still 0.5. This is true because the result of the second trial is in no way influenced by the result of the first trial. Therefore,

$$P(T_2|T_1) = P(T_2) = 0.5.$$

In general, if two events, A and B, are independent, then

$$P(A|B) = P(A).$$

On the other hand, if two events, A and B, are dependent, then,

$$P(A|B) \neq P(A).$$

In Table 2.1, the sex and academic major of the applicants are not independent of each other,[5] since

$$P(F|B) \neq P(F).$$

This is illustrated by the fact that (see Table 2.1)

$$P(F|B) = \frac{15}{70} \quad \text{and} \quad P(F) = \frac{30}{100}.$$

Joint probability

Two events, A and B, are said to be "joint" if they have sample points in common. Equation 2.7 can be used again to derive a joint probability. Thus, from

$$P(A|B) = \frac{P(A \cap B)}{P(B)},$$

by multiplying both sides by $P(B)$ and rearranging the product, we obtain

$$P(A \cap B) = P(A|B) \cdot P(B). \tag{2.8}$$

Equation 2.8 is a general form of joint probability. It is also referred to as a "multiplication rule" and enables us to compute the probability that two events will both occur. In Example 2.5, the probability that the company will hire a male who is also a psychology major is

$$P(M \cap P_s) = P(M) \times P(P_s|M)$$

$$= \frac{60}{100} \times \frac{5}{60} = \frac{5}{100}.$$

[5] This implies that more female students tend to major in psychology than do male students.

In this illustration, the probability that a male applicant is hired is 60 out of 100, or 6/10, and the probability that a psychology major will be selected from the male applicants is 5 of 60, or 5/60.

EXAMPLE 2.6 / What is the probability that a female applicant with a business major will be selected? The answer is 15/100 and is determined in the following manner:

$$P(F \cap B) = P(F) \times P(B|F) = \frac{40}{100} \times \frac{15}{40} = \frac{15}{100}.$$

It is relatively easy to prove that $P(F \cap B) = P(B \cap F)$,[6] for

$$P(B \cap F) = P(B) \times P(F|B)$$

$$= \frac{70}{100} \times \frac{15}{70} = \frac{15}{100}.$$

EXAMPLE 2.7 / When you roll a balanced die, what is the probability that either even numbers *or* numbers greater than 3 will appear?

SOLUTION: Let

$P(A)$ = probability that even numbers appear (II, IV, VI);

$P(B)$ = probability that numbers greater than 3 appear (IV, V, VI).

Then

$$P(A \cup B) = P(A) + P(B) - P(A \cap B),$$

where $P(A \cap B) \neq 0$. Since two events (IV and VI) were counted twice, once under $P(A)$ and once under $P(B)$, a subtraction is necessary to adjust for the portion that is duplicated. The area of overlap is, however, merely a joint event (A and B); this joint probability is illustrated in the Venn diagram of Fig. 2.4 as the shaded area. Therefore,

$$P(A) = P(\text{II}) + P(\text{IV}) + P(\text{VI})$$

$$= \frac{1}{6} + \frac{1}{6} + \frac{1}{6} = \frac{3}{6},$$

$$P(B) = P(\text{IV}) + P(\text{V}) + P(\text{VI})$$

$$= \frac{1}{6} + \frac{1}{6} + \frac{1}{6} = \frac{3}{6},$$

$$P(A \cap B) = P(B) \times P(A|B) = \frac{3}{6} \times \frac{4}{6} = \frac{2}{6}.$$

[6] In essence, we are saying that the probability of hiring a *female* candidate with a *business* major is the same as the probability of hiring a candidate with a *business* major who is also *female*. This identity of expression in a joint probability plays a vital role in the development of Bayes' theorem (see under the next section, "Bayes' Theorem").

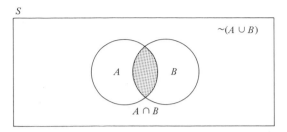

Figure 2.4 / Joint probability.

Accordingly,

$$P(A \cup B) = P(A) + P(B) - P(A \cap B)$$

$$= \frac{3}{6} + \frac{3}{6} - \frac{2}{6} = \frac{4}{6}$$

In the preceding conditional probability, $P(A|B)$ indicates the probability that an even number is selected, but the even numbers should be greater than III. There are only two even numbers (IV and VI) among the three possible numbers conditioned by B(IV, V, VI). Therefore, $P(A|B) = 4/6$.

If sets A and B were independent, $P(A)$ would replace $P(A|B)$ in Equation 2.8, and the equation would be stated as

$$P(A \cap B) = P(A) \times P(B). \tag{2.9}$$

EXAMPLE 2.8 / A reward of $100 was offered to anyone who tossed a "1" and a "3" in that order on two successive rolls of a die. What is the chance of receiving the reward?

The probability that the individual receives a "1" on the first roll is 1/6, and the probability that the individual receives a "3" on the second roll is also 1/6. The probability that these two sets both occur, with proper order, is

$$P(1 \cap 3) = P(1) \times P(3)$$

$$= \frac{1}{6} \times \frac{1}{6} = \frac{1}{36}.$$

It is interesting to note that if we were to compute and add the probabilities of all possible joint events in this situation, we would obtain 1.0. For example, in Table 2.1 there are four mutually exclusive events; i.e., $M \cap B$, $M \cap P_s$, $F \cap B$, and $F \cap P_s$. The sum of these joint probabilities is

$$P(M \cap B) + P(M \cap P_s) + P(F \cap B) + P(F \cap P_s)$$

$$= \frac{55}{100} + \frac{5}{100} + \frac{15}{100} + \frac{25}{100} = 1.0.$$

Marginal probability

The fact that the joint events are mutually exclusive provides the necessary condition for another important probability concept. For example, let us assume that we are interested in only the business major column in Table 2.1, ignoring the psychology major column. The probability of selecting an applicant with a business major is, of course, $70/100 = 0.7$. If we had analyzed the business major column a bit more carefully, we could have arrived at the same result by summing the joint probabilities contained in the business major column, or

$$P(B) = P(M \cap B) + P(F \cap B)$$

$$= \frac{55}{100} + \frac{15}{100}$$

$$= \frac{70}{100} = 0.7.$$

This is called a "marginal probability." In other words, the marginal probability is the sum of the probabilities of a mutually exclusive joint event, whenever one or more criteria of classifications is ignored; (e.g., in the preceding case, we ignored the classification of a psychology major).

Bayes' theorem

Up to this point we were concerned only with the *effect* of a problem, and used purely *deductive* reasoning to solve all the problems. Bayes' theorem uses *inductive* reasoning to determine the causes based on observed effects. This theorem was developed by Reverend Thomas Bayes (1702–1761), an English Presbyterian minister and mathematician.

Bayes' theorem—general form. Example 2.6 illustrated that $P(F \cap B) = P(B \cap F)$. In its most general form, a joint probability (Equation 2.8) can be expressed in two alternative forms:

$$P(A_j \cap B) = P(A_j) \times P(B|A_j),$$

or

$$P(B \cap A_j) = P(B) \times P(A_j|B),$$

where A_j = mutually exclusive events and $P(B) > 0$.
    Since $P(A_j \cap B) = P(B \cap A_j)$, it follows that

$$P(A_j) \times P(B|A_j) = P(B) \times P(A_j|B).$$

Solving for $P(A_j|B)$,

$$P(A_j|B) = \frac{P(A_j) \times P(B|A_j)}{P(B)}. \tag{2.10}$$

Since A_j are mutually exclusive events and $P(B) > 0$,

$$P(B) = P(A_1) \times P(B|A_1) + P(A_2) \times P(B|A_2)$$
$$+ \cdots + P(A_n) \times P(B|A_n)$$

$$= \sum_{i=1}^{n} P(A_i) \times P(B|A_i).$$

Accordingly, Equation 2.10 becomes

$$P(A_j|B) = \frac{P(A_j) \times P(B|A_j)}{\sum_{i=1}^{n} P(A_i) \times P(B|A_i)}. \tag{2.11}$$

Equation 2.11 is referred to as Bayes' theorem and $P(A_j|B)$ is called a "revised" probability, a "posterior" probability, or a "Bayesian" probability. Using this equation, it is possible to solve for $P(A_j|B)$, if $P(B|A_j)$ is given. In essence, the revised probability $P(A_j|B)$ is the ratio of the joint probability $P(A_j) \times P(B|A_j)$, to the marginal probability, $\sum_{i=1}^{n} P(A_i) \times P(B|A_i)$.

Criticism of Bayes' theorem. The validity of Bayes' theorem in the decision-making process under conditions of uncertainty usually arouses extensive discussion. The center of the controversy is the claim in Bayes' theorem that the prior information or probability, $[P(A_j)$ in Equation 2.11] can take any probability form, including subjective assessment concerning the existing state of nature. The opponents of Bayes' theorem, however, object to the idea of using any information derived from this type of subjective assessment. Their argument rests on the concept that the probabilities developed in the decision-making process should always be based on objective, long-run frequencies of occurrence, regardless of whether they are prior probabilities or conditional probabilities. It should be pointed out, however, that both opponents and proponents of Bayes' theorem agree with the mathematical reasoning inherent in Bayes' theorem.[7]

EXAMPLE 2.9 / The quality control chart of the Hickory Manufacturing Company in Peoria shows that 40% of the defective parts produced in a manufacturing plant were due to mechanical errors and that 60% of the defective parts produced were caused by operators' error. The defective parts caused by mechanical problems can be detected, with 90% accuracy

---

[7] For an excellent source of argument for and against Bayes' theorem, see J. D. Weber, *Historical Aspects of the Bayesian Controversy* (Tucson, Ariz.: University of Arizona Press, 1973).

rate, at an inspection station. The detection rate, however, drops to 50% if the defective parts are due to an error caused by a machine operator. Suppose that a defective part was detected at the inspection station. What is the probability that such a defective part was caused by a machine error?

SOLUTION: The tree diagram in Figure 2.5 illustrates the problem in question.

Let

$P(E_i)$ = probability that a defective part was produced, owing to either a mechanical error (E_m) or an error by a machine operator (E_o).

$P(DT)$ = probability that a defective part was detected at the inspection station.

$P(DT|E_i)$ = conditional probability that a defective part is detected, given that it was produced because of E_i reasons.

From the information given,

$$P(E_m) = 0.4 \qquad P(DT|E_m) = 0.9$$
$$P(E_o) = 0.6 \qquad P(DT|E_o) = 0.5$$

Mathematical approach. From Equation 2.11 and the information given above,

$$P(E_m|DT) = \frac{P(E_m \cap DT)}{P(DT)} = \frac{P(E_m) \cdot P(DT|E_m)}{P(DT)}$$

Figure 2.5 / Tree diagram for Example 2.9.

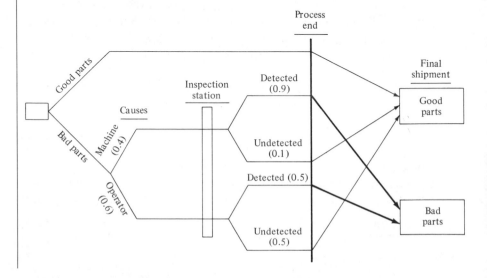

where

$$P(DT) = P(E_m)P(DT|E_m) + P(E_o)P(DT|E_o)$$
$$= 0.4 \times 0.9 + 0.6 \times 0.5 = 0.66.$$

$P(DT)$ is a marginal probability—the probability that defective parts will be detected at the inspection station. Therefore,

$$P(E_m|DT) = \frac{0.4 \times 0.9}{0.66} = 0.55.$$

Since the probability of selecting a defective part that was caused by mechanical error—provided it was already detected, $P(E_m|DT)$—and the probability of selecting a defective part that was caused by the operator's error, $P(E_o|DT)$, are mutually exclusive and totally exhaustive,

$$P(E_m|DT) + P(E_o|DT) = 1.0.$$

Therefore,

$$P(E_o|DT) = 1.0 - P(E_m|DT) = 1.0 - 0.55 = 0.45.$$

Table approach. There is a 55% chance that the defective part detected at the inspection station was caused by a mechanical error; it follows, therefore, that the probability is 45% that the defective part detected at the inspection station was caused by a machine operator error.

Table 2.2 / Causes of Defective Part

Causes (E_i)	Prior probability $P(E_i)$	Conditional probability $P(DT|E_i)$	Joint probability $P(E_i) \cdot P(DT|E_i)$	Revised probability $P(E_i|DT)$
Machine (m)	0.4	0.9	0.36	0.55
Operator (o)	0.6	0.5	0.30	0.45
	1.0		0.66*	1.00

* Marginal probability

EXAMPLE 2.10 / Using the same data given in Example 2.9, suppose that a customer returned a defective part which went undetected at the inspection station. What is the chance that the defective part was due to a machine operator error?

SOLUTION: From the given information,

$$P(-DT|E_m) = 1 - P(DT|E_m) = 1 - 0.9 = 0.1,$$

and

$$P(-DT|E_o) = 1 - P(DT|E_o) = 1 - 0.5 = 0.5,$$

where $P(-DT)$ = probability that the defective part went undetected at the inspection station. Therefore,

$$P(E_o|-DT) = \frac{P(E_o \cap -DT)}{P(-DT)} = \frac{P(E_o) \cdot P(-DT|E_o)}{P(-DT)}$$

where

$$P(-DT) = P(E_m) \cdot P(-DT|E_m) + P(E_o) \cdot P(-DT|E_o)$$
$$= 0.4 \times 0.1 + 0.6 \times 0.5$$
$$= 0.34.$$

Accordingly,

$$P(E_o|-DT) = \frac{0.6 \times 0.5}{0.34} = 0.88.$$

The probability is 88% that the undetected defective part was caused by a machine operator error. On the other hand, the probability that the undetected defective part was caused by a mechanical error is therefore $1 - 0.88 = 0.12$.

EXAMPLE 2.11 / Jim Reay is the production manager for a manufacturer of electronic parts. The firm produces items on a subcontractor basis for the Video Equipment Company. Currently Jim has four machines: A_1, A_2, A_3, and A_4, with each machine capable of producing 10,000 parts per week. Depending upon the age of the equipment, one machine must be replaced each year. It is generally true that the newer the machine, the fewer the number of defective parts it produces. A control chart maintained by the company reveals that the rate of defective parts produced by each machine was as follows: The rate by new machine A_1 is almost zero; by A_2, which is one year old, the rate is 1%; by A_3, which is two years old, the rate is 5%; and by A_4, which is three years old, the rate is 10%. At the end of its fourth year of operation, each machine is replaced with a new machine. Just prior to shipment, Jim selected one sample at random for examination purposes. The part he sampled turned out to be defective. What is the probability that the defective item was produced by machine A_2? Or, phrasing the question in a different form, what is the chance that the sample was produced by machine A_2, given that the item selected was a defective part?

SOLUTION:

Table approach. The solution to the problem in Example 2.11 is illustrated in Table 2.3, where D = defective parts.

Column (1) in Table 2.3 is the proportion of defective parts produced by each machine. Since each machine produces the same number of parts, the chance (prior probability) that the sample would have come from any one of these four machines is 1/4, or 0.25; this is shown in column (2).

Table 2.3 / Jim Reay Plant

Defective rate (A_i) (1)	Prior probability: $P(A_i)$ (2)	Conditional probability: $P(D\|A_i)$ (3)	Joint probability: $P(A_i) \cdot P(D\|A_i)$ (4) = (2) × (3)	Posterior probability: $P(A_i\|D)$ (5) = (4)/$\sum$(4)
$A_1(0\%)$	0.25	0.00	0.0000	0.0000
$A_2(1\%)$	0.25	0.01	0.0025	0.0625
$A_3(5\%)$	0.25	0.05	0.0125	0.3125
$A_4(10\%)$	0.25	0.10	0.0250	0.6250
	1.00		0.0400*	1.0000

* Marginal probability

Column (3) shows the conditional probability that a random sample selected is a defective part, given that it was produced by an A_i machine. For example, the conditional probability that a defective part is selected, given that it was produced by machine A_2, is 0.01, since machine A_2 is known to produce 1% defective parts. The rest of column (3) was likewise determined. Since the probability that any one part was produced by any machine is 0.25, as shown in column (2), the joint probability that a defective part will be selected and that the part was produced by machine A_2, is

$$P(A_2 \cap D) = P(A_2) \cdot P(D|A_2) = 0.25 \times 0.01 = 0.0250,$$

shown in column (4). The sum of the joint probabilities, or column (4), is the marginal probability. The posterior probability (Bayesian probability) is computed by dividing each of the joint probabilities in column (4) by the marginal probability (0.0400) as shown in column (5) or, in other words, by computing the ratio of the joint probability to the marginal probability. Therefore the answer to our original question is

$$P(A_2|D) = \frac{0.0025}{0.0400} = 0.0625, \quad \text{or } 6.25\%.$$

It is interesting to compare the list of prior probabilities contained in column (2) with the list of posterior probabilities in column (5) to see what effect, if any, the use of new information (sample result) had on the outcome of this problem. After the prior probability was modified by new information (column 3) to produce the revised or posterior probability (column 5), the probability of a defective part being produced by machine A_4 increased significantly from 0.10 to 0.625. Such an increase is not too difficult to understand. The number of parts produced by each machine remained the same, with each machine producing 25% of the total parts. The probability that we are interested in, however, is the one concerning which machine actually produced the defective part. As column (3) shows, machine A_1 did not produce any defective parts. Therefore, there is no chance (i.e., the probability is zero) that the defective part could have been produced by machine A_1. The chance that a defective part was produced by machine A_3 is, of course, higher than A_2, since machine A_3 is known to produce 5% defective parts, whereas machine A_2 produces only 1% defective parts, and so on.

Mathematical approach. Let

$P(D)$ = probability that a defective part is observed.

$P(A_i)$ = probability that a part is produced by each of the different machines.

$P(D|A_i)$ = probability that a defective part is selected, given that it is produced by a specific machine (A_i).

Then,

$$P(A_1) = 0.25 \qquad P(D|A_1) = 0.00$$
$$P(A_2) = 0.25 \qquad P(D|A_2) = 0.01$$
$$P(A_3) = 0.25 \qquad P(D|A_3) = 0.05$$
$$P(A_4) = 0.25 \qquad P(D|A_4) = 0.10$$

The probability that machine A_2 produces an item, given it is a defective part, is

$$P(A_2|D) = \frac{P(A_2 \cap D)}{P(D)} = \frac{P(A_2) \cdot P(D|A_2)}{P(D)},$$

where $P(D) > 0$. As defined above, $P(D)$ is the probability of selecting a defective part. In this problem, there are four machines that could produce a defective part: A_1, A_2, A_3, or A_4. Therefore, the probability that a defective part will be produced by these four machines is

$$P(D) = P(A_1) \times P(D|A_1) + P(A_2) \times P(D|A_2)$$
$$+ P(A_3) \times P(D|A_3) + P(A_4) \times P(D|A_4).$$

The probability that a part is produced by machine A_2, given that the sample was a defective part, is

$$P(A_2|D)$$

$$= \frac{P(A_2) \times P(D|A_2)}{P(A_1) \cdot P(D|A_1) + P(A_2) \cdot P(D|A_2) + P(A_3) \cdot P(D|A_3) + P(A_4) \cdot P(D|A_4)}$$

$$= \frac{0.25 \times 0.01}{(0.25 \times 0.0) + (0.25 \times 0.01) + (0.25 \times 0.05) + (0.25 \times 0.10)}$$

$$= \frac{0.0025}{0.0400} = 0.0625.$$

This answer is the same as the one computed in the second row of column (5) in Table 2.3. In the preceding equation, the value of the numerator (0.0025) is simply one of the joint probabilities computed in column (4) of Table 2.3, and the value of the denominator (0.0400) is the marginal probability.

What is the probability that a given part was produced by machine A_4, provided the sample was a defective part? The solution is

$$P(A_4|D) = \frac{P(A_4) \times P(D|A_4)}{P(D)} = \frac{0.25 \times 0.10}{0.040}$$

EXAMPLE 2.12 / This example provides a complete summary of the probability theory up to this point in our discussion.

A Super-8 movie projector produced by the Electro Company contains three basic tubes: A, B, and C. Tube B operates independently of tubes A and C, but tubes A and C are interdependent. Company records show that tube A works properly 70% of the time; tube B, 80% of the time; and tube C, 90% of the time. However, if tube C fails, there is an 85% chance that tube A will fail also. Assume that at least two parts must operate to enable the projector to function. What is the probability that the projector will function properly?

SOLUTION: Since at least two tubes must work properly, the following combinations of two tubes are possible, where $\sim$ indicating improper function:

$\quad P(A \cap B \cap C)$ $\quad$ (All three tubes function properly.)
$\quad P(A \cap B \cap \sim C)$ $\quad$ (Tubes A and B function properly.)
$\quad P(A \cap \sim B \cap C)$ $\quad$ (Tubes A and C function properly.)
$\quad P(\sim A \cap B \cap C)$ $\quad$ (Tubes B and C function properly.)

Given that tubes A and C are interconnected, the following conditional probabilities should be formulated, based on Equation 2.7:

$$P(A \cap B \cap C) = P(A) \times P(B) \times P(C|A),$$
$$P(A \cap B \cap \sim C) = P(A) \times P(B) \times P(\sim C|A),$$
$$P(A \cap \sim B \cap C) = P(A) \times P(\sim B) \times P(C|A),$$
$$P(\sim A \cap B \cap C) = P(\sim A) \times P(B) \times P(C|\sim A).$$

From the information given above:

$$P(A) = 0.70 \qquad P(\sim A) = 1.00 - 0.70 = 0.30$$
$$P(B) = 0.80 \qquad P(\sim B) = 1.00 - 0.80 = 0.20$$
$$P(C) = 0.90 \qquad P(\sim C) = 1.00 - 0.90 = 0.10$$
$$P(\sim A|\sim C) = 0.85 \qquad P(A|\sim C) = 1.00 - 0.85 = 0.15$$

35

The following additional conditional probabilities are then computed from the probabilities given above:

$P(C|A)$ = probability that tube C functions, given that tube A also functions properly.

$P(\sim C|A)$ = probability that tube C fails to function, given that tube A functions properly.

$P(C|\sim A)$ = probability that tube C functions properly, given that tube A fails to function.

From Equation 2.11,

$$P(\sim C|A) = \frac{P(\sim C) \times P(A|\sim C)}{P(A)} = \frac{0.1 \times 0.15}{0.7} = 0.02.$$

From Equation 2.6,

$$P(C|A) = 1.00 - P(\sim C|A) = 1.00 - 0.02 = 0.98,$$

and from Equation 2.6 and Equation 2.11,

$$P(C|\sim A) = 1.00 - P(\sim C|\sim A),$$

$$P(\sim C|\sim A) = \frac{P(\sim C) \times P(\sim A|\sim C)}{P(\sim A)}$$

$$= \frac{0.1 \times 0.85}{0.3} = 0.28$$

Then,

$$P(C|\sim A) = 1.00 - 0.28 = 0.72.$$

Therefore, the probability that at least two tubes will function properly is the sum of the following computations:

$$P(A \cap B \cap C) = P(A) \times P(B) \times P(C|A)$$
$$= 0.7 \times 0.8 \times 0.98 = 0.55$$

$$P(A \cap B \cap \sim C) = P(A) \times P(B) \times P(\sim C|A)$$
$$= 0.7 \times 0.8 \times 0.02 = 0.01$$

$$P(A \cap \sim B \cap C) = P(A) \times P(\sim B) \times P(C|A)$$
$$= 0.7 \times 0.2 \times 0.98 = 0.14$$

$$P(\sim A \cap B \cap C) = P(\sim A) \times P(B) \times P(C|\sim A)$$
$$= 0.3 \times 0.8 \times 0.72 = 0.17$$

or

$$0.55 + 0.01 + 0.14 + 0.17 = 0.87.$$

In other words, there is a 87% chance that the movie projection will function properly.

2.5 / Summary

In this chapter we have discussed the basic concepts of probability theory, and the use of probability concepts in business and other related fields. The following brief statements may be helpful in recalling the contents of this chapter:

(1) Because of the problem of uncertainty and subsequent partial information, the use of probability theory is necessary and inevitable in order to obtain an optimum decision.

(2) Probability is primarily based on any one of the three sources: (a) long-run relative frequency (objective probability), (b) theoretical analysis, and (c) subjective probability.

(3) The value of any particular probability is always greater than or equal to zero, and less than or equal to 1, or $0 \leq P(X) \leq 1.0$.

(4) If events A_1 and A_2 are mutually exclusive, the probability that one or the other will occur is the sum of their simple probabilities, or

$$P(A_1 \cup A_2) = P(A_1) + P(A_2).$$

(5) If events $A_1, A_2, A_3, \ldots A_n$ are mutually exclusive and exhaustive in sample space S, the sum of their simple probabilities is equal to 1.0, or

$$P(S) = P(A_1 \cup A_2 \cup A_3 \cup \cdots \cup A_n)$$
$$= P(A_1) + P(A_2) + P(A_3) + \cdots + P(A_n)$$
$$= 1.0;$$

or

$$P(S) = \sum_{i-1}^{n} P(A_i) = 1.0,$$

where $i = 0,1,2,\ldots n$.

(6) The conditional probability of A, given B, is equal to

$$P(A|B) = \frac{P(A \cap B)}{P(B)}.$$

(7) Two events, A and B, are said to be "joint" if they have sample points in common.

(8) The marginal probability is a summation of the probabilities of mutually exclusive joint events whenever one or more criteria of classifications are ignored.

(9) The Bayes' theorem is concerned with a *cause* based on an observed *effect*. Mathematically, this is stated as

$$P(A_j|B) = \frac{P(A_j) \times P(B|A_j)}{P(B)},$$

where the A_j are mutually exclusive events, and $P(B) > 0$.

EXERCISES

1. Discuss the significance of using probability theory in the decision-making process under conditions of uncertainty. How does probability theory contribute to the decision-making process? Name at least three practical business applications that are amenable to, or could be improved by, the use of probability theory.

2. Discuss the following terms:
   (a) set
   (b) sample space
   (c) empty set
   (d) marginal probability
   (e) statistically dependent event
   (f) statistically independent event
   (g) conditional probability
   (h) joint probability

3. How is Bayesian probability different from objective probability?

4. Let A represent an acceptable electronic part and D represent a defective part. Describe the set of outcomes that provides the maximum number of combinations for obtaining three acceptable parts in five samples.

5. Candidate S conducted two opinion polls concerning the outcome of a local election. One poll was conducted at the beginning of the campaign and the other was taken two days prior to the actual election day. Suppose that the results of polls taken between these two periods are described in terms of "increase," "decrease," and "no change" to show the voter preference for candidate S. Report the results of the two polls in terms of a sampling space and interpret each resulting subset.

6. In the following sets, mark whether the answer is true or false. If the answer is false, give the correct answer.
   (a) $\{A,B,D\} \cup \{A,C,E\} = \{A,B,C,D,E\}$.
   (b) $\{A,B\} \cap \{A,C\} = \{A,B,C\}$.
   (c) $\{A,B,C\} \cup \{A,B,D\} \cap \{A,E\} = \{A\}$.
   (d) $\{A,D,E\} \cap \{A,C,F\} = \{A,C\}$.

7. The ACME Vacuum Cleaner Company normally conducts the major portion of its advertising campaign by direct mail. The company calls the prospective buyers only after it receives an initial response via mail. The probability that, following a telephone response, a salesman will actually make a sale is 0.4. On a typical day, a salesman will follow up three telephone calls. Assuming that the sales are independent,
   (a) Compute the probability that a salesman will make all three sales.
   (b) Compute the probability that he will make at least one sale.
   (c) Compute the probability that he makes one sale.
   (d) Compute the probability that he does not make a sale.

8. What is the probability of selecting from a deck of cards two successive diamond-suit cards without replacing in the deck the first card you selected? Is this probability different from the probability of selecting diamonds on two successive draws, when the first card is replaced in the deck before making the second draw?

9. What is the probability of selecting a card with either a diamond or a picture from a deck of 52 cards? Give the full solution.

10. Of the employees of the Gardenstate Garment Company in New York City, 40% use their cars, 50% use the subway to commute to work, and 20% use both their cars and the subway.
    (a) If an employee says that he used a subway, what is the chance that he also uses a car?
    (b) What is the chance that employees use either a subway or a car?

11. John, a high school student in Denton, Texas, is considering delivering newspapers to the houses in his town. A preliminary study shows that occupants of 35% of the houses read only the Denton Star, 30% read only the Independent Press, an additional 20% read both newspapers, and 15% do not read a newspaper. If John's cousin reads the Independent Press, what is the probability that his cousin also reads the Denton Star?

12. In a public hearing regarding women's rights, the Judiciary Committee of the U.S. Congress decided to call 40 witnesses to testify before the Committee. The Committee chairman decided to invite witnesses based on sex and political party affiliation (i.e., Republican, Democratic, and Independent). The breakdown of witnesses was as follows: 10 males, 20 Democrats, 10 Independents, 8 female Republicans, and 5 male Independents.
    (a) What is the probability that a male will be called to testify if the person selected is an Independent?
    (b) Compute the probability of selecting a female who is, politically, an Independent.
    (c) Compute the probability of selecting a male who is a Democrat.
    (d) Is sex independent of political affiliation?

13. A survey of the graduate business school in a university shows how the members of the faculty subscribe to various professional journals. The survey yielded the following results (based on 100%):

American Economic Review (AER)	20%
Economic Journal (EJ)	15%
Southern Economic Journal (SEJ)	10%
None of the above three	5%
AER and EJ	20%
AER and SEJ	15%
EJ and SEJ	5%
All three	10%
	100%

(a) What is the probability of selecting a faculty member who subscribes to the AER?
(b) What is the probability of selecting a faculty member who subscribes to AER, given that he subscribes to AER *and* SEJ?
(c) What is the probability of selecting a faculty member who subscribes to AER and EJ, given that he subscribes also to SEJ?
(d) What is the probability of selecting a faculty member who subscribes to EJ and SEJ, given that he reads AER?
(e) Is the fact that a faculty member subscribes to the AER independent of the amount and type of subscribers to the EJ?
(f) A faculty member, selected at random, subscribed to both the AER and the EJ. What is the chance that he subscribes to the SEJ?

14. The Pierce-Acheson Investment Company maintains a standard policy of investing its capital into three different stock market exchanges: (1) 50% in the New York Stock Exchange (NYSE); (2) 30% in the American Stock Exchange (AE); and (3) the remaining 20% in the Over-The-Counter market (OTC). The results of company investments for the past fiscal year have just been published and distributed to the stockholders. These results were as follows: 60% of the investments in NYSE-traded stocks show a profit and 40% of the investment shows a loss; 50% of the

investments in AE-traded stocks show a profit and 50% show a loss; and only 30% of the investment in OTC-traded stocks show a profit, while 70% show losses. Assuming that the NYSE and the AE are designated as the major stock exchanges:

(a) What is the marginal probability of selecting a stock which shows (1) a price increase and (2) a price decrease?

(b) What is the probability of selecting a stock which shows a price decrease, given it is traded on a major stock exchange?

(c) A stock was selected at random and turned out to be one whose price increased from the previous year's value. What is the probability that it is traded on (1) the NYSE, (2) the AE, and (3) the OTC?

(d) Is the price behavioral pattern independent of the type of market in which the stock is traded? Test for two different price behavioral patterns for the three different markets.

15. The Jansen Technical Writing Company located in St. Louis has 200 employees. A detailed breakdown of the employees of the firm by sex and home residence is given in the following table.

	Residence	
Sex	Illinois	Missouri
Female	75	50
Male	40	35

(a) What is the marginal probability of selecting an employee who lives in Illinois?

(b) What is the joint probability of selecting a female employee who lives in Missouri?

(c) What is the conditional probability that an employee is a male, given that he lives in Illinois?

(d) An employee selected at random turned out to be a female. What is the chance that she lives in Illinois? Solve this problem using the conditional probability format (Equation 2.7) as well as the Bayesian formula (Equation 2.11).

(e) Is sex independent of place of residence?

16. Mr. Endejohn, who owns and operates a soft-drink bottling company, purchased a sophisticated filling machine that was designed to greatly reduce variable costs. This machine required an on-site employee constantly in attendance for inspection purposes. If the machine is inspected every day for possible readjustment, there is a 95% chance that the machine will fill the exact amount. However, if the machine is not inspected for several days, the efficiency rate drops to 40%. Unfortunately, the only employee who is qualified to inspect the machine has a bad reputation for absenteeism or for improper inspection of other plant equipments. During the past several reporting periods, the inspecting foreman has failed to inspect the equipment 30% of the time. Returning from a week-long trip, Mr. Endejohn found that the machine was not working properly.

(a) What is the chance that the foreman failed to properly inspect the machine?

(b) What is the probability that the foreman did inspect the machine properly?

(c) Assuming that the machine did work properly after Mr. Endejohn's return from his trip, what is the chance that the foreman failed to inspect the machine?

17. Small-business operators usually prepare their business income taxes by themselves. The record shows when these companies file tax returns, they do so properly 70% of the time. If the companies file improperly, there is a 90% chance that the internal revenue service (IRS) will call for an audit. However, the chance is 0.4 that the IRS will request an audit even though the companies have filed properly. One small company has just received a notice from the IRS for an audit. What is the chance that the company filed its income tax properly?

SUGGESTED READING

Clark, Charles T., and Schkade, Lawrence L. *Statistical Methods for Business Decisions.* Cincinnati: South-Western Publishing, 1969, chapter 6.

Dyckman, T. R., Smidt, S., and McAdams, A. K. *Management Decision Making under Uncertainty.* New York: Macmillan, 1969, chapters 1–3.

Ewart, P. J., Ford, J. S., and Lin, C. Y. *Probability for Statistical Decision Making.* Englewood Cliffs, N.J.: Prentice-Hall, 1974, chapters 2–5.

Freund, John E., and Williams, Frank J. *Elementary Business Statistics* (2d ed.). Englewood Cliffs, N.J.: Prentice-Hall, 1972, chapter 4.

Hays, William L., and Winkler, Robert L. *Statistics: Probability, Inference, and Decision.* New York: Holt, Rinehart and Winston, 1971, chapters 1–2.

Shook, Robert C., Highland, Harold J., and Highland, Esther H. *Probability Models with Business Applications.* Homewood, Ill.: Richard D. Irwin, 1969, chapter 3.

3 / A Survey of Probability Distribution (I)—Discrete Probability Function

3.1 / Purpose

Decision makers are often interested in determining not only the probability of obtaining one successful occurrence in any given trial, but also the probability of success in many, or a series of, trials. For example, while the credit manager of a bank is interested in the likelihood that one specific new customer may default on a loan, he is probably more concerned with estimating the probability that several loan accounts among N customers may be delinquent in their payments. Similarly, a quality control manager is probably more interested in the total number of defective products occurring among the next 100 items produced rather than merely knowing the probability that the next item produced will be defective. This broader type of analysis greatly expands the ability of a decision maker to plan adequately for future operations and to solve problems.

The determination of a series of probabilities for a given number of trials requires special computational techniques. The purpose of this chapter and Chapter 4 is to discuss these methods, especially as they relate to the decision-making process under conditions of uncertainty. We will discuss the general topics of discrete probability distributions in this chapter, and in Chapter 4 we will examine several continuous probability distributions.

3.2 / Random Variable and the Probability Function

The term "random variable" is used quite often in statistics, especially in the area of probability theory. A random variable is defined as a numerical quantity whose value is determined by the outcome of a random experiment. By using this definition, we can define a probability distribution as a systematic arrangement of possible values of a random variable and the corresponding probabilities for such values. Illustrated in the following table is a simple example of a probability distribution that describes all possible outcomes when a balanced die is rolled once. In the table, the value that a given random variable (X) assumes is determined by chance, and the chance is in turn computed by an appropriate probability function or model. For example, in the preceding illustration, we can express the probability of each event as

$$f(x) = \frac{x}{6} \quad \text{for } x = 1,2,3,4,5,6.$$

Random Variable, (x_i)	Probability, $P(X = x)$
1	1/6
2	1/6
3	1/6
4	1/6
5	1/6
6	1/6
	1.0

Mathematically, a function is described as follows: Given the set D, called a "domain," and the set R, called the "range," a function is a correspondence that associates each element of D with one and only one element in R. Therefore, if $D = \{1,2,3,4,5\}$, and $R = \{1,4,9,16,25\}$, then the correspondence or relationship of D to R is shown by the arrow,

$$D = 1, 2, 3, \quad 4, \quad 5$$
$$\downarrow \downarrow \downarrow \quad \downarrow \quad \downarrow$$
$$R = 1, 4, 9, 16, 25$$

or

$$f = \{(1,1), (2,4), (3,9), (4,16), (5,25)\}.$$

A probability function is therefore a special type of function in which the elements of the range are probability values; i.e., $R = \{f(x)\}$, and the domain is the sample space. Since a random variable must assume one of its values in the given sample space, the sum of the possible values of all random variables in a given sample space will always be equal to 1.0.

3.3 / Discrete Probability Distribution

A probability function is said to be discrete if the sample space or domain has a finite or countable infinite number of outcomes. Such experiments as rolling a die, tossing a coin, or selecting a red pencil at random from a jar containing six red pencils and six black pencils are examples of discrete probability functions.

The discrete probability function must satisfy the following conditions:

(1) $\quad 0 \le f(x) \le 1.0,$

(2) $\quad \sum f(x) = 1.0,$

for all X in the sample space. The first condition is a familiar one. As discussed in Chapter 2, the probability of any given random variable must be at least zero, but not greater than 1. The second condition merely states that at least one element in the sample space must occur. In terms of the language of set theory, the probability of the universal set must be equal to 1.

From the conditions described above, the following general probability functions can be derived:

$$P(a \leq X \leq b) = \sum_{x=a}^{b} f(x), \tag{3.1}$$

$$P(a \leq X \leq a) = P(X = a)$$
$$= \sum_{x=a} f(X) = f(a). \tag{3.2}$$

EXAMPLE 3.1 / There are 55 dimes in a jar with each being minted during the 10-year period from 1967 to 1976, inclusive. The distribution of coins is shown in Table 3.1. The probability function of the coin distribution is

$$f(x) = \frac{x}{55} \qquad \text{for } x = 1, 2, 3, \ldots 10,$$

$$f(x) = 0 \qquad \text{for all others.}$$

In the preceding illustration, the probability of obtaining a coin that was minted prior to 1969 is

$$P(1 \leq X \leq 2) = \sum_{x=1}^{2} \frac{x}{55} = \frac{1}{55} + \frac{2}{55} = \frac{3}{55},$$

Table 3.1 / Distribution of Coins

Year minted	Number of coins (x)	$f(x) = x/55$
1967	1	1/55
1968	2	2/55
1969	3	3/55
1970	4	4/55
1971	5	5/55
1972	6	6/55
1973	7	7/55
1974	8	8/55
1975	9	9/55
1976	10	10/55
	55	1.00

and the probability of selecting a coin that was minted after 1970 is

$$P(5 \leq X \leq 10) = \sum_{x=5}^{10} \frac{x}{55} = 1 - \sum_{x=1}^{4} \frac{x}{55}$$

$$= 1 - \left(\frac{1}{55} + \frac{2}{55} + \frac{3}{55} + \frac{4}{55} \right)$$

$$= 1 - \frac{10}{55} = \frac{45}{55}.$$

The probability of selecting a coin that was minted in 1972 is

$$P(6 \leq X \leq 6) = P(x = 6) = \frac{6}{55},$$

and so forth.

3.4 / Expected Value of a Random Variable

Suppose you flip a balanced coin four times. How many "tails" would you expect to appear? In this situation you would expect to receive an average of two tails. Of course this does not mean that you would always obtain exactly two tails in every four trials. In some series of trials, no tails may appear, while in others you may obtain only one tail, and so forth. But, in the long run, it is highly likely that two tails would appear more often than any of the other possible combinations. In other words, the occurrence of two tails is located in the center (mean) of the distribution whose values are 0,1,2,3, and 4 tails. This approximation of a mean actually represents the expected value of the random variable.

Mathematically, the expected value (E) of a random variable (X) is computed as follows:

$$E(X) = \sum_{i=1}^{n} P(X_i) \cdot X_i \qquad (i = 1,2, \ldots n), \qquad (3.3)$$

where X_i are random variables and $P(X_i)$ are the probabilities associated with the random variables (X_i).

Also the expected value of squared random variable (X^2) is

$$E(X^2) = \sum_{i=1}^{n} P(X_i) \cdot X_i^2. \qquad (3.4)$$

EXAMPLE 3.2 / Compute the expected value of receiving a tail when a balanced coin is flipped twice.

SOLUTION: Table 3.2 summarizes the solution to this problem. The first column represents the different number of tails that could occur in the two

Table 3.2 / Expected Value

Random variable (X_i), tail	Probability, $P(X_i)$	Expected value, $E = \sum P(X_i) \cdot X_i$
2	0.25	0.50
1	0.50	0.50
0	0.25	0
		$E(X) = \overline{1.0}$ tail

trials, while the second column represents the probability associated with the possible outcomes of the two trials.

In this example, the probability of getting two tails is $0.5 \times 0.5 = 0.25$. There are two ways of getting one tail: (T,H) and (H,T). Since these two sets are mutually exclusive, the probability of getting $1T$ is

$$P(T \cap H) + P(H \cap T) = (0.5 \times 0.5) + (0.5 \times 0.5) = 0.5.$$

Finally, the probability of getting no tails is the same as the probability of getting two heads. Therefore,

$$P(\text{no } T) = P(2H) = P(H \cap H) = 0.5 \times 0.5 = 0.25.$$

Table 3.2 shows that the expected value of this problem is one tail, which means that we would expect to get one tail more often than any of the other possible results; i.e., zero, or two tails.

EXAMPLE 3.3 / The Baxter Investment Company is considering whether it should invest in a gold mining venture in Canada or in an oil drilling expedition in Alaska. A preliminary report shows that the investment in mining will generate a $1,000,000 net profit if gold is discovered. If gold is not discovered, however, the company will lose $800,000. On the other hand, the company will either make a $1,500,000 net profit on the Alaskan venture or suffer a $1,000,000 loss, depending on whether or not oil is discovered.

(1) Compute the minimum probability of success which leads the company to undertake each venture.

(2) Suppose that a geologist's report estimates that there is a 70% chance that gold will be discovered and a 50% chance that oil will be discovered. Assuming that both investments require the same amount of initial capital and that, owing to insufficient capital, only one venture can be selected, which investment should the company undertake?

SOLUTION:

(1) Let P = probability of success,

q = probability of failure,

where $P + q = 1.0$. The probability of making no difference whether the company decides to undertake the venture (i.e., a break-even probability) is computed as follows for the Baxter problem:

(a) Mining venture:

$$1,000,000P + (-800,000)q = 0,$$
$$1,000,000P + (-800,000)(1 - P) = 0;$$

$$P = \frac{800,000}{1,800,000} = \frac{8}{18},$$

and

$$q = 1 - P = 1 - \frac{8}{18} = \frac{10}{18}.$$

(b) Oil venture:

$$1,500,000P + (-1,000,000)q = 0,$$
$$1,500,000P + (-1,000,000)(1 - P) = 0;$$

$$P = \frac{1,000,000}{2,500,000} = \frac{10}{25},$$

and

$$q = 1 - P = 1 - \frac{10}{25} = \frac{15}{25}.$$

Accordingly, if the estimated probabilities of success in the mining and oil ventures are greater than 10/18 and 15/25, respectively, the company should undertake the ventures, but which of the two ventures the company should undertake depends upon the expected value.

(2) The expected value for the Baxter Company will be computed by the use of Tables 3.3 and 3.4.

Tables 3.3 and 3.4 were constructed on the basis of the following logic: The company will make a $1,000,000 profit if it invests in the mining business and subsequently discovers gold. The chance of discovering gold is, of course, 0.70. If, on the other hand, the company fails to discover any gold,

Table 3.3 / Expected Return from the Mining Venture

Event (1)	Probability, $P(X_i)$ (2)	Payoff, (X_i), $ (3)	Expected value, $ (4) = \sum(2) \cdot (3)$
Success	0.7	1,000,000	700,000
Failure	0.3	− 800,000	− 240,000
		Expected return =	460,000

Table 3.4 / Expected Return from the Oil Venture

Event (1)	Probability, $P(X_i)$ (2)	Payoff, (X_i), \$ (3)	Expected value, \$ (4) $= \sum (2) \cdot (3)$
Success	0.5	1,500,000	750,000
Failure	0.5	$-1,000,000$	$-500,000$
		Expected return $=$	250,000

it will incur an \$800,000 loss. The chance that the company will not discover any gold is $1.00 - 0.70 = 0.30$. Therefore, the expected profit of a gold mining investment is

$$\$1,000,000 \times 0.70 + (-\$800,000) \times 0.30 = \$460,000.$$

The expected profit for the oil business venture is similarly computed:

$$\$1,500,000 \times 0.5 + (-\$1,000,000) \times 0.5 = \$250,000.$$

Since the expected profit is greater for the gold mining venture, the long-run optimum decision should be to invest in gold mining.

EXAMPLE 3.4 / Assume that you will receive a reward of \$10.00, \$5.00, or \$1.00 if you obtain three tails, two tails, or one tail, respectively, in three flips of a balanced coin. There will be no prize for zero tails, and a contest entrance fee of \$3.00 is required. Would you play the game?

SOLUTION: If all other subjective factors regarding the game were put aside (e.g., whether you can financially withstand the loss), then the only real factor in determining whether or not the game should be played depends upon its expected monetary outcome. If, for example, you expect to receive a payoff of more than the \$3.00 entrance fee, you certainly would like to play the game. The complete solution to this problem is given in Table 3.5.

Table 3.5 / Sample Game

Events (1)	Payoff, (X_i), \$ (2)	Probability, $P(X_i)$ (3)	Expected value, \$ $\sum P(X_i) \cdot X_i$ (4) $= \sum (2) \cdot (3)$
3 tails	10.00	0.125	1.250
2 tails	5.00	0.375	1.875
1 tail	1.00	0.375	0.375
0 tail	0.00	0.125	0.00
		1.000*	\$3.500 $=$ Expected return

* Students are urged to verify the probabilities contained in column (3) of this table.

Since the expected monetary return is greater than $3.00, you should play the game. The reader should be cautious, however, when analyzing the results of an expected monetary return such as the one computed above. The $3.50 does not imply that you will *always* earn a net gain of $3.50 − $3.00 = $0.50 if you play. You may receive $10.00 if you obtain three tails in three trials, or you may lose $2.00 if only one tail appears. What the expected monetary value implies is that if you play the game long enough, the outcome will result, on the average, in a total return of $3.50, which means that you will make a net average gain of $0.50.

EXAMPLE 3.5 / Compute $E(X^2)$, using the data in Table 3.5.

SOLUTION: By Equation 3.4,

$$E(X^2) = \sum P(X_i) \cdot X_i^2$$
$$= 0.125(10)^2 + 0.375(5)^2 + 0.375(1)^2$$
$$+ 0.125(0)^2 = \$22.25$$

Expected value for a linear function with a constant value

When a random variable is an integral part of a function that has a constant value, the expected value of this type of function is computed in a slightly different way from that used to compute a function without a constant value. In general, if the data in a problem exhibit a linear relationship in the form of

$$y = a + bX,$$

where y = monetary value, a = constant, b = constant, and X = random variable, then the expected value of y is

$$E(y) = E(a + bX)$$
$$= E(a) + E(bX)$$
$$= E(a) + bE(X)$$
$$= a + bE(X).$$

In general:
   (1) The expected value of the constant a is the constant itself:

$$E(a) = a. \tag{3.5}$$

   (2) The expected value of a random variable with a constant value is the product of the constant and the expected value of the random variable, or

$$E(bX) = bE(X). \tag{3.6}$$

EXAMPLE 3.6 / The Kirk Moving and Storage Company is considering the purchase of a new machine that is expected to save a considerable amount of operating costs associated with the current machine. The new machine costs $10,000 and is expected to save $0.50 per hour over the current machine, or

$$S = -\$10,000 + \$0.50X,$$

where S = amount of savings and X = number of machine hours to be used.

There is a considerable amount of uncertainty concerning the exact number of hours that the company will actually use the machine. The management of the company has expressed this uncertainty in terms of the following probability distribution.

Hours of use	Probability
10,000	0.1
20,000	0.3
30,000	0.5
40,000	0.1
	1.0

Should the company buy the machine? Compute the expected savings of the new machine.

SOLUTION: Two different approaches will be discussed as a means of solving the problem:

(1) *Table approach.* Table 3.6 illustrates a table solution to the Kirk problem. Columns (1), (2), and (3) are self-explanatory. Column (4) shows the gross savings associated with each possible event. For example, if the expected use of the new machine is 10,000 hours, the total savings would be $5,000 ($0.50 × 10,000). Column (5) contains the net savings; this is simply the difference between the fixed cost (column 3) and the gross savings (column 4) for each event. For example, if the machine is used for only 10,000 hours, the company actually faces a loss of $5,000 ($-\$10,000 + \$5,000$). If the company uses the machine for 40,000 hours, it will save $10,000 ($-\$10,000 + \$20,000$). Column (6) is the expected net savings; this is the product of columns (2) and (5).

Table 3.6 / Expected Savings for the Kirk Moving Company

Projected hours of use (1)	Probability (2)	Fixed cost, $ (3)	Gross savings, $ (4) = (1) × 0.5	Net savings, $ (5) = (3) − (4)	Expected net savings, $ (6) = $\sum$(2) × (5)
10,000	0.1	−10,000	5,000	− 5,000	− 500
20,000	0.3	−10,000	10,000	0	0
30,000	0.5	−10,000	15,000	5,000	2,500
40,000	0.1	−10,000	20,000	10,000	1,000
	1.0				3,000

51

Since the aggregate expected net savings is \$3,000, the company should buy the machine.

(2) *Mathematical approach.* From a given function of

$$S = -10,000 + 0.5X,$$

we can derive the following:

$$
\begin{aligned}
E(S) &= E(-10,000 + 0.5X) = E(-10,000) + E(0.5X) \\
&= -10,000 + 0.5E(X) \\
&= -10,000 + 0.5[\sum P(X_i) \cdot X_i] \quad \text{(from Eq. 3.3)} \\
&= -10,000 + 0.5(0.1 \times 10,000 + 0.3 \\
&\quad \times 20,000 + 0.5 \times 30,000 + 0.1 \times 40,000) \\
&= -10,000 + 0.5(26,000) = -10,000 + 13,000 \\
&= 3,000.
\end{aligned}
$$

Again, since the expected savings $E(S)$ shows a positive value of \$3,000, the company should buy the machine.

3.5 / Variance (*V*) of a Random Variable

The expected value (E) is a useful descriptive measurement for explaining a special aspect of a distribution. However, the expected value alone cannot comprehensively describe the characteristics of distribution, since two symmetrical distributions with the same expected values (average) may actually be quite different in nature, as illustrated in Figure 3.1.

Figure 3.1 / Two distributions with identical mean and different variance.

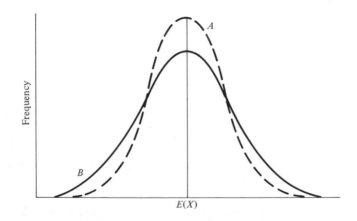

For the two curves (A and B) in Figure 3.1, suppose that the area of A is the same as that of B and that the expected values are also identical. However, it seems clear that the amount of dispersion (variance) around the mean(E) in the curve-A area seems to be larger than the dispersion (variance) in the curve-B area.

Dispersion reflects the "poorness" of a distribution and expresses how "far off" the expected value is from any given value of a random variable. Therefore, the main reasons for measuring the dispersion are to form a judgment about the *reliability* of the expected value and to determine the extent of the scatter so that steps may be taken to *control* the existing variation. For example, when a member of a track team has a record of 10 seconds for the 100-yard dash, but also has a wide variation of recorded times for the total races during the entire track season, the reliability of his performance is judged to be a rather poor one. Or, the variation of a patient's temperature is a very important indicator in the diagnosis (control) of a patient by the physician.

One way to express the dispersion σ of a given random variable is

$$\sigma = \sum (X_i - \mu) \cdot P(X_i),$$

where $P(X_i)$ is the probability of the random variable and μ is a mean of its distribution. However, the sum of the deviations from its own mean always yields zero; this is meaningless in any practical analysis. One possible way to solve this inconsistency is to square both sides of the equation; this yields

$$\sigma^2 = \sum (X_i - \mu)^2 P(X_i),$$

or, by Equation 3.3,

$$\sigma^2 = E(X_i - \mu)^2.$$

Accordingly, the variance of a random variable X is more logically expressed as

$$V(X) = \sigma^2 = E(X - \mu)^2. \tag{3.7}$$

Equation 3.7 can be further reduced to

$$V(X) = E(X^2) - [E(X)]^2. \tag{3.8}$$

The standard deviation σ of a random variable X then becomes

$$\sigma = \sqrt{V(X)} = \sqrt{E(X^2) - [E(X)]^2}. \tag{3.9}$$

For the derivation of this equation, see Note A at the end of this chapter.

EXAMPLE 3.7 / Let us compute the variance and standard deviation for the data given in Example 3.4. In that example, the expected value $E(X)$

was \$3.50. By Equations 3.8 and 3.9, and from Table 3.5,

$$V(X) = E(X^2) - [E(X)]^2,$$

where

$$\begin{aligned}
E(X^2) = \sum P(X_i) \cdot X_i{}^2 &= 0.125 \times (10)^2 \\
&+ 0.375 \times (5)^2 + 0.375 \times (1)^2 \\
&+ 0.375 \times (0)^2 = \$22.25,
\end{aligned}$$

and

$$[E(X)]^2 = (3.5)^2 = \$12.25$$

Therefore,

$$V(X) = \$22.25 - \$12.25 = \$10,$$

and the standard deviation is

$$\sigma = V(X) = \sqrt{10} = \$3.33.$$

In general, the smaller the standard deviation, the greater the chance of selecting a random variable whose value is relatively close to its expected value. Although in reality the variance is an adequate way of describing the degree of variability of the distribution, the standard deviation is more often used to express such deviation; the obvious reason for this is that the variance is a squared unit, and therefore it has a unit different from the expected value.

For example, the expected value in Example 3.7 is \$3.50, whereas the variance is expressed in squared dollar units. One way of refining this apparent discrepancy is to take the square root of the variance; this is shown in Equation 3.9.

Variance of random variable with a constant value

If a represents a constant real number and X is a random variable with an expected value of $E(X)$ and a variance of $\sigma^2(X)$, then the variance of $(X + a)$ is equal to $V(X)$, or

$$V(X + a) = V(X) = \sigma^2(X). \tag{3.10}$$

For the derivation of this equation see Note B at the end of this chapter.

In other words, the variance of a constant is zero, since the constant a takes only one value, with no deviation.

EXAMPLE 3.8 / Compute the variance of $(X + 3)$, where X is the random variable given in Table 3.5.

SOLUTION: By Equation 3.10,

$$V(X + 3) = \sigma^2(X) = E(X^2) - [E(X)]^2$$
$$= 22.25 - 12.25 = 10 \quad \text{(from Example 3.7)}.$$

It is apparent from Equation 3.10 that

$$V(aX) = a^2\sigma^2(X), \tag{3.11}$$

where X is the random variable and a is some constant with real number.

EXAMPLE 3.9 / Compute the variance of $3X$, where X is random variable given in Table 3.5.

SOLUTION: By Equation 3.11,

$$V(3X) = (3)^2\sigma^2(X)$$
$$= 9(10) \quad \text{(from Example 3.8)}$$
$$= 90.$$

3.6 / Survey of a Discrete Probability Distribution

In most practical business problems, the optimum decision under a condition of uncertainty is derived basically from the expected value; this value is in turn dependent upon the prevailing probability distribution in the problem being analyzed. Accordingly, the probability distribution is an important factor in the decision-making process under conditions of uncertainty.

Several types of probability distributions can be used; these types vary, depending primarily upon the nature of the problem being analyzed. This section examines various methods of determining discrete probability distributions and associated applications to practical business problems. The various forms of continuous probability distributions will be considered in Chapter 4.

Among the several types of discrete probability distributions, the most common types of distributions are the binomial, Poisson, hypergeometric, and Pascal distributions.

Binomial probability distribution

Assumption. The binomial probability has the following characteristics:

(1) The random variable has a discrete distribution.
(2) All events are independent.
(3) The outcome of the event is two-way mutually exclusive; e.g., yes *or* no; black *or* white, success *or* failure.
(4) The population probability remains constant until the operation is completed.

(5) The sample sizes[1] are usually small ($n \leq 30$). Many practical cases are amenable to the use of a binomial probability distribution. A few examples follow.

EXAMPLE 3.10 / Compute the probability of obtaining two tails when a balanced coin is tossed three times.

This problem contains all the prerequisites necessary for use of a binomial probability model, since (1) the random variable is discrete (i.e., there is no such probability that occurrence such as 1.6 or 0.9 tail will appear); (2) each trial is an independent event, since the outcome of the second trial has nothing to do with the outcome of the first trial; (3) the result of a trial is either tail (T) or head (H), with no other possibilities available; and (4) the population probability remains constant at $p = 0.5$. As previously illustrated in this chapter (Table 3.5), the answer to the present problem is $P(2T) = 0.375$.

In addition to the simple illustration given above, examples of other areas subject to probability analysis using binomial distribution techniques include: senatorial or presidential elections in a two-party campaign; new business ventures in which only two alternatives are feasible—success or failure as measured by certain predetermined criteria.

Binomial probability function. The binomial probability for a given problem can be computed as[2]:

$$P(X = x) = \binom{n}{x} p^x q^{n-x} = {}_nC_x p^x q^{n-x}$$

$$= \frac{n!}{(n-x)!\, x!} p^x q^{n-x}. \tag{3.12}$$

where

$n =$ sample size,

$X =$ random variable,

$x =$ any possible value that X may assume,

$P =$ population probability,

$q = 1 - p$,

$! =$ factorial.[3]

---

[1] This condition should not be overemphasized. In other words, the binomial distribution can be used when n is greater than 30. Experiments have proved, however, that when the sample size is greater than 30, both the Poisson and normal probability functions generate a close approximation to the binomial distribution. For a more detailed analysis, see Chapter 4.

[2] For a detailed explanation of the derivation of this function, see Note C at end of this chapter.

[3] Mathematically, the symbol ! is called a "factorial," signifying multiplication of a series of numbers in descending order, beginning with the number specified in the problem. Therefore, 6! is equivalent to $6 \times 5 \times 4 \times 3 \times 2 \times 1 = 720$. The only rule we have to keep in mind is that $0! = 1$. Appendix Table C provides the selective values of n and corresponding answers.

EXAMPLE 3.11 / Compute the probability of obtaining $3T$ (three tails) from four flips of a balanced coin. From the information given in the problem, $n = 4$, $p = 0.5$ (for a balanced coin), and $X = 3$.

$$P(X = 3) = \frac{4!}{(4 - 3)!\, 3!} (0.5)^3 (1 - 0.5)^{4-3}$$

$$= \frac{4 \times 3 \times 2 \times 1}{1 \times 3 \times 2 \times 1} (0.125)(0.5)$$

$$= 4 \times 0.0625 = 0.25.$$

Note that there are four possible combinations that can produce three tails from four trials. What are these four combinations?

EXAMPLE 3.12 / At a distance of 100 yards, Mr. Smith can hit the bull's-eye with a rifle shot 60% of the time. Compute the probability (1) that he will hit the bull's-eye three times in the next five shots, and (2) that he misses the bull's-eye with all five shots. The solution to this problem follows.

(1) Three bull's-eyes in five shots:

$n = 5$,

$X = 3$,

$P = 0.6$ (probability of success),

$q = 0.4$ (probability of failure),

$$P(X = 3) = \frac{5!}{(5 - 3)!\, 3!} (0.6)^3 \times (0.4)^{5-3}$$

$$= \frac{5 \times 4 \times 3 \times 2 \times 1}{2 \times 1 \times 3 \times 2 \times 1} (0.216) \times (0.16)$$

$$= 10(0.03456)$$

$$= 0.3456, \quad \text{or } 34.56\%.$$

(2) No bull's-eye in five shots:

$n = 5$,

$X = 0$,

$P = 0.6$,

$q = 0.4$,

$$P(X = 0) = \frac{5!}{(5 - 0)!\, 0!} (0.6)^0 \times (0.4)^5$$

$$= \frac{5 \times 4 \times 3 \times 2 \times 1}{5 \times 4 \times 3 \times 2 \times 1} (1) \times (0.01024)$$

$$= 0.01024, \quad \text{or } 1\%.$$

Binomial table. The computation of a binomial probability, using the probability function in Example 3.12, is relatively simple when the sample size is relatively small. This computational method also clearly provides the maximum number of different combinations that are possible in the solution. It has a severe computational disadvantage, however, in that when the sample size is relatively large (approximately 10 or greater), the computations are both very time consuming and subject to human error. Another disadvantage is that when the questions surrounding the problem are stated in cumulative terms, the derivation of the results requires not one but a series of computations. For example, if you are asked to express the probability that at least four tails will appear in seven trials of a balanced coin, it is actually necessary to compute the probability of occurrence of four tails, five tails, six tails and seven tails separately. Some of the disadvantages of this relatively time-consuming process are diminished by the use of a binomial probability table.

In order to avoid some of the obvious disadvantages of the computations of probabilities when using the binomial function, binomial probability tables have been prepared for selected sample sizes and population probabilities. There are two basic binomial tables: a table for individual terms and a table for cumulative terms. Appendix Table D.1 provides the individual probabilities, and Appendix Table D.2 contains the cumulative probabilities. The top row in each table indicates the selected population probabilities, beginning with 0.01 and continuing in irregular intervals to 0.50. The first column (n) gives the different sample sizes ranging from 1 to 30, and the second column (X) indicates the corresponding number of the random variable, with each random variable associated with a given sample size.

Individual terms

EXAMPLE 3.13 / Compute the probability of getting two tails in three flips of a balanced coin.

The solution to this or any binomial problem, using the binomial tables, requires a series of steps: First, locate column $n = 3$ in Appendix Table D.1. Second, pinpoint $X = 2$ under $n = 3$. Third, move across the line from $X = 2$, under $n = 3$, until $P = 0.5$ is reached. The value (and answer) that corresponds to these coordinates is 0.375. In other words,

$$P(X = 2|n = 3, P = 0.5) = 0.375$$

Use of the binomial table when P is greater than 0.50. As the preceding examples illustrate, Appendix Table D.1 is confined to probabilities that are less than or equal to 0.5. Assume, however, that you have an unbalanced coin. In this case, the probability of getting a tail (T) in any given trial may be 0.6, for example, and therefore the probability of obtaining a head (H) in any given trial is 0.4. Under these conditions, determine the probability of getting $2T$ in three trials.

Since the probability of obtaining a tail is 0.6, we are unable to use Table D.1, as currently constructed. The fact that we must have two tails from three trials, however, implies that we will also obtain one head. As a result, the probability of getting two tails is the same as the probability of receiving one head, or $P(2T + 1H) = P(1H + 2T)$. In order to use Table D.1, the problem is restated as follows:

$$P(2T|n = 3, P = 0.6) = P(1H|n = 3, q = 0.4)$$

or

$$P(X = 2|n = 3, P = 0.6) = P(X = 1|n = 3, q = 0.4).$$

As shown above, instead of computing the probability of receiving two tails (X), which has a given probability greater than 0.5, we could compute the probability of getting one head (X'); this latter variable has a probability of less than 0.5. The problem then becomes

$$P(X' = 1|n = 3, q = 0.4).$$

We can now return to Appendix Table D.1 as before.

Step 1. Locate $n = 3$.

Step 2. Pinpoint $X = 1$ under $n = 3$.

Step 3. Find $P = 0.4$.

Step 4. Move across the line from $X = 1$ until the line meets $P = 0.4$.

Answer: 0.432.

Therefore, $P(X = 2|n = 3, P = 0.6) = 0.432$. We can verify this answer by solving the problem with use of the binomial probability function (see Example 3.12). In this case, for two tails, we obtain the following result:

$$P(X = 2) = \frac{3!}{(3 - 2)! \, 2!} (0.6)^2(0.4)$$

$$= \frac{3 \times 2 \times 1}{1 \times 2 \times 1} (0.36)(0.4) = 3(0.144) = 0.432.$$

Similarly, for one head, the answer is the same:

$$P(X = 1) = \frac{3!}{(3 - 1)! \, 1!} (0.4)(0.6)^2$$

$$= \frac{3 \times 2 \times 1}{2 \times 1 \times 1} (0.4)(0.36) = 3(0.144) = 0.432.$$

Therefore,

$$P(2T|n = 3, P = 0.6) = P(1H|n = 3, q = 0.4),$$

or

$$P(X = 2|n = 3, P = 0.6) = P(X' = 1|n = 3, q = 0.4).$$

EXAMPLE 3.14 / Based on previous experience, baseball Team A in the Texas League usually wins 70% of the time it plays Team B in any given game. Team A is scheduled to play Team B 20 times this year. What is the chance that Team A will win 15 games over Team B this season?

SOLUTION:

$$P(X = 15|n = 20, p = 0.7) = P(X' = 5|n = 20, q = 0.3)$$
$$= 0.179, \quad \text{or almost } 18\%.$$

Cumulative terms. If a problem uses terms like "at least" or "more than" in statements such as, "What is the probability that at least one tail will appear out of three flips of a balanced coin," we have the basic condition required for the use of a cumulative probability function. The probability of getting at least one tail means in fact that we must sum the individual probabilities associated with receiving one tail, two tails, and three tails, or

$$P(\text{at least } 1T) = P(1T) + P(2T) + P(3T).$$

From Appendix Table D.1,

$$P(\text{at least } 1T) = 0.375 + 0.375 + 0.125$$
$$= 0.875.$$

The preceding solution becomes rather time consuming and complicated when the number of random variables is relatively large. Appendix Table D.2, "Binomial Probability Distribution—Cumulative Terms," was prepared to facilitate the computation when this situation exists. The table is constructed in a manner similar to Appendix Table D.1 and the same step is used to determine a desired probability.

EXAMPLE 3.15 / What is the probability of obtaining at least one tail when flipping a balanced coin three times?

SOLUTION:

$P(X \geq 1|n = 3, P = 0.5)$

Step 1. Locate $n = 3$ in Appendix Table D.2.

Step 2. Pinpoint $X = 1$ under $n = 3$.

Step 3. Find $P = 0.5$.

Step 4. Move across the line from $X = 1$ until it meets $P = 0.5$.

Answer: 0.875.

Appendix Table D.2 can also be used to solve problems that are stated in terms such as "not more than" or "at most." In order to use Appendix Table D.2 for solving this type, it is necessary to reformulate the statement of the problem.

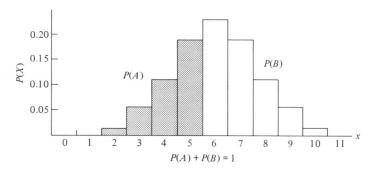

Figure 3.2 / $N = 12$, $P = 0.5$, $X \leq 5$.

EXAMPLE 3.16 / What is the probability that not more than 5 tails will appear among 12 tosses of a balanced coin?

SOLUTION: Using Appendix Table D.1, the problem may be solved in the following way:

$$P(X \leq 5|n = 12, P = 0.5) = P(X = 5) + P(X = 4) + P(X = 3)$$
$$+ P(X = 2) + P(X = 1) + P(X = 0)$$
$$= 0.193 + 0.121 + 0.054$$
$$+ 0.016 + 0.003 + 0^+ = 0.387.$$

Appendix Table D.2, however, is constructed in such a way that only questions stated in terms like "at least" or "more than" can be solved. Therefore, any question expressed in terms other than the types listed above must first be rearranged into an acceptable format before Appendix Table D.2 can be used.[4] Therefore, let $P(A)$ = probability that not more than five tails will appear, and $P(B)$ = probability that at least six tails appear among 12 tosses of a balanced coin. Then $P(A) = 1 - P(B)$, since $P(A) + P(B) = 1.0$. Accordingly,

$$P(A) = 1 - P(B) = 1 - P(X \geq 6|n = 12, P = 0.5)$$
$$= 1 - 0.613 = 0.387.$$

This solution is shown graphically in Figure 3.2, where $P(B)$ is the probability of obtaining other than $P(A)$. Since we are dealing with a discrete probability, $P(B)$ is described as the "probability that 6 or more tails will be obtained in 12 trials," or $P(B) = P(X \geq 6|n = 12, P = 0.5)$.

---

[4] As the students are aware, the distinction between the terms "at least," on the one hand, and "not more than," on the other hand, is rather subtle. The use of one or the other of these terms does, however, greatly affect the computational method and subsequent answer that is derived from the analysis of the problem. For this reason, it must be reemphasized that students should always read the problems and the instructions accompanying the statistical tables very carefully before attempting their actual use.

EXAMPLE 3.17 / Mr. Wildbrier, a junior in a business school, usually makes simple mistakes at an average of 5% of the time when he writes his examinations. What is the probability that he will make less than four simple mistakes in answering ten questions on the mid-term examination?

SOLUTION: Let $P(A) = P(X \leq 3|n = 10, P = 0.05)$, and $P(B) = 1 - P(A)$; then

$$P(A) = 1 - P(B)$$
$$= 1 - P(X \geq 4|n = 10, P = 0.05)$$
$$= 1 - 0.001 = 0.999.$$

In this problem, $P(B)$ is the chance that Mr. Wildbrier will make more than three mistakes; the probability is only 0.001. This relationship is illustrated in Figure 3.3.

EXAMPLE 3.18 / The Reflectorlite Battery Manufacturing Company claims that its batteries work 80% of the time, even in temperatures below $-30°F$. Official U.S. Weather Bureau records show that in January 1975, there were 18 days in which the temperature dropped below $-30°F$ in a sample northern city. What is the probability that the battery functioned properly for less than 11 days during the month of January?

SOLUTION: This problem can be restated in probability terms as follows:

$$P(X \leq 10|n = 18, P = 0.8).$$

Since the population probability of 0.8 is not in either Appendix Table D.1 or Table D.2, the best way to solve this problem, and still use the binomial tables, is to transform the question from one concerning the probability of

Figure 3.3 / $N = 10$, $P = 0.5$, $X \leq 3$.

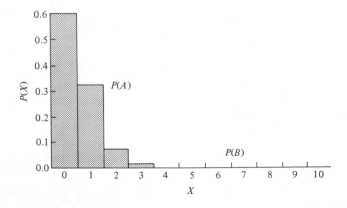

Table 3.7 / Reflectorlite Battery Company Number of Successes and Number of Failures

Number of days battery worked (X)	Number of days battery failed (X')	Total (n)
10	8	18
9	9	18
8	10	18
7	11	18
6	12	18
5	13	18
4	14	18
3	15	18
2	16	18
1	17	18
0	18	18
$P(X \leq 10\|n = 18, P = 0.8)$	$P(X' \geq 8\|n = 18, P = 0.2)$	

success (ignite) to one concerning the probability of failure (fail to ignite). One method of converting these data is illustrated in Table 3.7.

In other words, instead of solving for the probability of getting less than 11 days battery use in the total 18 days, it is easier to compute the probability of getting at least 8 days use, or

$$P(X \leq 10|n = 18, P = 0.8) = P(X' \geq 8|n = 18, q = 0.2)$$
$$= 0.901.$$

EXAMPLE 3.19 / Of the plastic bag products manufactured by the Jackson Company, 85% exceed the minimum consumer-safety requirements set by the government. What is the probability that at least 11 bags, but not more than 15 bags, meet the minimum government requirement out ot 20 randomly selected samples?

SOLUTION:

$P(11 \leq X \leq 15|n = 20, P = 0.85)$
$\quad = P(X \geq 11|n = 20, P = 0.85) - P(X \geq 16|n = 20, P = 0.85)$
$\quad = P(X' \leq 9|n = 20, q = 0.15) - P(X' \leq 4|n = 20, q = 0.15)$
$\quad = [1 - P(X' \geq 10|n = 20, q = 0.15)]$
$\quad\quad - [1 - P(X' \geq 5|n = 20, q = 0.15)]$
$\quad = (1 - 0) - (1 - 0.170)$
$\quad = 1 - 0.830$
$\quad = 0.170, \quad$ or 17%.

This problem could also be solved by using a different approach, as follows:

$$P(11 \leq X \leq 15 | n = 20, P = 0.85)$$
$$= P(X \leq 15 | n = 20, P = 0.85) - P(X \leq 10 | n = 20, P = 0.85)$$
$$= P(X' \geq 5 | n = 20, q = 0.15) - P(X' \geq 10 | n = 20, q = 0.15)$$
$$= 0.170 - 0^+ = 0.170.$$

Mean and variance of binomial probability distribution. When the random variables satisfy the binomial probability conditions, the mean(μ), variance (V), and standard deviation (S) of a binomial probability distribution can be computed by the use of the following equations[5]:

$$\mu = np, \tag{3.13}$$
$$V = npq, \tag{3.14}$$
$$S = \sqrt{V} = \sqrt{npq}. \tag{3.15}$$

EXAMPLE 3.20 / It was determined that only approximately 60% of the individuals who take the bar examination in a particular state actually pass the test. The State Board Committee which administers the bar exam received 30 requests to take the test in the current year. Compute the mean and standard deviation of the number of individuals who are expected to pass the exam.

Mean(μ) $= np = 30 \times 0.6 = 18,$

Variance(V) $= npq = 30 \times 0.6 \times 0.4 = 7.2,$

Standard deviation(σ) $= \sqrt{V} = \sqrt{7.2} = 2.7 \equiv 3.$

In other words, we should expect about 18 applicants to pass the exam, with a variation of 3 applicants.

Poisson probability distribution

When the population probability assumes a relatively small value in a binomial probability function, the resulting probability distribution for the possible values of a random variable X takes the form of a skewed distribution. In other words, when p is small and n is large, so that the value of np remains relatively constant, then the distribution of such a random variable forms a unique probability distribution called a "Poisson probability distribution."

[5] For the derivation of these functions, see Notes at the end of this chapter.

Assumption. The Poisson probability distribution should satisfy the following conditions:

(1) The random variable assumes only discrete values.
(2) All events are independent.
(3) The outcomes of all events are two-way mutually exclusive.
(4) The population probability remains stable.
(5) The sample sizes are rather large (e.g., $n \geq 100$).
(6) The size of the difference between the population probabilities of P and q should be large.

Poisson probability function. A Poisson distribution is expressed by the following Poisson probability function[6]:

$$\lim P(x|n,p) = \frac{(np)^X}{x!} e^{-np},$$

$$P(X = x|m) = \frac{(m)^X}{x!} e^{-m}, \tag{3.16}$$

where

$X = $ random variable,
$x = $ any possible value that X may assume,
$m = $ mean of the Poisson distribution; i.e., $m = np$,
$e = $ base of the natural log, or 2.718,
$n = $ sample size,
$p = $ population probability.

There are many practical cases where the use of a Poisson distribution is applicable; e.g., determining the number of accidents in the airline industry; the mortality rate for individuals insured by a life insurance company; the number of automobile accidents per 1,000 policyholders for an automobile insurance company; the incoming or outgoing telephone calls on a switchboard during a given time interval in a telephone company; and the rate of defective parts produced in a manufacturing company. All of these problems are similar in that the given events in each problem are characterized by a discrete random variable in a given time period or sample space; e.g., three accidents in a two-week time period, ten outgoing calls in every 2-minute interval, or five floor blemishes in an area of 3 square feet.

An illustration should help in understanding the nature of the Poisson probability distribution. Assume that machine A, which is owned by the General Dynamo Manufacturing Company, has a record of producing defective parts at a rate of 3% of its total production. What is the probability that exactly 5 defective parts are found in a random sample of 100 parts? This is

[6] For a mathematical derivation of this function, see Note at the end of this chapter.

essentially a Poisson probability problem because (1) the random variable is discrete, since we are stipulating exactly 5 parts, not 5.2 parts or 0.9 parts; (2) each event is independent, since the outcome of the previous selection has no influence over the outcome of the current event; (3) the outcomes of the events are mutually exclusive (i.e., we must obtain either a defective or acceptable part); (4) the population probability, $P = 0.03$, remains stable; (5) the sample of 100 is regarded as large; and (6) the difference between P and q is substantial (i.e., $P = 0.03$ versus $q = 0.97$).

Equation 3.16 is used to solve the example cited for the General Dynamo Company. Here:

$$X = 5 \text{ parts},$$
$$n = 100,$$
$$p = 0.03,$$
$$m = np = 100 \times 0.03 = 3.0.$$

Therefore,

$$P(X = 5) = \frac{(3)^5 \cdot (2.718)^{-3}}{5!} = 0.101.$$

In other words, the probability is about 10% that exactly five defective parts will be selected.

EXAMPLE 3.21 / A record of calls on a university telephone switchboard shows ten incoming calls every 5 minutes. What is the probability of receiving seven incoming calls during any 5-minute period?

SOLUTION:

$$X = 7,$$
$$m = 10,$$
$$P(X = 7) = \frac{(10)^7 \cdot (2.718)^{-10}}{7!} = 0.090, \qquad \text{or } 9\%.$$

There is a 9% chance that seven calls will be received during any 5-minute period.

Poisson table. From the two preceding illustrations we see that the Poisson probability function is a rather complicated and time-consuming method to estimate probability. To aid in the use of the Poisson function, statistical tables similar to those described to solve binomial problems were developed. Appendix Tables E.1 and E.2 contain probabilities applicable to the solution of individual and cumulative term Poisson distributions. The instructions concerning the use of these tables are similar to those for the binomial tables except that the use of the Poisson tables requires the use of the mean (m) of the Poisson distribution. This means that the mean should be computed prior to using the tables.

EXAMPLE 3.22 / The records of the Metro Taxi Cab Company show that on the average, three cabs are in the garage for service every month. Compute the probability that (1) exactly five cars, (2) at least four cars, and (3) not more than three cars will be in the garage for service in any given month.

SOLUTION:

(1) Exactly five cars:

$$P(X = 5|m = 3.0) = 0.101. \qquad \text{(from Appendix Table E.1)}$$

(2) At least four cars:

$$P(X \geq 4|m = 3.0) = 0.168. \qquad \text{(from Appendix Table E.2)}$$

(3) Not more than three cars:

$$P(X \leq 3|m = 3.0) = 1 - P(X \geq 4|m = 0.3) = 1 - 0.168$$
$$= 0.832. \qquad \text{(from Appendix Table E.2)}$$

EXAMPLE 3.23 / A quality-control chart in the Staypress Shirt Manufacturing Company shows that five shirts of each shipment of 10,000 shirts are returned to the company for exchange, owing to an imperfect finish on the cloth. The company recently forwarded two shipments to a New York customer. What is the probability that more than 10 shirts will be returned for exchange?

SOLUTION:

$$P = 0.0005 \text{ (since 5/10,000)},$$
$$n = 20,000 \text{ (2 shipments} \times 10,000 \text{ each)},$$
$$X \geq 11,$$
$$m = n \cdot p = (20,000) \cdot (0.0005) = 10.0.$$

Therefore,[7]

$$P(X \geq 11|m = 10.0) = 0.417.$$

Poisson approximation to the binomial probability distribution. The Poisson probability function can be used in lieu of a binomial probability function to estimate probabilities when the sample sizes become large ($n \geq 100$) and the population probability remains small. As we discussed in the section on the binomial probability function, two of the assumptions inherent in that function are that the sample sizes will usually be limited to a maximum of 30 and that the difference in the population probability between p (success) and q (failure) is not substantially large. As indicated above, when a given problem has a large sample size and a relatively small population probability

---

[7] If the mean of the Poisson distribution (m) is greater than 10.0, a normal probability function may be used to approximate the Poisson distribution. For further details, see Chapter 4.

(in Example 3.23, $n = 20,000$ and $p = 0.0005$), the Poisson distribution may be substituted for a binomial probability distribution, with the outcome yielding very similar results.

Tables 3.8 and 3.9 illustrate the Poisson approximation of a binomial probability when $p = 0.05$ and $n = 10$ (Table 3.8), and $p = 0.05$ and $n = 100$ (Table 3.9). As you will notice, the Poisson approximation yields a much closer approximation when the sample size increases from 10 to 100.

Table 3.8 / Poisson Approximation of a Binomial Probability when $p = 0.5$ and $n = 10$

Random variable, (X_i)	Poisson method, $P(X = x_i \vert m = 0.5)$	Binomial approximation, $P(X = x_i \vert n = 10, p = 0.05)$
0	0.607	0.599
1	0.303	0.315
2	0.076	0.075
3	0.013	0.010
4	0.001	0.001
5	0^+	0^+

Table 3.9 / Poisson Approximation of a Binomial Probability when $p = 0.05$ and $n = 100$

Random variable, (X_i)	Poisson method, $P(X = x_i \vert m = 5)$	Binomial approximation,* $P(X = x_i \vert n = 100, p = 0.05)$
0	0.007	0.006
1	0.034	0.031
2	0.084	0.081
3	0.140	0.140
4	0.176	0.178
5	0.176	0.180
6	0.146	0.150
7	0.104	0.106
8	0.065	0.065
9	0.036	0.035
10	0.018	0.017
11	0.008	0.007
12	0.003	0.003
13	0.001	0.001
14	0.001	0^+
≥ 15	0^+	0^+

* A sample size of 100 is not in Appendix Table D.1.

Mean and variance of a Poisson distribution. As previously discussed, the mean of a Poisson distribution is estimated by

$$m = np. \tag{3.17}$$

The variance(V) and the standard deviation(S) are estimated as

$$V = m, \tag{3.18}$$
$$S = \sqrt{V} = \sqrt{m}. \tag{3.19}$$

As you will notice, the variance is the same as the mean. This is not difficult to understand. Let us explain this situation by the use of an example.

As stated earlier in this section, a basic distinction between the Poisson distribution and the binomial distribution is that in the former, the population probability (p) is rather small, and therefore the value of q must be relatively large. Accordingly, if

$$n = 100,$$
$$p = 0.01,$$
$$q = 0.99,$$

then

$$m = np = 100 \times 0.01 = 1,$$
$$V = np = 100 \times 0.01 = 1,$$
$$S = \sqrt{V} = 1.0.$$

Assume, however, that the variance of the Poisson distribution is equal to npq, as is the case in the binomial distribution. Then,

$$V = npq = 100 \times 0.01 \times 0.99 = 0.99 \to 1.0.$$

In other words, as p approaches zero and therefore q approaches 1, the value of npq approaches the value of np.

EXAMPLE 3.24 / Three percent of the light bulbs produced by the Edison Light Company are known to be defective. A lot which contains 1,000 bulbs is ready to be shipped. Compute the mean and the variance of the defective bulbs using (1) a Poisson probability function, and (2) a binomial probability function.

(1) *Poisson Probability Function:*

Mean: $\quad m = np = 1,000 \times 0.03 = 30$

Variance: $\quad V = np = 1,000 \times 0.03 = 30$

Std. Deviation: $\quad S = \sqrt{V} = \sqrt{m} = \sqrt{30} = 5.48 \equiv 5.5.$

(2) *Binomial Probability Function:*

Mean: $\quad \mu = np = 1{,}000 \times 0.03 = 30$

Variance: $\quad V = npq = 1{,}000 \times 0.03 \times 0.97 = 29.1$

Std. Deviation $\quad \sigma = \sqrt{V} = \sqrt{29.1} = 5.39 \equiv 5.4.$

Once again, as illustrated above, when p is small and q approaches 1, the value of npq quickly approaches np.

Hypergeometric probability distribution

In the discussions of the binomial probability distribution, it was implicitly assumed that, after each trial, the sample selected was replaced so as to maintain a constant population density. However, if the samples are actually drawn from a finite sample space and are not replaced, the first and second conditions necessary in a binomial probability function (namely, stability of population probability and independence) are violated. As a result, the use of a binomial probability distribution is inappropriate for approximating the actual population probability distribution in these cases. When the conditions of a stability of population probability and independence are violated in the binomial probability function, a hypergeometric probability model is used to provide a closer approximation of the actual population probability.

Hypergeometric probability function. The probability of selecting x successes in n trials, given a finite population of N and K successes, is determined by

$$f(x) = P(X = x|n,K,N) = \frac{\dbinom{N-K}{n-x}\dbinom{K}{x}}{\dbinom{N}{n}}, \tag{3.20}$$

where

N = population (universe),

n = sample size, where $N \geq n$,

X = random variables

x = any possible value that X may assume

K = total number of successes in N, where $K \geq x$

EXAMPLE 3.25 / Suppose a box contains ten light bulbs, three of which are known to be defective. A random sample of three light bulbs is drawn from the box without replacement. Compute the probability that the random sample of three light bulbs contains: (1) no defective bulbs; (2) one defective bulb; (3) two defective bulbs; and (4) three defective bulbs.

SOLUTION: From the given information, where $N = 10$, $K = 3$, $n = 3$, and $X = i$, and where $i = 0,1,2,3$, we compute:

(1)

$$P(X = 0|n = 3, K = 3, N = 10) = \frac{\binom{10-3}{3-0}\binom{3}{0}}{\binom{10}{3}}$$

$$= 0.292,$$

or 29.2%.

(2)

$$P(X = 1|n = 3, K = 3, N = 10) = \frac{\binom{10-3}{3-1}\binom{3}{1}}{\binom{10}{3}}$$

$$= 0.525,$$

or 52.5%.

(3)

$$P(X = 2|n = 3, K = 3, N = 10) = \frac{\binom{10-3}{3-2}\binom{3}{2}}{\binom{10}{3}}$$

$$= 0.175,$$

or 17.5%.

(4)

$$P(X = 3|n = 3, K = 3, N = 10) = \frac{\binom{10-3}{3-3}\binom{3}{3}}{\binom{10}{3}}$$

$$= 0.008,$$

or 0.8%.

Since events (1) through (4) are both mutually exclusive and exhaustive, the sum of all these probabilities is equal to 1.0 (i.e., $0.292 + 0.525 + 0.175 + 0.008 = 1.0$).

The logic of the hypergeometric probability function is relatively easy to understand. For example, the maximum number of possible combinations that can occur when drawing three light bulbs from a population of ten is equal to the number of sets of combinations of ten light bulbs that can be taken three at a time, or

$$\binom{10}{3} = \frac{10!}{(10-3)!\,3!} = 120.$$

In order to determine the probability of drawing three acceptable bulbs (no defective parts), first determine the maximum number of ways that three good bulbs can be obtained. This is equal to

$$\binom{7}{3} = \frac{7!}{(7-3)!\,3!} = 35.$$

In other words, there are 35 different ways that three acceptable bulbs can be drawn from the available bulbs in the box. Since each of these 35 selections has an equal chance of being selected, the probability of drawing three acceptable bulbs and no defective parts is

$$P(X = 3) = \frac{\binom{7}{3}}{\binom{10}{3}} = \frac{35}{120} = 0.292.$$

The probability that two of the three samples will be acceptable (one defective) is computed in a similar manner. There are $\binom{7}{2} = 21$ ways of drawing two acceptable bulbs, and for each of these combinations there are $\binom{3}{1} = 3$ ways of combining one defective with the two acceptable bulbs. Accordingly, the total number of combinations that can be arranged to form two acceptable and one defective bulb is

$$\binom{7}{2}\binom{3}{1} = \left[\frac{7!}{(7-2)!\,2!}\right]\cdot\left[\frac{3!}{(3-1)!\,1!}\right] = 63.$$

Therefore, the probability that the three samples include two acceptable bulbs and one defective bulb is

$$P(X = 2) = \frac{\binom{7}{2}\binom{3}{1}}{\binom{10}{3}} = \frac{63}{120} = 0.525.$$

The probability of selecting one acceptable bulb (two defective items) from the three random samples is similarly computed as follows:

$$P(X = 1) = \frac{\binom{7}{1}\binom{3}{2}}{\binom{10}{3}} = \frac{21}{120} = 0.175.$$

Finally, the probability that none of the three samples is acceptable (three defective bulbs) is

$$P(X = 0) = \frac{\binom{7}{0}\binom{3}{3}}{\binom{10}{3}} = \frac{1}{120} = 0.008.$$

EXAMPLE 3.26 / The Tryharder Auto Rental Company in San Francisco has a total fleet of 20 passenger cars available for rental. Of these cars, 30% are not equipped with air conditioning. What is the probability that five of the next ten rental customers will receive cars without air conditioners?

SOLUTION: From the given information, $N = 20$, $K = 6(20 \times 0.3)$, $n = 10$, and $X = 5$. Then, by Equation 3.16,

$$P(X = 5|n = 10, K = 6, N = 20) = \frac{\binom{20 - 6}{10 - 5}\binom{6}{5}}{\binom{20}{10}}$$

$$= \frac{26,208}{184,756} = 0.142.$$

Binomial approximation of hypergeometric probability. The binomial probability distribution can often provide a good approximation of a hypergeometric probability, especially when the total population being considered is relatively large. This approximation can be used because the value of N in Equation 3.20 is relatively large, and therefore the probability that a trial will be a "success" changes only slightly for each successive trial. Table 3.10 illustrates the binomial approximation of a hypergeometric probability for the light bulb problem in Example 3.25. In this example, $N = 10$, $K = 3$, $n = 3$, and $X = x$, and where $x = 0,1,2,$ and 3.

Table 3.10 / Exact and Approximated Probability of a
Hypergeometric Probability

Number of defective bulbs, x_i (1)	Hypergeometric probability, $P(X = x_i|n,K,N)$ (2)	Binomial probability $P(X = x_i|n = 3, P = 0.3)$, (3)
0	0.292	0.343
1	0.525	0.441
2	0.175	0.189
3	0.008	0.027
	1.000	1.000

Pascal probability distribution

In a binomial probability distribution it was assumed that X successes were random variables and that n represented a fixed number of trials. In some situations, however, it is necessary to take samples continuously until a fixed number of X successes is observed. Under these circumstances, n becomes a random variable while the number of X successes remains fixed. This type of probability function is called a "Pascal probability function." Thus, the Pascal probability distribution is closely related to the binomial probability distribution. In fact, the Pascal probability distribution is often referred to as the "negative binomial probability distribution."

For example, suppose that, on the average, approval is granted to one of every five individuals who apply for a line of credit at a commercial bank. The credit manager wishes to know the probability that, on a typical day, he will have to evaluate only *five* applications in order to obtain two applicants who qualify for credit. An implied assumption is that the second approval must be made for the fifth applicant; otherwise, the credit manager would not have to evaluate five applications. Because of this assumption, this problem is different from the binomial probability distribution previously discussed. To solve it, let us designate as "A" those credit applications that are approved, and as "R" those that are rejected. Then,

Sequence	Probability
$A_1R_2R_3R_4A_5$	$(0.2)(0.8)(0.8)(0.8)(0.2) = 0.02048$
$R_1A_2R_3R_4A_5$	$(0.8)(0.2)(0.8)(0.8)(0.2) = 0.02048$
$R_1R_2A_3R_4A_5$	$(0.8)(0.8)(0.2)(0.8)(0.2) = 0.02048$
$R_1R_2R_3A_4A_5$	$(0.8)(0.8)(0.8)(0.2)(0.2) = \underline{0.02048}$
	0.08192

Since each of the five events is mutually exclusive, the probability that two credit approvals will be granted among five applications, with the second approval being the fifth applicant, is 0.08192.

Pascal probability function. The Pascal probability function can be derived directly from the binomial probability function. If n trials are required to obtain X successes, the $X - 1$ number of successes must have occurred on the $n - 1$ trials before the last trial. Furthermore, the last success must occur on the last trial; otherwise, the process of selection of success could not have been completed. Therefore, there are

$$\binom{n-1}{x-1} = {}_{(n-1)}C_{(X-1)}$$

number of combinations, and each combination has the following probabilities:

$$(\underbrace{p \cap p \cap p \cap \cdots p}_{X \text{ successes}})(\underbrace{q \cap q \cap q \cap \cdots q}_{n - X \text{ failures}}) = p^x q^{n-x}.$$

Accordingly, the probability that n trials are required to obtain X successes is determined as follows:

$$P(n|X = x, p) = \binom{n - 1}{X - 1} p^x q^{n-x} = {}_{(n-1)}C_{(x-1)} p^x q^{n-x}$$

$$= \frac{(n - 1)!}{(X - 1)!\,(n - X)!} p^x q^{n-x} \qquad (3.21)$$

Thus, by using the Pascal probability function to estimate the probability that the credit manager will approve the credit of the fifth applicant, we obtain the same result as before, or

$$P(n = 5|X = 2, p = 0.2) = \binom{5 - 1}{2 - 1}(0.2)^2(0.8)^{5-2}$$

$$= \frac{4!}{(4 - 1)!\,1!}(0.04)(0.512)$$

$$= 0.08192.$$

EXAMPLE 3.27 / Previous experience indicates that in the current production process in a factory, there is a 0.90 probability of producing an acceptable television tube. The production line will be halted and a complete inspection process initiated if three defective tubes of a sample of seven are produced before four acceptable tubes are produced. What is the probability that the production line will be halted?

SOLUTION: $n = 7$, $X = 4$, and $p = 0.9$. From Equation 3.21,

$$P(n = 7|X = 4, p = 0.9) = \binom{7 - 1}{4 - 1}(0.9)^4(0.1)^{7-4}$$

$$= \frac{6!}{(6 - 3)!\,3!}(0.6561)(0.001)$$

$$= 0.013122.$$

Accordingly, the probability that the production process is halted is 0.013122 or about 1.3%.

Binomial approximation of Pascal probability distribution. The computation of a Pascal probability, using Equation 3.21, is often time consuming. A close examination of the assumptions that are used to determine both the Pascal and binomial probabilities shows that a binomial probability function may actually be used to derive a Pascal probability.

From Equation 3.21, it follows that

$$P(n|X = x, p) = {}_{(n-1)}C_{(x-1)}P^x q^{n-x}$$
$$= [{}_{(n-1)}C_{(x-1)}P^{x-1}q^{n-x}] \cdot p \tag{3.22}$$

The first part of Equation 3.22 (inside the parentheses) is nothing other than $P(x|n,p)$, a binomial probability function, whose $n = n^* - 1$, and $X = X^* - 1$, where n^* and X^* are particular values of n (number of trials) and X (number of successes). The computation of $p(n|x,p)$ is therefore greatly simplified by the use of the binomial tables.

EXAMPLE 3.28 / On a typical day, the patient records at the St. Mary's hospital in Atlanta reveal that approximately 40% of the patients request a private room. There are usually five such rooms available on any one day. What is the probability that, for the next 15 patients admitted to the hospital, all five private rooms will be requested?

SOLUTION:

$$P(n = 15|X = 5, p = 0.4)$$
$$\text{(Pascal)}$$
$$= P(X = 4|n = 14, p = 0.4) \cdot (0.4)$$
$$\text{(Binomial)}$$
$$= 0.155 \times 0.4 = 0.0620.$$

By using Equation 3.21,

$$_{(n-1)}C_{(x-1)}p^x q^{n-x} = \frac{(n-1)!}{(n-x)!(x-1)!} p^x q^{n-x}$$

$$= \frac{(15-1)!}{(15-5)!(5-1)!}(0.4)^5(0.6)^{15-5}$$

$$= 0.06196, \quad \text{or } 0.0620.$$

Accordingly, there is a 6.2% chance that all private rooms will be requested on a typical day.

EXAMPLE 3.29 / The Safety Auto Service Company of Houston operates an emergency road service. This service covers all accidents that might occur in the metropolitan area. Company records show that 45% of all incoming calls require a tow truck. The company currently has 9 tow trucks available and 18 calls are currently waiting to be processed. What is the probability that the 18 calls will be completed before all the tow trucks are used?

SOLUTION:

$$P(n = 18|X = 9, p = 0.45) = P(X = 8|n = 17, p = 0.45) \cdot (0.45)$$
$$= 0.188 \times 0.45$$
$$= 0.0846.$$

EXAMPLE 3.30 / Solve Example 3.27, using the binomial probability approach.

SOLUTION[8]:

$$P(n = 7|X = 4, p = 0.9) = P(X = 3|n = 6, P = 0.9) \cdot (0.9)$$
$$= P(X' = 3|n = 6, q = 0.1) \cdot (0.9)$$
$$= 0.015 \times 0.9 = 0.0135.$$

3.7 / Summary

The main purpose of this chapter has been to survey various discrete probability functions (models). The following summary may be helpful in understanding the major subject areas discussed.

(1) A probability distribution is a systematic arrangement of possible values of random variables and the corresponding probabilities associated with each of these values.

(2) Probability distributions provide us with a basic tool, the expected monetary value, from which an optimum decision may be derived.

(3) An optimum decision is normally the one that provides the largest expected monetary value. Mathematically, the expected monetary value (E) is estimated as

$$E = \sum_{i=1}^{n} P(X_i) \cdot X_i.$$

(4) The expected monetary value is an average weighted by its probability of occurrence, and therefore always has a variance (V) or a standard deviation (σ). These values are computed as follows:

$$V(X) = E(X^2) - [E(X)]^2,$$
$$\sigma = \sqrt{V(X)} = \sqrt{E(X^2) - [E(X)]^2}.$$

(5) A probability distribution is basically either a discrete or a continuous distribution. The binomial, Poisson, hypergeometric, and Pascal probability functions all belong to the family of discrete probability distributions.

---

[8] The difference of answers between Example 3.27 (0.0131) and Example 3.30 (0.0135) is due to the rounding error.

(6) The binomial probability function has to satisfy the following conditions: (a) all events are both discrete in number and independent; (b) the population probability remains stable; (c) the outcome of a trial must be mutually exclusive (e.g., "yes" or "no"); and (d) the sample sizes are usually less than 30.

(7) The mean (μ) and standard deviation (σ) of a binomial distribution can be computed as

$$\mu = np,$$
$$\sigma = \sqrt{npq}.$$

(8) The Poisson distribution has to satisfy the following conditions: (a) It has discreteness and independence of events; (b) the population probability remains constant; (c) all outcomes are mutually exclusive; (d) the sample sizes are relatively large; and (e) the difference between p and q is substantial.

(9) When the selection of a random variable is made from a finite universe and without replacement, we have the additional conditions necessary for a hypergeometric function. The probability is computed by

$$P(X = x|n,K,N) = \frac{\binom{N - K}{n - x}\binom{K}{x}}{\binom{N}{n}}.$$

When the value of n is relatively large, the hypergeometric probability can be approximated by the use of a binomial probability function.

(10) If n trials are required to select a fixed number of X successes, the use of the Pascal probability is appropriate. This function is

$$P(N|X = x, p) = \binom{N - 1}{X - 1}p^x q^{n-x} = {}_{n-1}C_{x-1}p^x q^{n-x}$$
$$= \frac{(n - 1)!}{(x - 1)!\,(n - x)!}\, p^x q^{n-x}.$$

Unlike the binomial probability in which the number of successes is a random variable and the number of trials is fixed, in a Pascal probability the number of trials becomes a random variable and the number of successes is fixed. The Pascal probability can also be approximated, by the use of a binomial probability function, as

$${}_{n-1}C_{x-1}p^x q^{n-x} = \left[{}_{n-1}C_{x-1}p^{x-1}q^{n-x}\right] \cdot p.$$

EXERCISES

1. Discuss how the use of a probability distribution can be helpful in the decision-making process.

2. What is the probability of obtaining four heads in six trials of a balanced coin?

3. What is the probability of getting three tails in five trials of a coin when the population probability of receiving a tail is 0.6? Prove that the probability of getting three tails (in this unbalanced coin case) is the same as the probability of obtaining two heads in five trials.

4. There are 10 pencils in a blue box; 40% of the current supply of 10 pencils are red, and 60% of the pencils are black. What is the probability of selecting two red pencils with replacement and without replacement from six samples selected at random?

5. In the summer of 1969 when astronaut Neil Armstrong stepped onto the surface of the moon, it was determined through a nationwide sample that 95% of the adults in the United States watched this historic event on television.
   (a) What is the probability of selecting 7 adults in a group of 10 adults who watched the moon landing?
   (b) What is the chance that more than half of the people in a group of 12 watched the event?

6. Using Appendix Table D.1 and Table D.2, solve the following problems:
   (a) $P(X = 2|n = 5, p = 0.25)$.
   (b) $P(X = 2|n = 20, p = 0.45)$.
   (c) $P(X - 7|n = 10, p = 0.80)$.
   (d) $P(X = 13|n = 20, p = 0.99)$.
   (e) $P(X \le 6|n = 9, p = 0.5)$.
   (f) $P(X \ge 6|n = 13, p = 0.35)$.
   (g) $P(X \ge 5|n = 12, p = 0.80)$.
   (h) $P(5 \le X \le 11|n = 17, p = 0.40)$.
   (i) $P(X \le 10|n = 15, p = 0.95)$.
   (j) $P(3 \le X \le 9|n = 14, p = 0.65)$.

7. The Harbridge Appliance Store in Boston has a credit policy of not charging interest to customers who pay their bills within 90 days and 1.5% service charge per month on any unpaid balance over 90 days. Company records show that 40% of the customers usually pay their bills within 90 days. The store is in the process of sending bills to 12 customers. What is the probability that
   (a) none of the bills are paid within 90 days?
   (b) exactly half of the customers will pay within 90 days?
   (c) at least a third of the customers will take longer than 90 days for payment?

8. According to the production engineer of a firm that manufactures combustion engines, there will be an average of 1 defective part out of every 10 items produced when the company uses normal materials. However, the ratio of defective parts to total number of parts produced will decline to a 1 in 20 ratio if the company uses superior materials. On the other hand, the ratio of defective to total parts will increase to 1 in 5 if the company uses inferior materials. The engineer selected 15 samples on a given day and found 3 defective parts. What is the probability that the company used
   (a) superior materials?
   (b) normal materials?
   (c) inferior materials?

9. In the Stanley Cup playoffs, an Eastern Division team has a 65% chance of winning a playoff game, and a Western Division team has a 35% chance of winning a playoff game. Seven games are played, and any team that wins four games first is the winner.

(a) What is the chance that the playoffs last the full seven games and the Eastern Division team wins the title?

(b) What is the chance that the Eastern Division team defeats the Western Division team in four consecutive games?

(c) Compute the probability that the playoffs last six games and the Western Division team wins.

(d) What is the chance that the playoffs last seven games? How is this problem different from question (a)?

(e) Solve questions (a) and (c) by using the Pascal probability function.

(f) What is the chance that a Western Division team will win the playoff?

10. The quality control chart of the Battajer Auto Accessory Manufacturing Company shows that one car in every lot (100 cars) has a faulty ignition problem. The quality control manager inspected one lot at random just prior to their shipment.

(a) What is the chance that exactly three cars will have a faulty ignition problem?

(b) What is the probability that not more than two cars will have an ignition problem?

(c) What is the chance that at least one car, but not more than five cars will have an ignition problem?

11. Three airplanes arrive at Kennedy International Airport every 10-minute interval.

(a) What is the chance that the air traffic controllers guide more than five planes during a 10-minute time period?

(b) What is the probability that no airplanes will arrive for the entire 10 minutes?

12. Miss Jean Carothers, a secretary in a large corporate law office, makes an average of two typing mistakes per page. She is about to type a two-page letter for her boss.

(a) What is the chance that she will make more than five mistakes in that letter?

(b) What is the chance that she will make more than five mistakes on each page?

13. Using Appendix Table E.1 and Table E.2, solve the following problems:

(a) $P(X = 3|m = 0.4)$.   (b) $P(X = 8|m = 5.9)$.

(c) $P(X \geq 4|m = 1.2)$.   (d) $P(X \leq 6|m = 7.2)$.

(e) $P(3 \leq X \leq 9|m = 6.0)$.   (f) $P(4 \leq X \leq 6|m = 2.5)$.

14. Simon Hardware Store in Dallas maintains an average of 100 account receivables, of which 25 accounts usually have a balance in excess of $50. To verify this, the manager of the store selected a random sample of ten accounts without replacements.

(a) What is the probability that five of these accounts will have a balance in excess of $50.00?

(b) Recompute (a), using the binomial approximation, and compare the results.

(c) Using the binomial approximation, what is the probability that at least two of these accounts will have a balance in excess of $50?

15. The Minolta University in Ohio consists of six colleges, each of which sends three representatives to the University Council. Six faculty members are randomly selected to chair the Council for the year. What is the probability that a representative from the College of Business Administration will be selected as chairman?

16. The filling machine used by the Lo-Cal Soft Drink Company in Tulsa has maintained an industry standard of producing acceptable products 90% of the time. Every morning the chief quality control engineer selects ten bottles at random for examination. If four defective bottles appear before all ten samples are examined, the production line is halted and the entire production from the previous day must be

examined for potential defective bottles. What is the probability that he has to evaluate all ten samples to halt operations?

17. A stewardess on a major airline has observed that 60% of the passengers prefer seats next to a window. On a typical flight, there are only 5 window seats left in a total of 15 vacant seats. What is the probability that the last seat taken is a window seat?

SUGGESTED READING

Clark, Charles T., and Schkade, Lawrence L. *Statistical Methods for Business Decisions.* Cincinnati: South-Western Publishing, 1969, chapter 7.

Dyckman, T. R.; Smidt, S.; and McAdams, A. K. *Management Decision Making under Uncertainty.* New York: Macmillan, 1969, chapters 4 through 7.

Freund, John E., and Williams, Frank J. *Elementary Business Statistics* (2d ed.). Englewood Cliffs, N.J.: Prentice-Hall, 1972, chapter 7.

DeGroot, Morris H. *Optimal Statistical Decisions.* New York: McGraw-Hill, 1970, chapter 4.

Hays, William L., and Winkler, Robert L. *Statistics: Probability, Inference, and Decision.* New York: Holt, Rinehart and Winston, 1971, chapter 4.

Hickman, Edgar P., and Hilton, James G. *Probability and Statistical Analysis.* New York: Intext, 1971, chapters 4 through 6.

Sasaki, K. *Statistics for Modern Business Decision Making.* Belmont, Calif.: Wadsworth Publishing, 1969, chapter 3.

NOTES

A / Derivation of the Variance of Random Variable *X*

By Equation 3.7,

$$V(X) = \sigma^2 = E(X - \mu)^2$$
$$= E(X^2 - 2\mu X + \mu^2)$$
$$= E(X^2) - E(2\mu X) + E(\mu^2).$$

Since μ is a constant and $\mu = E(X)$,

$$V(X) = E(X^2) - 2\mu E(X) + E(\mu^2)$$
$$= E(X^2) - 2[E(X)]^2 + [E(X)]^2$$
$$= E(X^2) - [E(X)]^2.$$

B / Derivation of the Variance with a Constant Value

$$V(X + a) = E[(X + a) - E(X + a)]^2$$

But

$$E(X + a) = E(X) + a \qquad \text{(by Equation 3.5)}$$

Therefore,

$$V(X + a) = E[(X + a) - E(X) - a]^2$$
$$= E[X - E(X)]^2$$
$$= \sigma^2(X)$$

C / Derivation of the Binomial Probability Function

Suppose that there are three balls identified as B_1, B_2, and B_3. What is the number of different ways that two of three balls can be arranged?

If each ball is arranged with a different ball, the maximum number of different combinations are as follows: $(B_1 B_2)$, $(B_1 B_3)$, $(B_2 B_1)$, $(B_2 B_3)$, $(B_3 B_1)$, and $(B_3 B_2)$. Thus, there are six different ways that any two balls of a total of three balls can be arranged. In other words, there are $n! = 3! = 3 \times 2 \times 1 = 6$ different combinations.

In general, if n different items are selected x at a time, the maximum number of possible combinations can be expressed by

$$_nC_x = n(n - 1)(n - 2)(n - 3) \cdots (n - x + 1) = n!.$$

This equation can be greatly reduced by multiplying the right-hand side of the equation by $(n - x)! | (n - x)!$, or

$$_nC_x = \frac{n(n - 1)(n - 2)(n - 3) \cdots (n - x + 1)(n - x)!}{(n - x)!}.$$

But the numerator in the preceding equation is simply $n!$. Therefore,

$$_nC_x = \frac{n!}{(n - x)!} \tag{C.1}$$

Where there were three balls ($n = 3$) and we selected two balls at a time ($x = 2$), then

$$_3C_2 = \frac{3!}{(3 - 2)!}$$

$$= \frac{3 \times 2 \times 1}{1} = 6.$$

Equation C.1 considers the order of arrangement as an important factor. There are many cases, however, where the order of arrangement is not significant. For instance, if we want to know the number of combinations that can form three tails by tossing a balanced coin five times, we are not really concerned with how these three tails are arranged. All we really need to know is the number of different possible arrangements that can form three tails in five tosses.

If the order of arrangement is not important, Equation C.1 can be revised to simplify the computations. For example, if the order is not significant in the previous ball arrangement problem, then there are only three (instead of six) ways to arrange two balls among three samples; i.e., $(B_1 B_2)$, $(B_1 B_3)$, and $(B_2 B_3)$. However, if we evaluate these combinations more carefully, it can be seen that each combination actually has two permutations; these are:

Combination	Permutations
$B_1 B_2$	$(B_1 B_2), (B_2 B_1)$
$B_1 B_3$	$(B_1 B_3), (B_3 B_1)$
$B_2 B_3$	$(B_2 B_3), (B_3 B_2)$

If we divide the total permutations reflected in Equation C.1, or $n!|(n - x)!$, by the number of permutations per combination (i.e., $x!$), Equation C.1 becomes

$$_nC_x = \frac{n!}{(n - x)!\, x!}. \tag{C.2}$$

In our problem where $n = 3$ and $x = 2$,

$$_3C_2 = \frac{3!}{(3 - 2)!\, 2!}$$

$$= \frac{3 \times 2 \times 1}{1 \times 2 \times 1} = 3.$$

In other words, from a three-ball population, there are three ways to combine two balls. Let us consider another example. If we toss a balanced coin five times, there are ten ways to form three tails, as illustrated by

$$_5C_3 = \frac{5!}{(5 - 3)!\, 3!}$$

$$= \frac{5 \times 4 \times 3 \times 2 \times 1}{2 \times 1 \times 3 \times 2 \times 1} = 10.$$

Binomial probability function

Suppose you want to determine the probability of obtaining two tails when tossing a balanced coin three times. The first step is to determine the maximum number of different combinations that can be derived to form two tails out of three trials. The maximum number of ways in which two tails turn up in three tosses can be determined by the general rule used to compute combinations, or

$$_nC_x = \left(X = \frac{n}{x} \right) = \frac{n!}{(n-x)!\,x!},$$

where

n = sample size, or number of trials,

X = random variable,

x = any possible value that X may assume.

Therefore, in this problem,

$$\binom{3}{2} = \frac{3!}{(3-2)!\,2!} = \frac{(3)(2)(1)}{(1)(2)(1)} = 3.$$

In other words, there are three ways that two tails can be obtained in three trials. These are $(T_1 T_2 H_3)$, $(T_1 H_2 T_3)$, and $(H_1 T_2 T_3)$. The probability that any one of three combinations will actually occur, however, is determined by the rule of multiplication. For example, the probability that combination $T_1 T_2 H_3$ will occur is

$$P(T_1 \cap T_2 \cap H_3) = P(T_1) \cdot P(T_2) \cdot P(H_3)$$
$$= (0.5) \cdot (0.5) \cdot (0.5) = 0.125.$$

Since the three combinations are mutually exclusive, the probability that at least one of the combinations will occur is

$$P(T_1 \cap T_2 \cap H_3) + P(T_1 \cap H_2 \cap T_3) + P(H_1 \cap T_2 \cap T_3)$$
$$= 3(0.125)$$
$$= 0.375.$$

These results can be generalized as follows: Suppose in a two-expression test that if the probability of success is p, it follows that the probability of failure is $1 - p = q$. Therefore, the probability that X successes and $n - X$ failures will occur is computed by

$$(P \cap P \cap P \cap P \cdots \cap P)(q \cap q \cap q \cap q \cap \cdots \cap q) = p^x q^{n-x}.$$

Since there are $_nC_x$ possible combinations that can form X successes and $n - X$ failures, the probability that X successes and $n - X$ failures will actually occur is

$$P(X = x) = \binom{n}{x} p^x q^{n-x} = {}_nC_x p^x q^{n-x}$$

$$= \frac{n!}{(n-x)!\,x!} p^x q^{n-x}.$$

D.1 / Derivation of the Mean of Binomial Probability Function

Let X represent the number of successes in a given trial, with the probability of occurrence of X being P. Then the expected value of X in a single trial is

$$E(X) = 1(P) + 0(Q)$$
$$= 1(P) + 0(1 - P)$$
$$= P$$

where $q = 1 - p$.

Let us further consider that X_1 and X_2 represent the number of successes in two succeeding trials. Then the total possible number of successes in two trials can be represented as $X_1 + X_2$ with the following corresponding distribution:

Number of successes $(X_1 + X_2)$	Probability $P(X_1 + X_2)$
0	q^2
1	$2pq$
2	p^2

In this case, the expected value of $X_1 + X_2$ is

$$E(X_1 + X_2) = E(X_1) + E(X_2)$$
$$= p + p = 2p.$$

In general,

$$E(X_1 + X_2 + \cdots + X_n) = np$$

D.2 / Derivation of the Variance of Binomial Probability Function

By Equation 3.8,

$$V(X) = E(X^2) - [E(X)]^2.$$

However, as shown in Equation D.1,

$$E(X^2) = 1^2 p + 0^2 q = p \quad \text{and} \quad [E(X)]^2 = p^2.$$

Therefore,

$$V(X) = p - p^2 = p(1 - p) = pq.$$

The variance of $X_1 + X_2 + \cdots + X_n$ is

$$V(X_1 + X_2 + \cdots + X_n) = V(X_1) + \cdots + V(X_n)$$
$$= pq + \cdots + pq$$
$$= npq.$$

D.3 / Derivation of the Standard Deviation of Binomial Probability Function

By Equation 3.9,

$$S = \sqrt{V(X)} = \sqrt{npq}.$$

E / Derivation of the Poisson Distribution Function

Let

n = number of trials,

X = random variable,

x = any possible value that X may assume,

m = average number of successes per a given time span.

Then

$$p(X = x) = \frac{n!}{(n - x)!\, x!}\left(\frac{m}{n}\right)^x\left(1 - \frac{m}{n}\right)^{n-x}.$$

We must now find the limit that $p(x)$ approaches as n becomes large. This is done in the following manner:

$$\lim_{n \to \infty} p(x) = \lim_{n \to \infty} \frac{n!}{(n - x)!\, x!}\left(\frac{m}{n}\right)^x\left(1 - \frac{m}{n}\right)^{n-x}$$

$$= \lim_{n \to \infty} \frac{n!}{(n - x)!\, x!\, n^x}\, m^x\left(1 - \frac{m}{n}\right)^n\left(1 - \frac{m}{n}\right)^{-x}$$

$$= \frac{m^x}{x!}\lim_{n \to \infty}\frac{n!}{(n - x)!\, n^x}$$

$$\times \lim_{n \to \infty}\left(1 - \frac{m}{n}\right)^n \cdot \lim_{n \to \infty}\left(1 - \frac{m}{n}\right)^{-x}.$$

Each of the three equalities may be evaluated separately, as follows:

(a) $\quad \lim_{n \to \infty} \dfrac{n!}{(n - x)!\, n^x} = \lim_{n \to \infty} \dfrac{n(n - 1)(n - 2)\cdots(n - x + 1)}{n \cdot n \cdot n \cdot n \cdots n}$

$$= \frac{n}{n}\lim_{n \to \infty}\left(1 - \frac{1}{n}\right)\lim_{n \to \infty}\left(1 - \frac{2}{n}\right)\cdots\lim_{n \to \infty}\left(1 - \frac{x - 1}{n}\right)$$

$$= 1 \cdot 1 \cdots 1 = 1.$$

(b) $\quad \lim_{n \to \infty}\left(1 + \dfrac{-m}{n}\right)^n = e^{-m},$

since

$$\lim_{n \to \infty} \left(1 + \frac{1}{n}\right)^n = e$$

and

$$\lim_{n \to \infty} \left(1 + \frac{a}{n}\right)^n = e^a.$$

(c) $\lim_{n \to \infty} \left(1 - \frac{m}{n}\right)^{-x} = 1^{-x} = 1.$

Accordingly,

$$P(X) = \frac{m^x}{x!} (1)(e^{-m})(1)$$

$$= \frac{(m)^x e^{-m}}{x!}.$$

4 / Survey of Probability Distribution (II)—Continuous Probability Function

4.1 / Purpose

The probability distribution functions discussed in Chapter 3 were all based on a discrete random variable. In this chapter we discuss the various forms of the continuous probability function.

A probability function is said to be continuous if the values of the random variable are continuous at every point in the sample space or domain. In other words, any outcome (X) is allowed to occur within some interval, $a \leq X \leq b$. The continuous probability function must satisfy the following conditions:

(1) $f(X) > 0$ for all possible values of X.
(2) The probability of an interval is the same as the area cut off by that interval under the curve for the probability densities; i.e., $P(a \leq X \leq b) = \int_a^b f(x)\, dx = F(b) - F(a)$, where $b \geq a$ (see Figure 4.1).
(3) The total area under the curve, $f(X)$, is equal to 1; i.e., $\int_D f(x = 1)$, where D is domain.

From the above conditions, it follows that

$$P(a \leq X \leq b) = F(b) - F(a), \tag{4.1}$$
$$P(a \leq X \leq a) = F(a) - F(a) = 0. \tag{4.2}$$

Figure 4.1 / Continuous probability distribution.

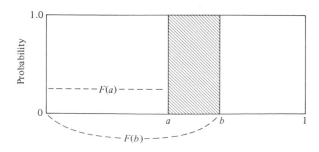

Equation 4.1 states that the probability associated with a random variable X is equal to the difference of area between the cumulative probability of b, $F(b)$, and the cumulative probability of a, $F(a)$, where $a \leq b$; this is illustrated graphically in Figure 4.1. Equation 4.2 implies that the probability associated with a particular continuous random variable is zero. The reasoning behind this assumption is as follows:

It was noted in Equation 4.1 that the probability of a continuous random variable X, $P(a \leq X \leq b)$, is determined by the area between the two cumulative values bounded by a and b on the X axis. In the case where there is only a single value, however, the difference in the area between a and b is only a line without width. Accordingly, the probability associated with such a line is extremely small, or almost zero when compared with the infinite (uncountable) number of possible values. Therefore, the probability of occurrence of a specific random variable in a continuous sample space, for all practical purpose, is zero; i.e., $P(a \leq X \leq 0) = \int_a^a f(x)\,dx = 0$.

Only the three more common types of continuous distributions are discussed in this chapter; these include the uniform, exponential, and normal probability distributions.

4.2 / Uniform Probability Distribution

Uniform probability function

The simplest form of continuous probability distribution is the uniform probability distribution. If the probability of occurrence for any event that lies between two random variables is identical, the basic condition required for a uniform distribution is met. This type of distribution is illustrated graphically in Figure 4.2, where the probability of occurrence of any random variables bounded by A and B is assumed to be uniform with a probability

Figure 4.2 / Uniform probability distribution.

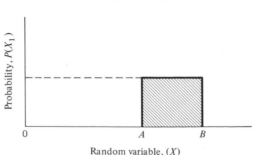

Random variable, (X)

of $P(X_1)$. In general, the uniform probability function is determined as follows:

$$P(X = x) = \frac{1}{B - A} \quad \text{if } A \leq X \leq B,$$

$$P(X) = 0 \quad \text{for all other variables.}$$

(4.3)

For proof of this function, see Note A at the end of this chapter.

EXAMPLE 4.1 / Records concerning the amount of water used by a local chemical company during the normal working hours show a uniform usage rate of 100 to 200 gallons per hour. Find the probability of the company's exact hourly demand for water.

SOLUTION: From Equation 4.3,

$$P(100 \leq X \leq 200) = \frac{1}{200 - 100} = 0.01, \quad (100 \leq X \leq 200),$$

and $P(X) = 0$ for all other X. The uniform probability for this problem is shown in Figure 4.3.

Cumulative terms

A cumulative density function can be derived from Equation 4.3. Suppose that A and B are two continuous random variables, where $A \leq X \leq B$, and A^* and B^* represent other continuous random variables within the bounds of A and B, where $A^* \leq X \leq B^*$. Accordingly, it follows that $A \leq A^*$ and $B \geq B^*$. In this case the cumulative probability function between random variables A^* and B^* is determined by

$$P(A^* \leq X \leq B^*) = P(X \leq B^*) - P(X \leq A^*),$$

and therefore

$$P(X = x) = \frac{B^* - A^*}{B - A} \quad \text{if } A \leq X \leq B \text{ and } A^* \leq X \leq B^*, \quad (4.4)$$

and $P(X) = 0$ for all other X.

Figure 4.3 / Hourly demand for water consumption.

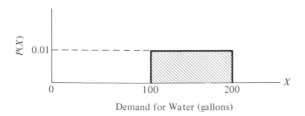

Demand for Water (gallons)

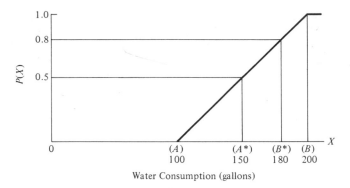

Figure 4.4 / Cumulative uniform probability for
water consumption.

EXAMPLE 4.2 / In Example 4.1, find the probability that the demand for
water is between 150 and 180 gallons per hour.

SOLUTION: By Equation 4.4, where $A = 100$, $B = 200$, $A^* = 150$, and
$B^* = 180$,

$$P(150 \leq X \leq 180) = \frac{180 - 150}{200 - 100} = 0.3.$$

In other words, since the probability of determining the exact demand for
1 gallon of water within the range of 100 to 200 is 0.01, and the distribution
is uniformly distributed, the probability of determining the demand for
30 gallons of water is $0.01 \times 30 = 0.3$.

The cumulative probability can also be computed from a cumulative
uniform probability chart. According to Figure 4.4, the cumulative probability that the demand for water is as high as 150 gallons per hour is 0.5,
and the probability that the demand will be as great as 180 gallons per
hour is 0.8. Therefore, the probability that the demand for water lies between 150 and 180 gallons per hour is $0.8 - 0.5 = 0.3$.

Mean and variance of a
uniform probability distribution

The mean and variance of a uniform probability distribution are computed
as follows[1]:

$$\text{Mean:} \quad E(X) = \frac{a + b}{2} \tag{4.5}$$

$$\text{Variance:} \quad V(X) = \frac{(b - a)^2}{12} \tag{4.6}$$

[1] For a mathematical proof of these models, see Note B at the end of this chapter.

EXAMPLE 4.3 / Compute the mean and variance for the water consumption data given in Example 4.1.

SOLUTION:

$$E(X) = \frac{100 + 200}{2} = 150 \text{ gallons per hour,}$$

$$V(X) = \frac{(200 - 100)^2}{12} = 833.3 \text{ gallons per hour.}$$

4.3 / Exponential Probability Distribution

As previously explained in Chapter 3 under the "Poisson Probability Distribution" in Section 3.6, the number of successes (X) is a random variable, and the sample space is fixed. In this type of distribution, the only parameter to be satisfied is the expected number of successes in a given space. In an exponential probability, the sample space becomes a random variable and the number of successes (X) is fixed. Since the sample space is considered to be continuous, the exponential probability distribution is normally categorized as a continuous probability distribution.

Exponential probability-density function

The exponential probability has the following probability function:

$$P(t) = \lambda e^{-\lambda t} \qquad \text{(for } 0 \leq t \leq \infty\text{)}$$

and (4.7)

$$P(t) = 0 \qquad \text{(for all other } t\text{)}$$

where

λ = average number of successes in a given space or time = np. The reciprocal of λ (or $1/\lambda$) is the average time between arrivals.[2]

t = random variable; in an exponential probability function, this equals the length of time or space between successive events.

e = 2.718.

The general shape of the exponential probability distribution is illustrated in Figure 4.5, where the probability between random variables A and B (where $B \geq A$) is shown as the shaded area.[3]

---

[2] When the Poisson distribution discussed in Chapter 3 has no direct connection with the binomial distribution, m is usually replaced by λ (lambda).

[3] The area under an exponential distribution can also be expressed as $P(a \leq t \leq b) = \int_a^b \lambda e^{-\lambda t} \, dt$.

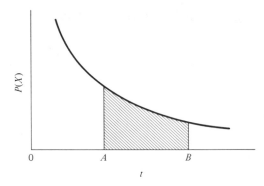

Figure 4.5 / General form of the exponential
probability density function.

Since the exponential probability function is a continuous probability
distribution, the computation of a given probability is greatly facilitated by
the use of a cumulative probability function; this function is expressed in the
following manner[4]:

$$P(t > a) = 1 - e^{-\lambda a} \tag{4.8}$$

where a is any given value of t. Equation 4.8 is based on logic similar to that
used to construct a cumulative probability distribution for a binomial probability density function in Chapter 3.

EXAMPLE 4.4 / Suppose that we have a situation where $\lambda = 1$, so that the
associated density function becomes

$$P(t) = e^{-t} \qquad (0 \leq t \leq \infty),$$

and

$$P(t) = 0 \qquad \text{(for all other } t\text{)}.$$

Compute $P(0 \leq t \leq 3)$.

SOLUTION: By Equation 4.8,

$$P(0 \leq t \leq 3) = 1 - e^{-3}$$
$$= 1 - (2.71828)^{-3}$$
$$= 1 - 0.0498 = 0.9502.$$

This result is illustrated graphically in Figure 4.6.

The computation of $e^{-\lambda t}$ can become very time consuming, however,
especially when the values of λ and t are relatively large. Appendix Table F
was prepared to facilitate this computation.

[4] For a mathematical proof of this equation, see Note C at end of this chapter.

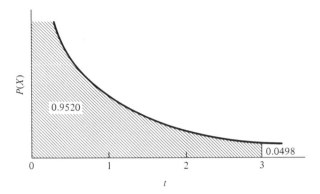

Figure 4.6 / $P(0 \leq t \leq 3)$.

In Appendix Table F, the value of λ^t is located in column (1) and the corresponding value of $e^{-\lambda t}$ is contained in column (2). Therefore, in Example 4.4 (where $\lambda^t = 3$), $e^{-3} = 0.0498$.

EXAMPLE 4.5 / Given the exponential density function

$$P(t) = 3e^{-0.5t} \qquad (0 \leq t \leq \infty),$$

and

$$P(t) = 0 \qquad \text{(for all other } t\text{)},$$

find $P(2 \leq t \leq 3)$.

SOLUTION: The solution to this problem is determined as follows:

$$P(2 \leq t \leq 3) = P(t \leq 3) - P(t \leq 2),$$
$$P(t \leq 3) = 1 - e^{-0.5 \times 3} = 1 - e^{-1.5}$$
$$= 1 - 0.2231 = 0.7769,$$
$$P(t \leq 2) = 1 - e^{-0.5 \times 2} = 1 - e^{-1.0}$$
$$= 1 - 0.3679 = 0.6321.$$

Therefore,

$$P(2 \leq t \leq 3) = 0.7769 - 0.6321 = 0.1448.$$

The result is shown in Figure 4.7.

EXAMPLE 4.6 / The motor of the average washing machine in a coin-operated launderette breaks down once every six months. The number of failures is assumed to be a Poisson distribution. Suppose that the motor has just been repaired. What is the probability that it will fail again within a year?

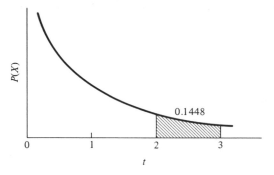

Figure 4.7 / $P(2 \leq t \leq 3)$.

SOLUTION: By Equation 4.8, where $1/\lambda = 0.5$ year (so that $\lambda = 2$) and $t = 1$,

$$P(0 \leq t \leq 1) = 1 - e^{-2 \times 1} = 1 - 0.1353 = 0.8647.$$

The probability that the motor will perform satisfactorily for at least one year, therefore, must be 0.1353, or 13.53%, as shown in Figure 4.8.

Mean and variance of exponential probability function

Let t be an exponentially distributed random variable whose density function is

$$P(t) = \lambda e^{-\lambda t} \qquad (0 \leq t \leq \infty),$$

and

$$P(t) = 0 \qquad \text{(for all other } t\text{)},$$

Figure 4.8 / Probability distribution of motor failures.

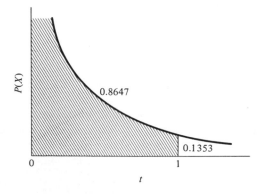

then the mean (E) and the variance (V) are determined as follows:[5]

$$E(t) = \frac{1}{\lambda} \tag{4.9}$$

$$V(t) = \frac{1}{\lambda^2} \tag{4.10}$$

EXAMPLE 4.7 / The average useful life of the transmission for a certain automobile is five years. The manufacturer will replace free of charge any transmission that fails within the first two years of operation. What is the probability that the transmission in a new car will have to be replaced free of charge?

SOLUTION: From Equation 4.9, with $E(t) = 5$,

$$E(t) = \frac{1}{\lambda} \quad \text{and} \quad \lambda = \frac{1}{E(t)}.$$

Then

$$\lambda = \frac{1}{5} = 0.2 \text{ failure per year.}$$

Accordingly, the density function of t is

$$P(t) = 0.2e^{-0.2t} \quad (0 \le t \le \infty),$$

and

$$P(t) = 0 \quad \text{(for all other } t\text{).}$$

Using Equation 4.8,

$$
\begin{aligned}
P(0 \le t \le 2) &= 1 - e^{-0.2 \times 2} \\
&= 1 - e^{-0.4} \\
&= 1 - 0.6703 \quad \text{(from Appendix Table F)} \\
&= 0.3297.
\end{aligned}
$$

In other words, the company will normally have to replace during the first two years approximately 33 transmissions of every 100 new cars that are sold.

4.4 / Normal Probability Distribution

General description

By far the most important and useful probability distribution is the normal probability distribution. This type of distribution is applicable when the

---

[5] For a mathematical proof of this function, see Note D at the end of this chapter.

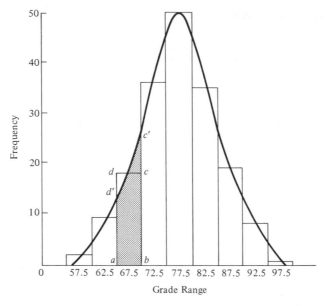

Figure 4.9 / Average grade distribution.

distribution of frequencies is symmetrical around the mean (μ). An example of the type of data used in a normal probability distribution is shown in Table 4.1, and Figure 4.9; this example concerns the grade distribution for the students in a college final examination.

Table 4.1 shows the distribution of grades for a class of 178 students in a college final examination. Figure 4.9 plots the given frequency distribution

Table 4.1 / Grade Distribution for
178 College Students

Range	Midpoint (X_i)	Frequency
55–59	57.5	2
60–64	62.5	9
65–69	67.5	18
70–74	72.5	36
75–79	77.5	50
80–84	82.5	35
85–89	87.5	19
90–94	92.5	8
95–99	97.5	1
	Total	178

in the form of a histogram. The areas contained within each rectangle represent the proportion (probability) of selecting students in each class, starting with the range 55–59 and ending with the distribution class of 95–99. If we draw a continuous line that passes through the midpoint of each rectangle in Figure 4.9, the total area under the continuous curve is approximately the same as the sum of the area of each rectangle. In other words, if the histogram is approximated by a continuous curve, the frequencies (probabilities) of each class can be computed by examining the corresponding area under the curve. Therefore, the probability of selecting those students whose grade averages range, for example, from 65 to 69, could be represented by either the rectangle *abcd* or by the shaded area bounded by *abc'd'*.

Normal probability density function

A normal probability distribution has the following probability density function:

$$P(X = x) = \frac{1}{\sigma\sqrt{2\pi}} e^{-\frac{1}{2}\left(\frac{x-\mu}{\sigma}\right)^2} \qquad (-\infty \leq X \leq \infty) \qquad (4.11)$$

where

X = random variable (not necessarily an integer),

x = *any given value that X may assume,*

μ = mean of the distribution,

σ = standard deviation,

π = constant (3.141),

e = constant (2.718).

As Equation 4.11 shows, the area under the normal curve can be determined by its mean (μ) and standard deviation (σ). Since the area (probability) under the curve depends on the mean and standard deviation, we would get a different curve and therefore a different probability for each different mean and standard deviation. For example, Figure 4.10 illustrates the frequency distributions for two different business statistics classes in a college, Class A and Class B. The means and standard deviations for these two classes are $\mu_A = 70$ with $\sigma_A = 10$; and $\mu_B = 85$ with $\sigma_B = 17$. In this case, the probability of selecting those students who scored between 65 and 70, for example, are obviously not the same for the two different distributions.

It is extremely difficult, if not impossible, to arrange all conceivable pairs of μ and σ to determine all possible probabilities that might occur under a normal curve. One possible solution to this problem is to transform the standard deviation contained in the X scale into a common unit, called the

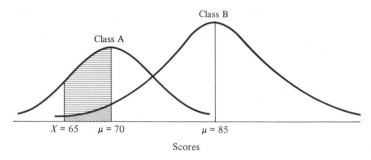

Figure 4.10 / Two normal curves with a different μ and σ.

"standard normal deviation" (Z). This can be accomplished in the following manner:

$$Z = \frac{X - \mu}{\sigma}. \qquad (4.12)$$

The Z scale is illustrated in Figure 4.11.

One distinct advantage of a conversion from the X scale to the Z scale is that, regardless of the value of μ and σ under the normal curve, the area (probability) between two points under the curve will be identical for two different curves as long as the standard normal deviate (Z) is the same in the two curves. For example, suppose that a company produces two different products, A and B. The values of mean (μ) of products A and B are 10 ounces and 20 ounces, respectively. The standard deviation (σ) of the corresponding products are 0.5 ounce and 2 ounces, respectively. In this case, it is shown in Figure 4.12 that the probability of selecting from A a product that weighs from 9.5 ounces to 10 ounces is the same as the probability of selecting a

Figure 4.11 / Probability under a normal distribution.

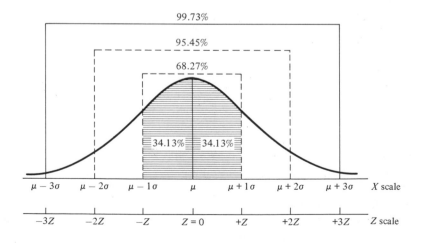

100

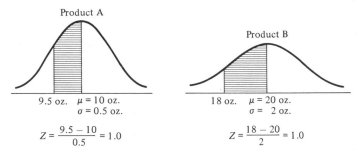

Figure 4.12 / Mean and standard deviation for two different products.

product from B which weighs from 18 ounces to 20 ounces, since the Z values for products A and B are identical. In the two charts shown in Figure 4.12, the shaded area (probability) in product A is the same as that of product B, although the values of μ and σ are different in the two distributions.

Using Equation 4.11, it can be shown that if the random variable X deviates by one standard deviation in both directions from the mean, the probability between the random variable X and μ under the curve is approximately 68.27% (34.13% on each side), as shown in Figure 4.11. If the random variable X deviates by two standard deviations from the mean (μ), the probability under the two points X and μ is approximately 95.45% (47.725% on each side), and if the random variable X deviates by three standard deviations from the mean (μ), the probability under the curve between two points is approximately 99.73% (49.865% on each side). It is clear from Figure 4.11 that the probability outside the three standard deviations from the mean is $100\% - 99.73\% = 0.27\%$ (or $50\% - 49.865\% = 0.135\%$ for each side). Table 4.2 shows these selected values of standard deviations with their corresponding Z values and probabilities.

The probabilities in Table 4.2 imply that if a random variable X is selected from any normally distributed frequency, the chances are approximately 68% that the random variable X will fall within $\mu \pm 1\sigma$; the chances are 95% that the random variable will fall within $\mu \pm 1.96\sigma$, and so on.

Table 4.2 / Probability under a Normal Curve for a Selected Number of Standard Deviations

If X deviates from μ by	Z value	Probability under the curve
$\mu \pm 1\sigma$	± 1.0	68.27% (34.13% on each side)
$\mu \pm 1.96\sigma$	± 1.96	95.0% (47.5% on each side)
$\mu \pm 2\sigma$	± 2.0	95.45% (47.725% on each side)
$\mu \pm 3\sigma$	± 3.0	99.73% (49.865% on each side)

Normal probability table. There is an infinite range of Z values, the standard normal deviation (e.g., 0.1, 0.05, 1.65, 2.91), other than those given in Table 4.2. A fairly complete list of these values, along with their corresponding probabilities, is given in Appendix Table G.

The first column in Appendix Table G represents the value of Z. The Z value at the mean (μ) is, of course, zero, with a corresponding standard deviation of 1.[6] As can be seen in Appendix Table G, the Z values start at 0.0 (i.e., the mean itself) and increase to 5.0, at which point almost 100% of the total area is contained under the curve. It must be remembered, however, that the probabilities contained in Appendix Table G are based on only one side of the curve. Therefore, if $Z = \pm 1.26$, the probability under the curve is 0.7924, or 39.62% on each side.

EXAMPLE 4.8 / The Goodstone Tire Manufacturing Company produces 2 million polybelted tires each year. Based on results from previous experiments, it was determined that each tire lasts an average of 40,000 miles, with a standard deviation of 10,000 miles. It is the company's policy that any tire which lasts less than 28,000 miles will be recalled and a refund of $5.00 per tire will be given to the customer. Assuming that the average tire has a normal distribution,

(1) What is the chance that the Goodstone Company will produce a tire which will last at least 55,000 miles? How many of these long-mileage tires will be produced each year?

(2) How many tires having an average life of 25,000 to 35,000 miles will be manufactured?

(3) What is the total expected annual cost of refunds paid for short-mileage tires?

(4) Assume that you bought a tire which is in the top 5% as far as quality is concerned. How many miles would you expect this tire to last?

SOLUTION:

$\mu = 40,000$ miles,

$\sigma = 10,000$ miles.

(1) Probability and number of tires expected to last at least 50,000 miles: Let $P(S) = P(X \geq 55,000)$, or the probability that a tire will last at least 55,000 miles. Therefore, $P(\sim S) = P(40,000 \leq X \leq 55,000)$, or the probability that a tire has an average life of 40,000 to 55,000 miles.

In order to compute $P(S)$, we first have to determine the corresponding Z value at $X = 55,000$. This is computed in the following

---

[6] For a mathematical proof of this concept, see Note E at the end of this chapter.

manner:

$$Z(X = 55,000) = \frac{55,000 - 40,000}{10,000}$$

$$= 1.5 \qquad \text{(from Equation 4.12)}.$$

From Appendix Table G:

$$P(Z = 1.5) = 0.4332$$

which is the probability of the tire lasting from 40,000 to 55,000 miles or $P(\sim S)$. Therefore,

$$P(S) = P(z \geq 1.5) = 0.5000 - 0.4332$$
$$= 0.0668, \qquad \text{or } 6.68\%$$

Accordingly,

$$2,000,000 \text{ tires} \times 6.68\% = 133,600 \text{ tires}$$

which will last at least 55,000 miles. This solution is shown graphically in Figure 4.13.

(2) Number of tires expected to last between 25,000 and 35,000 miles: In order to compute the probability of a tire life of 25,000 to 35,000 miles, we must first compute the probability of occurrence of the random variable ($X = 35,000$ miles) from the mean (μ) and then subtract this result from the probability of the occurrence of another random variable ($X = 25,000$) from the mean (μ):

$$Z(X = 35,000) = \frac{35,000 - 40,000}{10,000} = -0.5,$$

$$Z(X = 25,000) = \frac{25,000 - 40,000}{10,000} = -1.5,$$

$$P(25,000 \leq X \leq 35,000) = P(Z = -1.5) - P(Z = -0.5)$$
$$= 0.4332 - 0.1915$$
$$= 0.2417.$$

Figure 4.13 / $X \geq 55,000$.

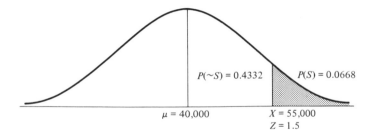

103

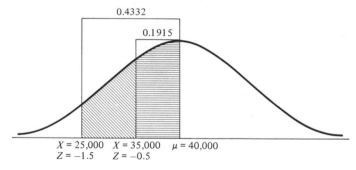

Figure 4.14 / $25,000 \leq X \leq 35,000$.

Accordingly, $2,000,000 \times 0.2417 = 483,400$ tires should average 25,000 and 35,000 miles of use. This is displayed in Figure 4.14.

(3) Similarly, the probability of a tire life of less than 28,000 miles is illustrated in Figure 4.15, where $P(S) = 0.5000 - P(\sim S)$. For $P(\sim S)$,

$$Z(X = 28,000) = \frac{28,000 - 40,000}{10,000} = -1.2,$$

$$P(\sim S) = P(Z = -1.2) = 0.3849,$$

and

$$P(S) = 0.5000 - 0.3849 = 0.1151.$$

Therefore, the probability is 0.1151, or 11.51%, that a tire which lasts less than 28,000 miles will be produced; the total cost of refunds to customers would be

$$2,000,000 \times 0.1151 \times \$5.00 = \$1,151,000.$$

Figure 4.15 / $X \leq 28,000$.

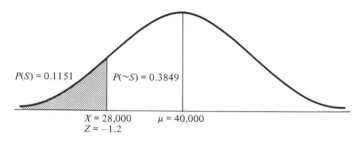

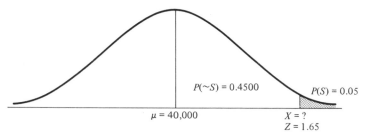

$P(\sim S) = 0.4500$

$P(S) = 0.05$

$\mu = 40,000$

$X = ?$
$Z = 1.65$

Figure 4.16 / Miles expected for top-grade tire.

(4) Number of miles expected for the top-grade tire: This problem is illustrated in Figure 4.16. From the information given in the figure,[7]

$$P(S) = 0.05 \quad \text{and} \quad P(\sim S) = 0.4500.$$

Appendix Table G shows that the Z value of 1.65 will contain 45% area under the curve. Therefore,

$$1.65 = \frac{X - 40,000}{10,000},$$

$$X = 40,000 + 10,000 \times 1.65$$
$$= 56,500 \text{ miles.}$$

In other words, you would expect the best-grade tire to last for at least 56,500 miles.

Normal frequency function

Any normal curve can be constructed by computing the ordinate of the curve in terms of the expected frequencies provided by the normal frequency function. Mathematically, the ordinate is computed as follows:

$$Y_i = \frac{Ni}{\sigma\sqrt{2\pi}} e^{\left[\left(-\frac{1}{2}\right)\left(\frac{x-u}{\sigma}\right)^2\right]} \tag{4.13}$$

[7] In Figure 4.16, the probability $P(\sim S) = 0.45$ is given, and we must compute the value of Z. Therefore, instead of proceeding from the Z value to the probability, as in the usual case, we must develop the solution to this problem in the reverse direction; i.e., from the probabili† to the Z value. There is, however, no Z value that corresponds to a probability of 0.45. Under these circumstances, the best solution is to select a Z value that comes closest to 0.4500; this value is 1.65. Incidentally, there are two Z values close to a probability of 0.45; these include $Z = 1.65$, which has a 0.4505 probability, and $Z = 1.64$, which has a 0.4494 probability. If the problem being analyzed requires a more exact answer, any number of procedures can be used to interpolate a closer approximation.

where

Y_i = ordinate (expected frequencies) at a given value of X,

N = finite population size grouped into intervals,

i = value of the class width.

Equation 4.13 can be further reduced to

$$Y_i = \frac{Ni}{\sigma}(W), \tag{4.14}$$

where

$$W = \frac{1}{\sqrt{2\pi}} e^{\left[\left(-\frac{1}{2}\right)\left(\frac{X-u}{\sigma}\right)^2\right]}$$

The maximum value of the ordinate (Y_{max}), or the maximum height, of the normal frequency curve occurs at the mean (μ) and is determined by

$$Y_{max} = \frac{Ni}{\sigma\sqrt{2\pi}} e^{\left[\left(-\frac{1}{2}\right)\left(\frac{0}{\sigma}\right)^2\right]}$$

$$= \frac{Ni}{\sigma\sqrt{2\pi}} \tag{4.15}$$

This is true because $X - u = 0$ at mean and $e^{-0} = 1$.

The computation of ordinates is greatly simplified by the use of Appendix Table H; this table was developed to provide various values of W. Accordingly, any desired frequency can be computed merely by multiplying Ni/σ by the corresponding value(s) of W in Appendix Table H.

EXAMPLE 4.9 / Let us use the grade distribution case presented in Table 4.1 to illustrate the construction of a normal frequency distribution.

Columns (1) and (2) in Table 4.3 were reproduced from Table 4.1, and the expected frequencies contained in column (6) were estimated by using Equation 4.15. Since the values of μ and σ are not given, we can use the sample mean ($\bar{X}$) and standard deviation (S) to approximate these values. The sample mean and associated standard deviation are estimated, respectively, by

$$\bar{X} = \frac{\sum f X_i}{n} \tag{4.16}$$

$$S = \sqrt{\frac{\sum f X_i^2 - (\sum f X_i)^2/n}{n-1}} \tag{4.17}$$

where, in this example (from Table 4.3), $n = 178$, $\Sigma fX_i = 13,765.0$, and $\Sigma fX_i^2 = 1,074,968.50$. Therefore,

$$\bar{X} = \frac{13,765.0}{178} = 77.33,$$

$$S = \sqrt{\frac{1,074,968.50 - (13,765.0)^2/178}{178 - 1}} = 7.7.$$

SOLUTION: In this example, the expected frequency of 20.6 contained in column (6), Table 4.3, with $X = 67.5$ was computed as follows: Using Equations 4.12 and 4.13, where $N = 178$, $i = 5$, $\bar{X} = 77.33$, and $S = 7.7$, the Z value is determined by

$$Z = \frac{67.5 - 77.3}{7.7} = -1.27.$$

Appendix Table H shows that the value of W that corresponds to $Z = -1.27$ is 0.1781. Accordingly, the expected frequency is

$$Y_i = \frac{178 \times 5}{7.7} \times 0.1781 = 20.58 = 20.6.$$

The other expected frequencies in column (6) of Table 4.3 were similarly computed. Both the observed (column 2) and expected frequencies (column 6) are depicted in Figure 4.17. In this example, the data contained in Figure 4.17 indicate that there is a close relationship between the observed and expected frequencies.

Table 4.3 / Observed and Expected Frequency for Grade Distribution

Midpoint (X) (1)	Observed frequency (f) (2)	(fX) (3)	(X²) (4)	(fX²) (5)	Expected frequency (Yᵢ) (6)
57.5	2	115.0	3,306.25	6,612.50	1.7
62.5	9	562.5	3,906.25	35,156.25	7.3
67.5	18	1,215.0	4,556.25	82,012.50	20.6
72.5	36	2,610.0	5,256.25	189,225.00	38.0
77.5	50	3,875.0	6,006.25	300,312.50	46.1
82.5	35	2,887.5	6,806.25	238,218.75	36.9
87.5	19	1,662.5	7,656.25	145,468.75	19.3
92.5	8	740.0	8,556.25	68,456.00	6.6
97.5	1	97.5	9,506.25	9,506.25	1.5
	178	13,765.0		1,074,968.50	178.0

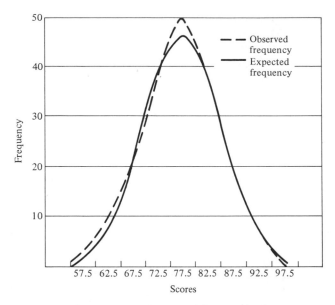

Figure 4.17 / Observed and expected frequencies for grade distribution.

Normal approximation of the binomial probability distribution

In Chapter 3 we discussed the fact that the use of the binomial probability function is generally limited to those cases where the sample size is not more than 30. When we increase the sample size, however, the shape of the associated probability curve tends to become symmetrical as shown in Figure 4.18. It is clear from the figure that as the sample size approaches 50, the curve becomes smoother until it finally becomes almost a perfectly normal distribution. Therefore, as the sample size increases, the normal probability function provides a close approximation of the binomial probability. As a result, the normal approximation of binomial probability provides a convenient tool to solve binomial probability problems when the sample sizes are relatively large.

EXAMPLE 4.10 / About 50% of the employees of the New York Power and Gas Company commute to work from New Jersey. If a sample of ten employees is selected at random, what is the probability that eight of those selected are commuters?

This problem has all the characteristics of a binomial probability, since (1) the event is an independent; (2) the probability distribution is discrete;

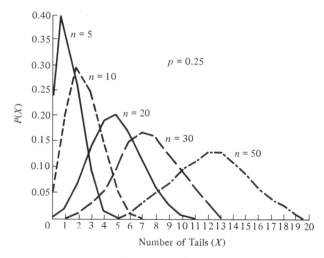

Figure 4.18 / Probability distributions showing the expected numbers of tails when the probability of receiving a given tail is 0.25.

(3) the population probability remains stable ($p = 0.5$); and (4) there are only two possible alternatives, i.e., either "commuter" or "noncommuter."

SOLUTION: Let us solve this problem in two different ways, first using the binomial probability function, and then using the normal probability density function.

Binomial probability approach: $P(X = 8/n = 10, p = 0.5) = 0.044$ (from Appendix Table D.1). That is, the probability that 8 of 10 employees, selected at random, are commuters is 4.4%. This result is shown as rectangle *abcd* in Figure 4.19.

Normal probability approach: In order to use the normal probability method, the mean and standard deviation must be estimated. From Equations 3.13 and 3.15,

$$\text{Mean } (\mu) = np = 10 \times 0.5 = 5.$$
$$\text{Standard deviation } (\sigma) = \sqrt{npq} = \sqrt{10 \times 0.5 \times 0.5}$$
$$= \sqrt{2.5} = 1.58.$$

Accordingly, the probability that exactly 8 employees are commuters is computed by subtracting the probability that 7.5 or less employees are commuters from the probability that 8.5 or less employees are commuters, or

109

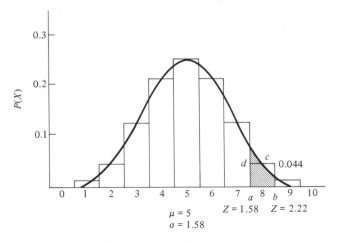

Figure 4.19 / Normal approximation of binomial probabilities.

$$P(X = 8|n = 10, p = 0.5) = P\left(Z = \frac{8.5 - 5}{1.58}\right) - P\left(Z = \frac{7.5 - 5}{1.58}\right)$$

$$= P(Z = 2.22) - P(Z = 1.58)$$

$$= 0.4861 - 0.4429 \qquad \text{(from Appendix Table G)}$$

$$= 0.0432.$$

The result is shown graphically by the shaded area in Figure 4.19. This probability of 0.0432 compares favorably with that of 0.044 computed by the binomial approach.

EXAMPLE 4.11 / About 70% of the freshmen students at Harvard University live outside the immediate Cambridge area. Assuming that next year there will be 1,000 freshmen, find the probability that (1) at least 750 students will be out-of-town residents, and (2) not more than 680 students will be out-of-town residents.

SOLUTION: From Equations 3.13 and 3.15,

$$\mu = np = 1,000 \times 0.7 = 700$$
$$\sigma = \sqrt{npq} = \sqrt{1,000 \times 0.7 \times 0.3}$$
$$= \sqrt{210} = 14.5.$$

(1) The probability that at least 750 freshmen are nonresidents is computed as follows: Let $P(S) =$ probability that at least 750 freshmen are

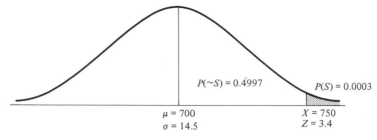

Figure 4.20 / $P(X \geq 750/n = 1,000, P = 0.7)$.

nonresidents and $P(\sim S) = 1 - P(S)$. Then

$$P(S) = 0.5 - P(\sim S)$$

$$= 0.5000 - P\left(Z = \frac{750 - 700}{14.5}\right)$$

$$= 0.5000 - P(Z = 3.4)$$

$$= 0.5000 - 0.4997 \quad \text{(from Appendix Table G)}$$

$$= 0.0003, \quad \text{or } 0.03\%.$$

This result is shown in Figure 4.20.

(2) Similarly, the probability that 680 students or less are nonresidents is computed as follows:

$$P(S) = 0.5000 - P(\sim S)$$

$$= 0.5000 - P\left(Z = \frac{680 - 700}{14.5}\right)$$

$$= 0.5000 - P(Z = -1.38)$$

$$= 0.5000 - 0.4162$$

$$= 0.0838, \quad \text{or } 8.38\%.$$

Figure 4.21 illustrates this result.

Figure 4.21 / $P(X \leq 680/n = 1.000, P = 0.7)$.

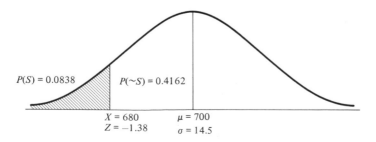

Normal approximation of the Poisson probability

If the mean of the Poisson probability distribution (m) is greater than 10.0, the normal probability density function also provides a close approximation of the Poisson probability.

In a Poisson probability function, the average number of successes for a given time interval or space is expressed as m; this is analogous to the variable μ in a normal probability distribution. Furthermore, the standard deviation of a Poisson distribution, expressed as S, is analogous to σ in a normal probability distribution. Accordingly, a normal approximation of a Poisson probability can be solved by

$$Z = \frac{m - \mu}{S}$$

EXAMPLE 4.12 / An average of three customers arrive every 5 minutes at the checkout counter in a grocery store. What is the probability that exactly five customers will arrive during a 5-minute interval?

(1) Poisson Probability Approach:

$$P(X = 5 | m = 3) = 0.101. \quad \text{(from Appendix Table E.1)}.$$

(2) Normal Probability Approach:

$$m = 3,$$
$$X = 5,$$
$$S = \sqrt{m} = \sqrt{3} = 1.73.$$

Therefore, the probability that exactly five customers will arrive is

$$P(X = 5 | m = 3) = P\left(Z = \frac{5.5 - 3}{1.73}\right) - P\left(\frac{4.5 - 3}{1.73}\right)$$
$$= P(Z = 1.44) - P(Z = 0.86)$$
$$= 0.4251 - 0.3051 \quad \text{(from Appendix Table G)}$$
$$= 0.1200.$$

In this case, there is only approximately a 2% difference between the results obtained by using the normal probability and those obtained by the Poisson probability approach. The comparison is illustrated in Figure 4.22.

4.5 / Summary

In this chapter, we are concerned with an explanation of the various forms of continuous probability distributions. This type of distribution arises when the random variable (X) in question can take the value of any number,

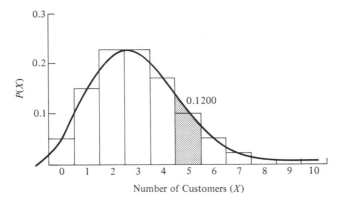

Figure 4.22 / Normal approximation to Poisson probabilities.

including an integer within a specified interval. Following is a summary of the contents of this chapter.

(1) If the probability of the occurrence of a random variable between two specified limits is identical, it is called a "uniform probability" and has the following probability function:

$$P(X = x) = \frac{1}{B - A} \qquad (\text{if } A \leq X \leq B)$$

$$P(X) = 0 \qquad (\text{for all other } X).$$

(2) When the sample size of a Poisson probability distribution becomes a random variable and the number of successes (X) is fixed, we have the necessary conditions for an exponential probability distribution. The density function associated with an exponential probability is denoted as

$$P(t) = \lambda e^{-\lambda t} \qquad (\text{for } 0 \leq t \leq \infty)$$
$$P(t) = 0 \qquad (\text{for all other } t).$$

(3) By far the most important and useful probability distribution is the normal probability distribution. This type of distribution can be used when the distribution of observed frequencies is symmetrical around the mean. The density function of a normal probability distribution is expressed as

$$P(X) = \frac{1}{\sigma\sqrt{2\pi}} e^{\left[\left(-\frac{1}{2}\right)\left(\frac{x - \mu}{\sigma}\right)^2\right]}$$

(4) A normal curve can be estimated by determining the ordinates at the given value of X. The normal frequency is determined by

$$Y_i = \frac{Ni}{\sigma\sqrt{2\pi}} e^{\left[\left(-\frac{1}{2}\right)\left(\frac{x - \mu}{\sigma}\right)^2\right]}$$

(5) Under certain conditions a normal density function can be used to approximate either the binomial (when $n > 30$) or the Poisson probability functions (when $m > 10$).

EXERCISES

1. Discuss the advantages and disadvantages of using continuous distribution over discrete distribution.

2. Explain the differences and similarities between
   (a) a Poisson and an exponential probability distribution.
   (b) a normal and a binomial probability distribution.

3. At the cafeteria of a university, approximately 50 to 150 students arrive at the check counter between 7 and 8 A.M., and approximately 100 to 250 arrive between 5 and 6 P.M. Construct the probability density function and compute the probability that at least 120 students arrive at the check counter each morning and evening. (Assume uniform probability function.)

4. Given a value of $\lambda = 2$, solve the following exponential probabilities:
   (a) $P(t \leq 1.5)$.   (b) $P(t \leq 3.2)$.
   (c) $P(2.1 \leq t \leq 3.5)$.   (d) $P(t \leq 2.0)$.

5. The arrival of airplanes at the Kennedy International Airport is assumed to be a Poisson probability distribution, with six planes arriving every hour. Suppose that an airplane just arrived. Let t be the random variable associated with the elapsed time between any two successive arrivals.
   (a) Derive the probability density function.
   (b) What is the probability that another plane will arrive within the next half-hour?
   (c) What is the probability that another airplane will arrive between the next half-hour and 45 minutes?

6. The International Office Machine Company manufactures low-price copy machines. The motor in the most popular model has a maximum life span of 4 years. If the motor breaks down within one year, the company will replace the entire copy machine free of charge, and if the motor breaks down within two years of purchase, the company will repair the motor free of charge. According to the current timetable, it takes an average of 2 hours to repair the motor. The company just sold 400 copy machines to a group of local banks.
   (a) Compute the probability and number of copy machines that will be replaced free of charge.
   (b) What is the probability that a copy machine will be repaired free of charge? How many man-hours will be required for such repair in the next two years?

7. A supplier of the Sicox Sign Manufacturing Company in Atlantic City visits the company an average of once every two months for business consultations. The number of visits is assumed to be a Poisson probability distribution. Suppose that the salesman has just completed a visit.
   (a) Express the above problem in terms of a probability density function.
   (b) What is the probability that the salesman will visit the company within the next two months?
   (c) What is the probability that the company will need the salesman within the next 15 days?

(d) What is the probability that the company will need the salesman between the next 15 days and 2 months?

(e) What is the probability that the company will not need the salesman for three months?

8. The Edison Light Bulb Company claims that its medium-priced light bulb product normally lasts 240 hours, with a standard deviation of 60 hours. The company makes two million such light bulbs per year.

(a) What is the chance that a light bulb, chosen at random, will last at least 300 hours?

(b) What is the chance that any bulb lasts less than 200 hours?

(c) How many of the light bulbs produced each year will last at least 350 hours?

9. The Wilson Drug Store sells an average 250 copies of *Outdoor Life* magazine per week. There is currently a 50-copy variation on either side of the mean distribution, and the number of copies sold approximates a normal distribution.

(a) What is the probability that at least 300 copies of *Outdoor Life* will be sold in a given week?

(b) How many copies should be ordered to ensure that magazines are in stock 95% of the time?

(c) Is there any chance that the drugstore will sell less than 100 copies per week?

10. Using Appendix Table G, solve the following problems.

(a) $P(X \geq 25 | \mu = 20, \sigma = 5)$. (b) $P(40 \leq X \leq 50 | \mu = 30, \sigma = 10)$.
(c) $P(X \leq 110 | \mu = 75, \sigma = 20)$. (d) $P(X \leq 85 | \mu = 150, \sigma = 35)$.
(e) $P(X \geq 250 | \mu = 300, \sigma = 60)$. (f) $P(70 \leq X \leq 120 | \mu = 150, \sigma = 70)$.
(g) $P(20 \leq X \leq 110 | \mu = 60, \sigma = 25)$. (h) $P(X \geq 16 | \mu = 25, \sigma = 4)$.

11. The Radiant Beauty Cosmetic Company sells its products only through individual salesmen. It has been shown that if a salesman visits the customer and explains the uniqueness of the various cosmetic products, the chances are 70% that the customer will buy at least one Radiant Beauty product. Each salesman makes an average of 300 customer calls per month.

(a) Compute the probability that at least 250 customers would buy some cosmetic products during any given month from each salesman.

(b) Studies have shown that, in order to stay in business, any salesman should convince at least 200 new customers per month to buy Radiant Beauty products. What is the chance that the average salesman will stay in business?

(c) If any salesman fails to convince at least 180 customers per month, he must seek new employment because he cannot recover his daily expenses. What is the chance that a dealer will not make this 180 quota and therefore quit his job?

12. An average of five customers arrive at the checkout counter in a grocery store every 5 minutes. The manager decided to open an additional counter to relieve the congestion and attract more customers. He intended to open the counter if more than seven customers arrived in an average 5-minute period. On the other hand, he also decided to close the existing counter if less than three customers arrived in the average 5-minute period. (Use normal approximation of Poisson probability.)

(a) What is the probability that the manager will open an additional counter?

(b) What is the probability that he will close the existing counter?

(c) Compute the (a) and (b) by using the Poisson tables in the Appendix and compare.

13. The state of Missouri badly needs medical doctors who are willing to work in rural and urban poverty areas. The state needs at least 70 more doctors next year for these areas. According to the record of the State Board Director and Missouri Medical Society, on the average, 60% of the doctors who applied for the Board test passed, and 70% of those who passed the Board examination remained in Missouri to practice medicine. Only 30% of those who remained in the state were willing, however, to work in rural or poverty areas. If the state fails to retain at least 55 doctors for rural and poverty areas, it will lose millions of dollars in aid from the federal government. Five hundred applicants for the Board examinations have been received thus far this year.
    (a) What is the chance that at least 300 applicants will pass the examination?
    (b) What is the probability that not more than 280 of the applicants will pass the test?
    (c) What is the probability that at least 230 doctors who passed the exam will remain and practice medicine in the state of Missouri?
    (d) What is the chance that the state of Missouri will retain at least 80 doctors next year from those who pass the Board exam?
    (e) Is there any chance that the state of Missouri may lose its current federal grant?

SUGGESTED READING

Clark, Charles T., and Schkade, Lawrence L. *Statistical Methods for Business Decisions.* Cincinnati: South-Western Publishing, 1969, chapter 8.

Dyckman, T. R., Smidt, S., and McAdams, A. K. *Management Decision Making under Uncertainty.* New York: Macmillan, 1969, chapter 8.

Freund, John E., and Williams, Frank J. *Elementary Business Statistics* (2d ed.). Englewood Cliffs, N.J.: Prentice-Hall, 1972, chapter 7.

DeGroot, Morris H. *Optimal Statistical Decisions.* New York: McGraw-Hill, 1970, chapter 4.

Hays, William L., and Winkler, Robert L. *Statistics: Probability, Inference, and Decision.* New York: Holt, Rinehart and Winston, 1971, chapter 4.

Hickman, Edgar P., and Hilton, James G. *Probability and Statistical Analysis.* New York: Intext, 1971, chapter 7.

Sasaki, K. *Statistics for Modern Business Decision Making.* Belmont, Calif.: Wadsworth Publishing, 1969, chapter 4.

NOTES

A / Derivation of Uniform Probability Function

Since the distribution is uniform, the density function must be constant (k) over the given interval a and b:

$$P(X) = \begin{cases} k & \text{(if } a \leq X \leq b) \\ 0 & \text{(for all other values of } X). \end{cases}$$

It is also true that

$$\int_{-\infty}^{\infty} P(X)\, dx = 1.0,$$

since the sum of all probabilities is equal to 1. Therefore

$$\int_{-\infty}^{\infty} P(X)\, dx = \int_{a}^{b} k\, dx = k \int_{a}^{b} dx = k(b - a),$$

$$k(b - a) = 1$$

$$k = \frac{1}{b - a}.$$

B / Derivation of Mean and Variance of Exponential Probability Function

For the mean,

$$E(X) = \int_{-\infty}^{\infty} X \cdot P(X)\, dx$$

$$= \int_{a}^{b} \frac{X}{b - a}\, dx = \frac{1}{b - a} \int_{a}^{b} X \cdot dx$$

$$= \frac{1}{b - a} \int_{a}^{b} \left[\frac{X^2}{2}\right] = \frac{1}{b - a}\left(\frac{b^2 - a^2}{2}\right)$$

$$= \frac{b^2 - a^2}{2(b - a)} = \frac{(b - a)(b + a)}{2(b - a)}$$

$$= \frac{b + a}{2}.$$

For the variance,

$$V(X) = E(X)^2 - [E(X)]^2 \qquad \text{(from Equation 3.8)}$$

$$E(X^2) = \int_{-\infty}^{\infty} X^2 \cdot P(X)\, dx = \int_{a}^{b} \frac{X^2}{b - a}\, dx$$

$$= \frac{1}{b - a} \int_{a}^{b} X^2\, dx$$

$$= \frac{b^3 - a^3}{3(b - a)} = \frac{(b - a)(b^2 + ab + a^2)}{3(b - a)}$$

$$= \frac{b^2 + ab + a^2}{3}$$

Therefore,

$$V(X) = E(X^2) - [E(X)]^2$$

$$= \frac{b^2 + ab + a^2}{3} - \left(\frac{b + a}{2}\right)^2$$

$$= \frac{4b^2 + 4ab + 4a^2 - 3b^2 - 6ab - 3a^2}{12}$$

$$= \frac{b^2 - 2ab + a^2}{12}$$

$$= \frac{(b - a)^2}{12}.$$

C / Derivation of Exponential Probability Function

Let $P(0 \le t \le k)$ be the exponentially distributed random variable. Then the area between 0 and k is given by

$$P(0 \le t \le k) = \int_0^k \lambda e^{-\lambda t}\, dt$$

$$= 1 - e^{-\lambda t}$$

$$= 1 - e^{\lambda t}.$$

D / Derivation of Mean and Variance of Exponential Probability Function

For mean,

$$E(t) = \int_0^\infty tme^{-mt}\, dt$$

$$= -te^{-mt} - \int_0^\infty - e^{-mt}\, dt$$

$$= \left[-te^{-mt} - \frac{e^{-mt}}{m}\right]_0^\infty$$

$$= \left[(0) - \left(\frac{-1}{m}\right)\right] = \frac{1}{m} \equiv \frac{1}{\lambda}.$$

For variance,

$$V(t) = E\left[\left(t - \frac{1}{m}\right)^2\right] = E(t^2) - \left(\frac{1}{m}\right)^2.$$

But,

$$E(t)^2 = \int_0^\infty t^2 m e^{-mt}\, dt = m \int_0^\infty t^2 e^{-mt}\, dt$$

$$= m \left[\left(\frac{t^2 e^{-mt}}{-m} \right) \Big|_0^\infty + \int_0^\infty \frac{2}{m} t e^{-mt}\, dt \right]$$

$$= m \left[0 + \frac{2}{m} \int_0^\infty t e^{-mt}\, dt \right]$$

$$= 2 \int_0^\infty t e^{-mt}\, dt = 2 \left(\frac{1}{m^2} \right) = 2 \left(\frac{1}{m} \right)^2.$$

Therefore,

$$V(t) = 2 \left(\frac{1}{m} \right)^2 - \left(\frac{1}{m} \right)^2 = \frac{1}{m^2} = \frac{1}{\lambda^2}.$$

E / Derivation of Mean and Variance of Standard Normal Deviate (Z)

For mean,

From $Z = \dfrac{X - \mu}{\sigma} = \dfrac{1}{\sigma}(X - \mu)$,

$$E(Z) = \frac{1}{\sigma}\left[E(X - \mu) \right] = \frac{1}{\sigma}\left[E(X) - \mu \right]$$

$$= \frac{1}{\sigma}(\mu - \mu) = 0.$$

For variance,

$$V(Z) = V\left[\frac{1}{\sigma}(X - \mu) \right] = \frac{1}{\sigma^2}\left[V(X - \mu) \right]$$

$$= \frac{1}{\sigma^2}\left[V(X) - V(\mu) \right] = \frac{1}{\sigma^2}\left[V(X) - 0 \right]$$

$$= \frac{1}{\sigma^2}\left[V(X) \right] = \frac{\sigma^2}{\sigma^2} = 1.$$

Decision-Making Process under Conditions of Uncertainty

5 / Decision-Making Process with and without Prior Information under a Discrete Random Variable Assumption

5.1 / Purpose

In Chapters 2 through 4, we discussed the basic concepts of probability theory and associated distributions. In this chapter (and the subsequent four chapters), we discuss the decision-making process under conditions of uncertainty. In the following discussions, it will become readily apparent that the concepts of probability theory and distributions play a vital role in deriving an optimum decision under these circumstances.

We have already mentioned that uncertainty concerning future events is the biggest obstacle confronting an individual in a decision-making role. In fact, the correct diagnosis of the states of nature of a problem concerning future events is often the most important step toward making an optimum decision under conditions of uncertainty.

The main purpose of this chapter is to discuss in detail the decision theory that is based on prior information with a discrete random variable. Accordingly, the following assumptions will be made in this chapter: (1) The state of nature in the problem being analyzed has a discrete random variable; (2) the terminal decision is derived from prior information only (i.e., without the benefit of new or additional sampling information); and (3) the utility of money to the decision maker is a linear function of the amount of money available. These assumptions will be gradually relaxed in subsequent chapters. At the end of this chapter we briefly discuss the decision-making process under conditions of uncertainty with no available prior information.

5.2 / Decision-Making Process under Conditions of Uncertainty

In this section we consider several cases that involve the three assumptions listed above.

Case with binomial assumptions

Consider the following data.

> EXAMPLE 5.1 / At a distance of 100 yards, Mr. Winchester hits the target an average of 60% of the time. In a local rifle contest, each contestant will

attempt three shots and receive the following reward, depending upon the number of targets that he hits: $100 for all three targets; $50 for two targets; and $10 for one target. If a contestant fails to hit at least one target, he must pay $50. Should Mr. Winchester enter the contest?

The data in this problem satisfy the assumptions necessary to use a binomial probability function (see Chapter 3). The advantages of classifying decision problems as binomial, Poisson, or any other type of probability distribution is that the likelihood (probability) that each event will occur can be generated by its probability functions. The following solution will illustrate how a binomial probability function is used to solve the rifle-contest problem.

SOLUTION: Table 5.1 summarizes the decision-making procedures in this problem, where $n = 3$ and $p = 0.6$. The probabilities in the second column were generated by using the binomial probability function, which was discussed in Section 3.6. Since the expected return from the rifle contest is positive ($+\$42.88$), Mr. Winchester should play the game.

Case with Poisson assumptions

The following case illustrates the solution of a problem using the Poisson probability function.

EXAMPLE 5.2 / Mr. Spotanski is in charge of stock control and inventory replenishment in a small local grocery store. He recently noticed that the sales of one particular brand of vegetable had a Poisson distribution with a mean of two cases per day. The grocery buys the vegetable for $50 per case and sells the product for $100 per case. Since the vegetable is perishable, unsold cases cannot be returned for a refund, and additional orders placed for delivery on the same day are unprofitable because of excessive transportation and handling costs. What is the maximum number of cases that Mr. Spotanski should order to minimize potential losses?

SOLUTION: Since $m = 2$, Appendix Table E.1 provides us with the pertinent probability distribution for this problem, as shown in Table 5.2.

Table 5.1 / 100-Yard Rifle Contest

Success events	Probability: $P(X = x_i\|n = 3, p = 0.6)$	Monetary rewards, $ (X_i)	$E = \sum P(X_i) \cdot X_i$, $
3	0.216	100	21.60
2	0.432	50	21.60
1	0.288	10	2.88
0	0.064	− 50	− 3.20
	1.000		42.88*

* Expected return.

Table 5.2 / Profit Matrix for the Spotanski Grocery Store

Cases sold	Probability ($m = 2$)	Profit or loss/cases purchased per day							
		1	2	3	4	5	6	7	8
0	0.135	−50	−100	−150	−200	−250	−300	−350	−400
1	0.271	50	0	−50	−100	−150	−200	−250	−300
2	0.271	50	100	50	0	−50	−100	−150	−200
3	0.180	50	100	150	100	50	0	−50	−100
4	0.090	50	100	150	200	150	100	50	0
5	0.036	50	100	150	200	250	200	150	100
6	0.012	50	100	150	200	250	300	250	200
7	0.003	50	100	150	200	250	300	350	300
8	0.001	50	100	150	200	250	300	350	400
	1.000								
Expected net profit or loss, $		36.50	45.90	28.05	−7.70	−52.20	−100.80	−150.35	−204.50

According to Table 5.2, there is almost no chance that the store will sell more than eight cases of vegetables on any given day. The profit matrix in this table was computed as follows: If the grocery orders one case and is unable to sell the product, the store faces a net loss of $50. If the store orders six cases and sells only four, the net profit would be $100, i.e., ($100 × 4 cases) − ($50 × 6 cases); and so on. The expected net profit is then computed by $E = \sum P(X_i) \cdot X_i$.

According to the solution shown above, the store could order up to three cases without incurring any loss. Of course the optimum decision is to order two cases, with an associated expected profit of $45.90 per day.

Two-action case

Consider a problem that does not require us to make any assumptions concerning the likelihood of the state of nature. First, we discuss a case with two alternative courses of action; this is followed by a discussion of a case involving multiactions.

EXAMPLE 5.3 / Suppose that the Clayton Construction Company in St. Louis was awarded a contract from the city highway department to build a bridge across the Mississippi River. The contract price is $1,400,000. If the company builds the bridge, the cost would be approximately $1,000,000. If, however, the company decides to subcontract the construction, the cost would be $800,000; this subcontract cost is valid only if the construction is completed on schedule. After a careful study of all potential subcontractors, the company concluded that there is a 1/5 chance that any subcontractor selected would experience one of the following types of strikes during the contract period[1]: (1) a wildcat strike (C_1); (2) a limited strike (C_2); and (3) a prolonged strike with a provision stipulating that the bridge would be completed prior to the strike (C_3). If C_1 occurs, the total subcontract cost is estimated to be $1,800,000; the cost under option C_2 is $1,100,000, and the cost of C_3 is estimated at $900,000. A preliminary report indicates that there is a 30%, 50%, and 20% chance that events C_1, C_2, and C_3 will occur, respectively. What is the optimum decision for the Clayton Company?

SOLUTION: This problem can be solved using two different approaches. The first approach involves selecting the decision that yields the largest expected profit (profit analysis). The second approach requires selecting the decision that produces the lowest expected cost (cost analysis). In solving this problem, using either method, we let

A_1 = the decision that the company builds the bridge itself,

A_2 = the decision that the company decides to hire subcontractors,

[1] This probability estimate could be based on either subjective judgment or on objective fact, but not on probability functions.

S_0 = the condition that the company uses subcontract labor and that there are no strikes by the subcontractors,

S_1 = the condition that the company uses subcontract labor but that there are strikes by the subcontractors.

Profit analysis. Figure 5.1 illustrates the decision-tree method of arriving at an optimum decision under conditions of uncertainty. This chart was constructed in the following manner: In the case of A_1, the net profit would be $400,000, since the contract price is $1,400,000 and the estimated cost is $1,000,000. If the company decided to delegate the actual construction of the bridge to subcontractors (A_2), two possible conditions should be considered: (1) the subcontractors will not be plagued with strikes (S_0), and (2) they will be involved with some type of strike (S_1). The net profit under option S_0 would be $600,000, on ($1,400,000 contract price) − ($800,000 subcontract cost). The net profit of $600,000 materializes, however, only if there is no strike. Furthermore, since the probability is 80% that there will be no strike, the expected profit under S_0 is $600,000 × 0.8 = $480,000.

In the case of S_1, there are three possibilities. If C_1 occurs, the company actually faces a loss of $400,000 because the cost of building the bridge under this option would be $1,800,000 and the contract price is only $1,400,000. Since the probability of C_1 is only 0.3, the expected loss for C_1 is (− $400,000) × 0.3 = $120,000. The expected profit for strike situations C_2 and C_3 are similarly computed as C_1, and are $150,000 and $100,000, respectively. Since C_1, C_2, and C_3 are mutually exclusive (i.e., if C_1 occurs, C_2 and C_3 will not occur), the expected value for S_1 is the sum of the projected profits for C_1, C_2, and C_3, or − $120,000 + $100,000 + $150,000 − $130,000. The $130,000 is the expected profit that the company can logically forecast if the contract is awarded to subcontractors *and* the subcontractors experience some type of strike. But, as specified in the problem, there is only a 20% chance that a subcontractor will suffer a strike

Figure 5.1 / Expected profit for the Clayton Company (unit = $1,000).

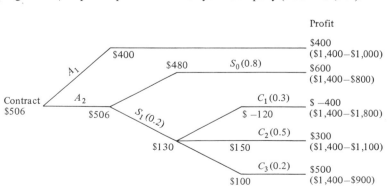

(S_1). Therefore, the expected profit for events A_2 and S_1 is $130,000 \times 0.2 =$ $26,000. Since events S_0 and S_1 are mutually exclusive, the expected profit from the alternative decision (A_2) is the sum of S_0 and S_1, or

$480,000 + $26,000 = $506,000.

As illustrated in the tree diagram, the expected profit from decision A_1 is $400,000. Therefore, the optimum decision is to give the contract to a subcontractor (A_2); this would raise the maximum expected profit from $400,000 to $506,000.

Cost analysis. The same problem can be analyzed by using costs as the primary variable. In this case, the company should choose the decision that produces the least expected cost. Figure 5.2 contains a tree diagram that shows the least-cost approach to solving the Clayton problem.

The same basic procedures apply when analyzing a problem using either the cost or profit approach. The expected cost under the three possible strike situations is $1,270,000; this result was derived as follows:

$1,800,000 \times 0.3 + 1,100,000 \times 0.5 + 900,000 \times 0.2 = $1,270,000.

There is, however, only a 20% chance that the subcontractors will actually encounter a strike; this would result in a final expected cost of $254,000 ($1,270,000 \times 0.2$).

The expected cost under option S_0 is $640,000 ($800,000 \times 0.8$). Because events S_0 and S_1 are mutually exclusive, the expected cost for option A_2 (hire subcontractors) is $254,000 + $640,000 = $894,000. Since the expected cost for option A_1 (no subcontractors) is $1,000,000, the optimum decision is therefore A_2, as this alternative has the least expected cost.

Multiaction problem

In the preceding example we discussed a case that contained only two choices, A_1 and A_2. There are many instances in the real world, however,

Figure 5.2 / Expected cost for the Clayton Company (unit = $1,000).

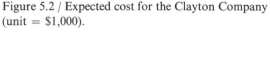
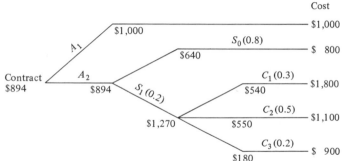

where there are more than two choices or alternatives. Consider the following case.

EXAMPLE 5.4 / Mr. Jansen, the owner of The Jansen Magazine Stand, has encountered great difficulty in determining the optimum number of certain magazines to order for sale at his stand. If he orders more magazines than he can sell, the unsold copies cannot be returned, and he must dispose of them. In this case, Mr. Jansen incurs a loss equal to a predetermined reduced cost of the magazine. On the other hand, if he orders fewer copies than he actually needs, he cannot place additional orders or receive extra copies because each order must be placed at least one week in advance. According to business records that Mr. Jansen has kept for the past 100 weeks, one magazine called *Sports Review*, which sells for $1.00 per copy and costs Mr. Jansen $0.50 per copy, shows a constant loss because of poor forecasts of demand.

After investigating the record of the past 100 normal weeks' sales, he found that the distribution of sales for the *Sports Review* was as shown in Table 5.3. According to these data, there was no single week in which more than 8,000 or less than 4,000 copies of *Sports Review* were sold. In the absence of better information, Mr. Jansen decided to use the information (number of weeks) as an indication of the likelihood for each event (weekly sales). The resulting probabilities are shown in the last column in Table 5.3.

Through the terms of a prior wholesale contract, Mr. Jansen can return any unsold copies of *Sports Review* for a refund under the following formula: Up to the first 500 copies, the refund is $0.30 for each unsold copy; 501 to 1,000 copies entitles Mr. Jansen to a refund of $0.20 for each copy; and for over 1,000 copies, the refund is $0.10 each.

(1) What is the optimum decision; i.e., how many copies of *Sports Review* should Mr. Jansen order?
(2) Compute the expected maximum profit.

SOLUTION: This problem contains several alternative choices; i.e., order any amount from 4,000 to 8,000 copies per week. The optimum decision in this

Table 5.3 / Demand for *Sports Review*

Number of copies sold per week (X_i)	Number of weeks	Probability $P(X_i)$
4,000	10	0.10
5,000	25	0.25
6,000	35	0.35
7,000	20	0.20
8,000	10	0.10
	100	1.00

Table 5.4 / Net Profit Matrix for *Sports Review*

No. of copies sold (X_i)	Probability $P(X_i)$	No. of copies ordered/profit or loss, $				
		4,000	5,000	6,000	7,000	8,000
4,000	0.10	2,000	1,750	1,350	950	550
5,000	0.25	2,000	2,500	2,250	1,850	1,450
6,000	0.35	2,000	2,500	3,000	2,750	2,350
7,000	0.20	2,000	2,500	3,000	3,500	3,250
8,000	0.10	2,000	2,500	3,000	3,500	4,000
Expected net profit or loss, $		2,000	2,425	2,647.50*	2,570	2,290

* Optimum decision.

case is to choose that course of action which yields the largest expected monetary return.

In order to illustrate the solution to this problem, the profit matrix in Table 5.4 was developed.

The net profit matrix in Table 5.4 was computed on the assumption that the amounts of net profit vary, depending on the size of the demand for the magazine in relation to the number of copies ordered. Assume, for example, that Mr. Jansen decided to order 4,000 copies of *Sports Review* and subsequently sold all 4,000 copies of the magazine. From Table 5.4 he would make a profit of $2,000:

Total revenue = $1.00 × 4,000 = $4,000
Total cost = $0.50 × 4,000 = $2,000
Net profit = $2,000

If the actual demand were 5,000 copies, and he had ordered only 4,000 copies, there is nothing Mr. Jansen could do to obtain additional copies to meet this demand because he could not get additional copies in time to sell them during the week. In this case the net profit remains at $2,000. As a matter of fact, even if the demand for the magazine were 8,000 copies, he would still make only a $2,000 net profit because he could not change his original supply of 4,000 copies!

Now assume that Mr. Jansen decided to order an original supply of 5,000 copies and that the subsequent actual demand was only 4,000 copies. The net profit in this case would be $1,750:

Sales:	$1.00 × 4,000 = $4,000	
Refund (up to 500 copies):	$0.3 × 500 = $ 150	
Refund (501–1,000 copies):	$0.2 × 500 = $ 100	
Total revenue		$4,250
Total cost ($0.50 × 5,000)		$2,500
Net profit		$1,750

If the actual demand were 5,000 *and* Mr. Jansen had originally ordered 5,000 copies of *Sports Review*, the net profit would be $2,500, since total revenue would be $5,000 ($1.00 × 5,000) with an associated total cost of $2,500 ($0.50 × 5,000). Following the same logic as discussed previously, the net profit would remain at $2,500, even if demand exceeded 5,000 copies. The net profits for the remaining alternative courses of action are similarly computed.

From the data given above, it can be seen that net expected profit is a conditional concept. For example, in the magazine case the amount of net profits is dependent upon the size of the actual demand and the number of magazines (supply) originally ordered. The net profit of $1,750, for example, occurs at only one point of intersection of the supply and demand curves, that point being where 4,000 copies were sold and 5,000 copies were ordered. The chance that the actual demand will be 4,000 copies is, however, only 10%, as shown in column (2) of Table 5.4. In other words, each net profit is conditioned by its likelihood or probability of occurrence, and accordingly the expected net profit should be adjusted (or weighted) by the probability of the occurrence of the events. The expected profit when Mr. Jansen orders 4,000 copies should therefore be equal to

$$E(4,000) = \$2,000 \times 0.1 + \$2,000 \times 0.25 + \$2,000$$
$$\times 0.35 + \$2,000 \times 0.20 + \$2,000 \times 0.1$$
$$= \$2,000.$$

The expected profits for the other possible course of actions are:

$$E(5,000) = \$1,750 \times 0.1 + \$2,500 \times 0.25 + \$2,500 \times 0.35$$
$$+ \$2,500 \times 0.20 + \$2,500 \times 0.1$$
$$= \$2,425.$$

$$E(6,000) = \$1,350 \times 0.1 + \$2,250 \times 0.25 + \$3,000 \times 0.35$$
$$+ \$3,000 \times 0.20 + \$300 \times 0.1$$
$$= \$2,647.50.$$

$$E(7,000) = \$950 \times 0.1 + \$1,850 \times 0.25 + \$2,750 \times 0.35$$
$$+ \$3,500 \times 0.20 + \$3,500 \times 0.1$$
$$= \$2,570.$$

$$E(8,000) = \$550 \times 0.1 + \$1,450 \times 0.25 + \$2,350 \times 0.35$$
$$+ \$3,250 \times 0.20 + \$4,000 \times 0.1$$
$$= \$2,290.$$

The optimum decision in this case is to order 6,000 copies; under this option, the expected profit would be $2,647.50. The reader should be cautioned, however, concerning the interpretation of this expected profit of $2,647.50. As previously discussed in Chapter 3, the expected profit of $2,647.50 *does not* explicitly state that Mr. Jansen will always make a net profit of $2,647.50 from weekly sales of 6,000 copies. Sometimes only 4,000 copies will be sold; on the other occasions, 8,000 copies may be demanded. What the expected net profit of $2,647.50 does imply is that if Mr. Jansen repeatedly orders 6,000 copies per week over an extended period of time, he will make an average profit of $2,647.50 per week, *in the long run*.

5.3 / Cost of Uncertainty

Expected Opportunity Loss (EOL)

Any decision, whether it is optimum or suboptimum, faces a possible opportunity loss unless the decision is based on perfect information; this latter situation is, of course, quite rare in the real world. This is true because information is constantly changing, day by day and even hour by hour. Therefore, the optimum decision reached during one given time period and under one set of circumstances often becomes suboptimal as time passes and as conditions or the states of nature change. The optimum decision at any given time, therefore, should be the one that uses existing information in such a way as to generate the highest monetary reward or largest utility. In other words, the optimum decision under conditions of uncertainty is the one that yields the largest economic reward and/or the decision that produces the lowest economic penalty or loss.

It may seem contradictory or inconsistent to explain or state a theory of optimization in terms of penalties or potential economic losses. Such economic losses or penalties are possible, however, because any optimum decision is usually based on imperfect information. If the decision were based on perfect information, there would, of course, be no economic penalty, since the risk of failure would not be a realistic alternative. Accordingly, the size of any economic loss due to imperfect information is equal to the difference between the expected monetary value of the optimum decision under conditions of certainty (perfect information) and the decision that is derived under the more realistic condition of uncertainty (imperfect information).

Let us return to the *Sports Review* case (Table 5.4). Under conditions of uncertainty, the optimum decision was to order 6,000 copies of *Sports Review*; the expected net profit from this decision was $2,647.50. Mr. Jansen's strategy probably would have been far different, however, if he had the advantage of perfect information. For example, instead of ordering 6,000 copies of *Sports Review* every week, Mr. Jansen would have ordered a different number of copies each week had he known in advance the actual size of the demand for the magazine; this optimum order size would, of course, be equal to the size of the actual demand for the week. Table 5.5 illustrates a probable strategy that Mr. Jansen might follow under conditions of *certainty*.

In other words, if Mr. Jansen knew in advance that the number of copies to be sold (actual demand) would be 4,000, he would certainly order 4,000

Table 5.5 / Net Profit Matrix for Jansen's Stand with Perfect Information

		No. of copies ordered/profit or loss, $				
No. of copies sold	Probability	4,000	5,000	6,000	7,000	8,000
4,000	0.10	2,000	0	0	0	0
5,000	0.25	0	2,500	0	0	0
6,000	0.35	0	0	3,000	0	0
7,000	0.20	0	0	0	3,500	0
8,000	0.10	0	0	0	0	4,000
	1.00					

copies and make a $2,000 profit ($4,000 − $2,000); he would order 5,000 copies with a subsequent profit of $2,500, if he knew in advance that he could sell exactly 5,000 copies; and so on.

The probability that weekly demands of exactly 4,000, 5,000, 6,000, 7,000, and 8,000 copies will occur are 0.10, 0.25, 0.35, 0.20, and 0.10, respectively. The expected profit under conditions of *perfect information* is therefore

$$E(X) = 0.10 \times \$2,000 + 0.25 \times \$2,500 + 0.35 \times \$3,000$$
$$+ 0.20 \times \$3,500 + 0.1 \times \$4,000$$
$$= \$2,975.$$

The amount of economic loss or expected opportunity loss due to *imperfect information* is therefore $327.50 ($2,975 − $2,647.50).[2]

Table 5.6 illustrates another way of deriving expected opportunity losses (EOL) for different possible courses of action that Mr Jansen could take.

Table 5.6 / Opportunity-Loss Matrix for Jansen's Stand

	Probability $P(X_i)$	No. of copies ordered/opportunity loss, $				
No. of copies sold		4,000	5,000	6,000	7,000	8,000
4,000	0.10	0	250	650	1,050	1,450
5,000	0.25	500	0	250	650	1,050
6,000	0.35	1,000	500	0	250	650
7,000	0.20	1,500	1,000	500	0	250
8,000	0.10	2,000	1,500	1,000	500	0
EOL		975	550	327.50*	405	685

* Expected value of perfect information.

[2] It should be noted that, in addition to the fact that there is an expected opportunity loss of $327.50, an analysis of this loss is extremely useful. The interpretation and application of this $327.50 loss will be discussed at the end of this section.

Table 5.6 demonstrates how much Mr. Jansen will suffer from making a suboptimum decision. For instance, if Mr. Jansen ordered 4,000 copies of *Sports Review* and the actual demand for the magazine was 4,000 copies, then he would have made a correct decision and, as a result, would suffer no opportunity loss. If the demand was 5,000 copies, he should have ordered 5,000 copies, thus making a $2,500 net profit and a corresponding zero opportunity loss. Since he ordered only 4,000 copies (due to insufficient information), he suffered an opportunity loss of $500 [($2,500 maximum profit) − ($2,000 actual profit)]. This opportunity loss increases as the difference between the number of copies ordered and the number of copies demanded becomes larger.

The remainder of Table 5.6 was computed in a similar manner. For instance, assume that Mr. Jansen ordered 7,000 copies, but actual demand was only 4,000 copies. If he had known that the demand would be only 4,000 copies, he would have ordered 4,000 copies, thus making a $2,000 net profit. Due to imperfect information, however, he actually ordered 7,000 copies and made a net profit of only $950 (see Table 5.4). The opportunity loss, therefore, would be $1,050 ($2,000 − $950). If, on the other hand, the actual demand was 7,000 copies, he would have made a correct decision. The opportunity loss in this latter case is, of course, zero.

Carrying the example one step further, Mr. Jansen would suffer a $2,000 opportunity loss if the actual demand were 8,000 copies and he ordered only 4,000 copies. The chance that this level of demand will occur is, however, only 10%. In other words, in order to portray a realistic picture of any projected opportunity loss, the results should be adjusted, or weighted, by its respective probabilities of occurrence in order to arrive at the final expected opportunity loss. When each alternative is weighted by its probability of occurrence, the expected opportunity loss (EOL) for the various courses of action in the Jansen problem become

$$
\begin{aligned}
\text{EOL(4,000 copies)} =\ & \$0 \times 0.1 + \$500 \times 0.25 \\
& + \$1,000 \times 0.35 + \$1,500 \times 0.20 \\
& + \$2,000 \times 0.1 \\
=\ & \$975.
\end{aligned}
$$

$$
\begin{aligned}
\text{EOL(5,000 copies)} =\ & \$250 \times 0.1 + \$0 \times 0.25 + \$500 \times 0.35 \\
& + \$1,000 \times 0.20 + \$1,500 \times 0.1 \\
=\ & \$550.
\end{aligned}
$$

$$
\begin{aligned}
\text{EOL(6,000 copies)} =\ & \$650 \times 0.1 + \$250 \times 0.25 + \$0 \times 0.35 \\
& + \$500 \times 0.20 + \$1,000 \times 0.1 \\
=\ & \$327.50.
\end{aligned}
$$

$$
\begin{aligned}
\text{EOL(7,000 copies)} =\ & \$1,050 \times 0.1 + \$650 \times 0.25 \\
& + \$250 \times 0.35 + \$0 \times 0.2 + \$500 \times 0.1 \\
=\ & \$405.
\end{aligned}
$$

$$\text{EOL(8,000 copies)} = \$1,450 \times 0.1 + \$1,050 \times 0.25$$
$$+ \$650 \times 0.35 + \$250 \times 0.2 + \$0 \times 0.1$$
$$= \$685.$$

Table 5.7 was prepared to present several interesting facets of the EOL concept. First, the sum of the expected opportunity losses (column 3) and the corresponding expected profit (column 2) in each course of action is always equal to the amount of maximum profit that would have been obtained under conditions of *certainty* ($2,975); this is shown in column (4) of the table.

It is not too difficult to understand why the sums of columns (2) and (3) in Table 5.7 are equal to the maximum profit derived under conditions of perfect information (column 4). When any suboptimum decision is made (e.g., ordering 4,000 copies when the actual demand is 6,000 copies), the expected profit is, of course, smaller, and therefore the opportunity loss is greater. In other words, the difference in expected profit between an optimum decision and any suboptimum decision is actually equal to the difference between the expected opportunity losses of the optimum decision and the suboptimum decision.

For example, consider the case where 8,000 copies are ordered. The maximum profit under uncertainty is $2,290 and the EOL is $685. The difference in profit where 6,000 copies are ordered (optimum decision) and where 8,000 copies are ordered (suboptimum) is $357.50 ($2,647.50 − $2,290). The difference in EOL between these two decisions is, of course, equal to $357.50 ($685 − $327.50). It follows that, by definition, the optimum decision always yields both the highest profit and the lowest EOL. This is illustrated in Table 5.7, where the EOL of $327.50 for a 6,000-copy order (an optimum decision) is the smallest in the group.[3]

Table 5.7 / Maximum Profit under Conditions of Certainty

No. of copies ordered (1)	Max. profit under uncertainty*, $ (2)	Expected opportunity loss†, $ (3)	Max. profit under certainty, $ (4) = (2) + (3)
4,000	2,000.00	975.00	2,975.00
5,000	2,425.00	550.00	2,975.00
6,000‡	2,647.50	327.50	2,975.00
7,000	2,570.00	405.00	2,975.00
8,000	2,290.00	685.00	2,975.00

* From Table 5.4.
† From Table 5.6.
‡ Optimum decision under conditions of uncertainty.

[3] The fact that the optimum decision, by definition, yields the lowest expected opportunity loss provides us with another approach to deriving an optimum decision. In other words, the optimum decision is equal to the lowest EOL among several alternatives.

It might seem odd that the optimum decision of 6,000 copies still carries an expected opportunity loss of $327.50. We must keep in mind the assumption, however, that this optimum decision is still made under conditions of uncertainty. Therefore, although an order of 6,000 copies is the optimum decision in our example, nevertheless it still carries some degree of uncertainty and subsequent opportunity loss.

Expected Value of Perfect Information (EVPI)

The decision maker often wonders whether he should make his terminal decision based solely on prior information or wait until further information is collected and analyzed for a better decision. If the decision problem carries a great deal of uncertainty, the decision maker might prefer to use additional information. The expected value of perfect information (EVPI) measures the degree of uncertainty that the decision maker faces in a given problem. Actually, the EVPI is the lowest amount of expected opportunity loss. This value represents the maximum expenditure that management would be willing to spend to obtain additional or better information. For example, in the *Sports Review* magazine case, the maximum profit under perfect information is $2,975.00 (Table 5.7), whereas the maximum profit *under conditions of uncertainty* is $2,647.50 (Table 5.4). It is reasonable, therefore, to assume that the only reason that management is unable to obtain the highest profit ($2,975) is the existence of uncertainty. Accordingly, the cost of obtaining better information with which to improve profit (or reduce losses) is the difference between these two maximum profits or,

[Maximum profit with perfect information ($2,975.00)]
 − [Maximum profit under uncertainty ($2,647.50)]
   = [cost of perfect information (EVPI = $327.50)].   (5.1)

In other words, it would not be profitable for the company to spend more than $327.50 to obtain additional information. Let us illustrate this point through an example.

Assume that management decides to spend $427.50 to improve its information base; this amount is $100 more than the maximum required (EVPI). This increase of $100 would eventually reduce the expected profit by a like amount, or from $2,647.50 to $2,547.50 (Table 5.4). But, as shown in Table 5.4, this expected profit of $2,547.50 for a 6,000-copy order would then not be an optimum decision, since an order of 7,000 copies would yield an expected profit of $2,570. Therefore, the implicit rules of profit maximization dictate that management cannot spend more than the EVPI to improve profit.[4]

EXAMPLE 5.5 / The Nadar Safety Auto Parts Company manufactures electronic parts for passenger cars. The company subsequently sells these parts in lots of 10,000 parts each. Furthermore, the company has a policy of

---

[4] By definition, the amount of EVPI is, of course, equal to the lowest EOL in the problem.

Table 5.8 / The Nadar Auto Company

Rating	Proportion of defective items	Frequency (days)
Excellent	0.02	35
Good	0.05	30
Acceptable	0.10	20
Fair	0.15	10
Poor	0.20	5
		100

inspecting each lot before it is actually shipped to the retailer. Five categories of parts deficiency have been established for the quality-control inspection. Each inspection category represents the percentage of defective items contained in each lot; this is shown in Table 5.8.

Since the company has received some complaints from its customers about the quality of the products, the Quality Control Department has initiated an investigation of the conditions of the machines producing the parts in question. The daily inspection charts for the past 100 inspections show the breakdown of the inspection ratings; these data are summarized in column (3) of Table 5.8. The Nadar management is considering two possible courses of action to correct the problem: (1) shut down the entire plant operation and thoroughly inspect each machine, or (2) continue production as it now exists, but offer the customer a refund for defective items that are discovered and subsequently returned. The first alternative will cost the company a total of $600, while the second alternative will cost the company $1.00 for each defective item that is returned. The Engineering Department estimates that inspecting and overhauling any deficient machine would significantly improve the machine's performance for a given period of time; subsequent to the given period of time, machine quality might again decline to the pre-overhaul performance level.

An important variable to consider in this problem is the high economic penalty for an incorrect decision; i.e., the company will lose customers if it continues to produce a high rate of defective goods. On the other hand, profits will decline and a great deal of production time will be lost if the company decides to shut down all machinery unnecessarily, especially if the production of defective items on most machines is found to be low.

(1) What is the optimum decision for the company?
(2) What is the expected cost?
(3) What is the EVPI?

SOLUTION: Let A = "Excellent," B = "Good," C = "Acceptable," D = "Fair," and E = "Poor." Then the decision matrix (Table 5.9) can be constructed.

Table 5.9 / Decision Matrix for the Nadar Parts Co.

Rating	Defective rate (1)	Probability (2)	Cost, $		Opportunity loss, $	
			Inspect (3)	Refund (4)	Inspect (5)	Refund (6)
A	0.02	0.35	600	200	400	0
B	0.05	0.30	600	500	100	0
C	0.10	0.20	600	1,000	0	400
D	0.15	0.10	600	1,500	0	900
E	0.20	0.05	600	2,000	0	1,400
		1.00	600*	670*	170†	240

* Expected cost.
† EVPI.

The cost of inspection is $600 regardless of the proportion of defective parts produced; this is shown in column (3) of Table 5.9. The cost of refunding, however, will depend upon which lot is produced. For example, assume that lot A is produced. Since lot A is known to contain 2% defective parts, based on a total lot size of 10,000 parts, the number of defective parts in lot A is $10,000 \times 0.02 = 200$. Since each defective part costs the firm $1.00 in refunds, the total cost of refunds for lot A would be $200 \times \$1 = \$200$. The refund values for the remaining lots are

B: $10,000 \times 0.05 \times \$1.00 = \$500,$
C: $10,000 \times 0.10 \times \$1.00 = \$1,000,$
D: $10,000 \times 0.15 \times \$1.00 = \$1,500,$
E: $10,000 \times 0.20 \times \$1.00 = \$2,000.$

The probability that any machine will produce lot A, B,C,D, or E is, as shown in column (2) of Table 5.9, 0.35, 0.30, 0.20, 0.10, and 0.05, respectively. Therefore, the expected cost of refunds to customers for defective parts would be

$$\text{Expected cost} = \$200 \times 0.35 + \$500 \times 0.30$$
$$+ \ \$1,000 \times 0.20 + \$1,500 \times 0.1$$
$$+ \ \$2,000 \times 0.05$$
$$= \$670.$$

Since the expected cost of inspecting the machines is $600 and the expected cost of refunds is $670, the optimum decision would be to shut down and inspect the equipment.

Computing EVPI. Let us assume that the machine produces items that mainly fall into the categories of lots A and B. In this case, the optimum decision would be alternative 2 (rely on refunds), since the total projected costs for refunds are $200 for lot A and $500 for lot B, and both figures are less than the $600, which is the cost of an inspection. In this case, due to the uncertainty, if the management of the Nadar Company selected the option of *inspection*, the cost would be $400 more than necessary for lot A and $100 more than necessary for lot B. On the other hand, if the machine produces mainly lots C, D, and E, the optimum decision under these circumstances would be to shut down and inspect, since the cost of inspection is cheaper than the corresponding refund costs. In this latter case, if the decision maker selected the refund option, he would pay $400 ($1,000 − $600), $900 ($1,500 − $600), and $1,400 ($2,000 − $600) more than necessary for lots C, D, and E, respectively. These figures are shown in columns (5) and (6) in Table 5.9.

As can be seen in column (2) of Table 5.9, the probability that each of the different lots will appear is different. Therefore, the expected opportunity losses for the two different decisions are

$$\text{EOL(inspection)} = \$400 \times 0.35 + \$100 \times 0.30$$
$$= \$170.$$

$$\text{EOL(refund)} = \$400 \times 0.20 + \$900 \times 0.10 + \$1,400 \times 0.05$$
$$= \$240.$$

The EVPI is therefore $170, since it is the lowest EOL between the two alternative decisions. As before, this EVPI represents (1) the cost of uncertainty, (2) the value of perfect information, or (3) the maximum cost that management would be willing to spend to reduce uncertainty.

The EVPI may be computed by using Equation 5.1. According to Equation 5.1,

EVPI = (maximum expected value under conditions of certainty)
 − (maximum expected value under conditions of uncertainty).

Since the problem in question is concerned with cost, Equation 5.1 can be modified as follows:

EVPI = [(least expected value under certainty)
 − (least expected value under uncertainty)],

where, from Table 5.9, the least expected value under uncertainty = $600, and the least expected value under certainty = $200 × 0.35 + 500 × 0.30 + $600 × 0.2 + $600 × 0.05 = $430. Therefore

EVPI = ($430 − $600) = $170.

5.4 / Decision-Making Process with No Prior Information

The decision-making process discussed in the preceding section was based on the assumption that the decision maker had a prior knowledge of the state of nature (events) in the problem being analyzed. There are some instances, however, where the decision maker either has no prior knowledge of the probability distribution, or for a variety of reasons may choose not to use the known distribution(s), even though the distributions are available to him. Under these circumstances, the decision-making process is quite different from the processes that we have discussed previously in this chapter. The purposes of this section are to study the decision-making process under conditions of uncertainty without the use of prior information.

Decision criteria with no prior information

The following decision criteria may be used in the absence of, or as a substitute for, prior knowledge of appropriate distributions:

(1) Criterion of pessimism: This criterion is also known as the maximin, or the Wald criterion.
(2) Criterion of optimism, or the maximax, or the Hurwicz criterion.
(3) Criterion of regrets, also known as the minimax or the Savage criterion.
(4) Laplace criterion, also known as the criterion of insufficient reason, or the Bayes' criterion.

An explanation of each criterion is given in the subsequent sections.

Criterion of maximin or minimax. The maximin criterion, a unique form of decision-making tool, was developed by Abraham Wald. It is used in the following manner: First, determine the several choices that are available in a problem. Then list the minimum returns that can be expected from each course of action. Finally, from this list, choose the course of action that produces the highest minimum return; in other words, maximize the minimum return. It should be obvious that this is essentially a pessimistic decision criterion; i.e., what we are actually saying is to prepare a list of the worst possible outcomes and then select the best possible alternative from these choices. If, on the other hand, the situation is concerned with an emphasis on costs rather than on profits, the decision-making process would be concerned with minimizing the largest potential cost from each course of action, or in this case, a minimax strategy.

In either case (i.e., maximin or minimax), the situation described above represents a conservative decision strategy. This type of strategy is not as unrealistic as it may appear at first glance. Suppose, for example, that you are in the process of establishing a business and are faced with a minimum amount of available capital to be used for operating expenses. In this case, while it would be desirable to establish estimates and corresponding operating conditions to meet the most optimistic forecasts (in terms of such items as sales and revenue), a more realistic approach (in view of the limited resources available) would be to plan expenditures and to estimate subsequent profits using more pessimistic assumptions. Example 5.6 will be used to illustrate the types of decision-making process we have been discussing.

EXAMPLE 5.6 / Mr. Wallstreet, an investment banker, is considering whether or not to invest in the stock market. Specifically, he is interested in investing in three categories of stocks: recreation (R); utilities (U); and the automobile industry (A). The expected returns from each of these categories will differ, depending upon the economic conditions that exist during the next two years; i.e., whether the economy is expected to expand (E), to remain in its present conditions (S), or to decline (D). Furthermore, due to a limited amount of funds available for investment purposes, Mr. Wallstreet feels that it will be most profitable to invest his money only in one group of stocks—either R,U, or A.

The estimates in Table 5.10 concerning the anticipated percentage rate of return that Mr. Wallstreet could expect to receive on his investment were supplied by Mr. Wallstreet's broker.

SOLUTION: If Mr. Wallstreet is a born pessimist, he would look into the return matrix, determine the worst possible outcome from each course of action, and then select the largest return from among the three worst returns. In this case, the least return from each strategy is 1% from R, 4% from U, and 3% from A. In other words, Table 5.10 indicates that an investment in a utility stock yields the largest return from among the three least returns available in the problem. The decision would be the same if the values in the matrix were to represent losses rather than profits. In this case, Mr. Wallstreet would determine the largest loss from each course of action (10%, 7%, and 8% from R, U, and A, respectively), and then select the least loss from among these three largest losses. The least loss is 7%; this outcome is also the utility stock.

Table 5.10 / Maximin Criterion

States of nature	Estimated rate of return (%)		
	R	U	A
E	10	7	8
S	4	5	6
D	1	4	3
Minimum return	1	4	3
Maximum of minimum		4	

Criterion of maximax. When using this decision criterion, the decision maker chooses the largest return from the highest yield obtained from each course of action. Let us slightly revise the data contained in Example 5.6 to illustrate the maximax decision strategy.

EXAMPLE 5.7 / Suppose that Mr. Wallstreet, instead of being a pessimist, is really an optimist by nature. Based upon this assumption, Mr. Wallstreet would evaluate each course of action with the hope that the best possible outcome would occur from the available alternatives; this process is shown in Table 5.11.

SOLUTION: Being an optimist, Mr. Wallstreet would choose the largest return from the three highest returns in each category; in this case, the alternative that yields the maximum result is an investment in recreation stocks (10%).

Criterion of minimax regrets. The decision-making process using the strategy of minimax regrets as the major-decision criterion is basically the same as the solution to a problem using the opportunity-loss approach discussed in Section 5.3. As previously mentioned, the optimum decision is, by definition, always the one that yields the least expected opportunity loss. According to the regret criterion, the decision maker evaluates what should have been done to maximize return, and, based upon the results of this analysis, "regrets" if the decision that was previously made was not the right decision. Accordingly, the next step is to determine the amount of regret or opportunity loss associated with each course of action and to select the alternative which yields the least amount of regret (opportunity loss). Once again, we will illustrate this type of strategy by the use of the Wallstreet investment problem.

EXAMPLE 5.8 / Table 5.12 was constructed to show a regret approach to the Wallstreet problem. As shown, had Mr. Wallstreet known that the economy would expand (i.e., state-of-nature E) during the next two years,

Table 5.11 / Maximax Criterion

	Estimated rate of return (%)		
States of nature	R	U	A
E	10	7	8
S	4	5	6
D	1	4	3
Maximum returns	10	7	8
Maximum of maximum return	*10*		

Table 5.12 / Minimax Regret Criterion

	A Course of action			Maximum profit	B Opportunity loss		
States of nature	R	U	A		R	U	A
E	10	7	8	10	0	3	2
S	4	5	6	6	2	1	0
D	1	4	3	4	3	0	2
Maximum opportunity loss					3	3	2
Least opportunity loss							2

he would have invested in the recreation stock because this alternative provides the largest return (10%) among the available choices. On the other hand, if he had actually invested in utility or automobile stocks, the returns would have been 7% and 8%, respectively, and the associated amounts of regret (opportunity loss) in this case would be, therefore, 3% (10% − 7%) and 2% (10% − 8%); this is shown in part B of Table 5.12. If Mr. Wallstreet knew, however, that the economy would remain constant (state of nature S), he would have invested in the automobile stocks; this alternative would be selected because it provides the highest return (6%) under the prevailing economic environment. The opportunity losses for the other alternatives (i.e., recreation stocks and utility stocks) are 2% (6% − 4%) and 1% (6% − 5%), respectively. The amounts of regret for a declining economy (state of nature D) are similarly computed.

SOLUTION: In our example, the largest regrets (opportunity losses) for each investment (course of action) are 3% for recreation (R), 3% for utilities (U), and 2% for automobile (A) stocks. Therefore, the optimum decision is to invest in automobile stocks, since this option provides the least regret (opportunity loss) among the three alternatives; this is also shown in Table 5.12 (part B). Because the ultimate selection of an optimum decision is based upon the least opportunity loss, the regrets (or Savage) criterion is sometime referred to as a "minimax" criterion. This minimax criterion is, of course, different from the Wald minimax criterion.

The Bayes' criterion. When the decision maker has no information concerning the states of nature, a common assumption is made that the likelihoods of occurrence for each event are equal. This decision criterion was first implicitly suggested by Pierre Laplace. According to this criterion, if there is no logical reason to assume that one state of nature is more likely to occur than any other state of nature, the decision maker should consider

Table 5.13 / Bayes' Criterion

States of nature	Probability	Estimated rate of return (10%)		
		R	U	A
E	1/3	10	7	8
S	1/3	4	5	6
D	1/3	1	4	3
Expected returns		15/3	16/3	17/3
Optimum decision				17/3

(or assume) that the probabilities of occurrence of any of the states of nature are the same.[5] According to this criterion, the decision-making process would then emulate one similar to the decision-making process under conditions of risk rather than a situation under conditions of uncertainty. The optimum decision in this case is the alternative that yields the largest expected monetary value.

In our previous example, suppose that the likelihoods of occurrence for each state of the economy—i.e., expansion (E), status quo (S), and decline (D)—are equal, or 1/3 each. The expected return for each course of action is then easily determined as shown in Table 5.13. According to this criterion, therefore, the optimum decision is to invest in the automobile stocks.

5.5 / Summary

The main purpose of this chapter has been to illustrate the decision-making process under conditions of uncertainty. The decision-making process was based on the following assumptions: (1) the states of nature have a discrete random variable; (2) all decisions are derived from, or based upon, only the prior information that is given in the problem; and (3) there is a linear utility of money. At the end of the chapter, we removed one of the assumptions (assumption 2) in order to analyze the decision-making process under conditions of uncertainty where no prior information exists. The following summary should be helpful to the student in understanding the contents of this chapter:

(1) Expected monetary value theory, or simply EMV theory, is a basic tool used in the decision-making process for deriving an optimum decision under conditions of uncertainty.

[5] Since the decision maker assigned a subjective (or judgmental) probability concerning the state of nature, it is called a "Bayes' criterion."

(2) The optimum decision under conditions of uncertainty is defined as the largest expected profit or the lowest expected loss among at least two alternatives.

(3) The optimum decision under conditions of uncertainty is always accompanied by the least expected opportunity loss. In general, the higher the expected profit, the lower the expected opportunity loss.

(4) The least EOL is called the expected value of perfect information, or EVPI.

(5) The EVPI plus the expected profit of the optimum decision under conditions of uncertainty are always equal to the maximum profit under conditions of certainty.

(6) EVPI represents (1) the cost of uncertainty, (2) the maximum amount of money that should be spent to reduce uncertainty, or (3) the value of perfect information.

(7) When prior information concerning the state of nature is either unavailable or considered inapplicable in the analysis of the problem, the following decision criteria may be used to develop an optimum decision under conditions of uncertainty (depending upon, of course, the attitude of the decision maker toward risk and the existing states of nature): maximin, maximax, minimax regret, and the criterion of insufficient reason.

(8) The maximin criterion, or criterion of pessimism, is developed by determining the minimum return from each course of action and then selecting the largest return among these several minimum returns.

(9) A decision strategy opposite to that of maximin is the maximax criterion. This strategy is concerned with the determination of the largest return from among several courses of action, followed by selection of the largest result from the various available alternatives.

(10) When using the minimax regret criterion, the decision maker selects as an optimum decision that course of action which yields the least regret (opportunity loss) from among several alternatives.

(11) When there are no clear-cut reasons for assuming that one specific state of nature is more likely to occur than any other, the decision maker may consider that the likelihoods of occurrence of either (all) of the states of nature are the same. According to this criterion, the decision-making process changes to one of risk from one of uncertainty.

EXERCISES

1. Explain EVPI in business language. How does EVPI differ from advertising expenses?

* 2. "EVPI is the difference between the expected monetary value of an optimum decision under conditions of uncertainty and under conditions of certainty." Discuss the implication of the preceding statement.

3. Assume that the following table is a profit matrix with its corresponding probability distribution.

Number of units demanded	Prob., %	Number of units ordered/profit or loss, $			
		1,000	2,000	3,000	4,000
1,000	0.1	$100	$ 60	$ 20	−$ 20
2,000	0.1	100	120	80	20
3,000	0.5	100	120	140	80
4,000	0.3	100	100	140	160

(a) Compute the optimum decision.
(b) Compute the cost of uncertainty.

4. A newsstand at the Grand Central Railway Station buys the *Financial Journal* at 10 cents a copy and sells it for 15 cents a copy. Unsold newspapers may be returned for a refund of 5 cents a copy. From past experience the projected daily *Journal* sales and their corresponding probabilities of occurrence are as follows:

Daily copy sales	Probability
500	0.05
1,000	0.15
1,500	0.30
2,000	0.25
2,500	0.20
3,000	0.05

(a) How many copies should the newsstand order each day?
(b) If unsold copies cannot be returned (i.e., unsold copies are useless), what is the optimal order each day?
(c) What is the total cost of uncertainty for (a) and (b)?

5. In the past, a bicycle manufacturer has sold his product exclusively in the domestic market, but is now planning to introduce his product in the world market. His research staff has provided the following market information and corresponding probabilities for this venture:

Possible market shares, %	Probability
10	0.3
5	0.5
1	0.2

In addition, his accounting department has estimated that the net payoff for this venture would be a $20 million profit for a 10% share of the market, $5 million profit for a 5% share of the market, and a $10 million loss for a 1% share of the market.

(a) Under these conditions, should the manufacturer enter the world market?

(b) What is the cost of uncertainty?

6. The Phi Delta Nebulum fraternity house is planning a beer party after the annual New Year's Day football game. The game is considered a social event, and therefore the fraternity wants to cover all costs although a large profit is not required. Five hundred advanced tickets have already been sold at $1.00 per ticket, and the students have projected a maximum attendance of 1,000. According to the record of similar events in previous years, the possible admissions and corresponding likelihoods are as follows:

Admissions	Likelihoods
500	0.05
600	0.15
700	0.20
800	0.30
900	0.20
1,000	0.10

From past experience it has been shown that one keg of beer is adequate for 50 students in the course of a football game. In addition, the price of $20 per keg is valid only if purchased prior to January 1, and no refund is given for unused beer. If additional beer is required, the price on the day of the game is increased to $30 per keg because of delivery cost on a holiday.

Determine the optimum number of kegs to order.

7. The Pulitzer Printing Company is considering replacing its present copying machine with a more efficient one. Currently, there are two types of machines on the market, the model QX100 and the model XL50. The former costs $5,000, but would save $2.00 per hour over the present machine expenditures; the latter costs $3,000, but would save only $1.00 per hour over the present current machine costs.

The exact hours of usage for one of these reproducing machines in the future are somewhat uncertain because of the tight competition in the printing industry. The management of Pulitzer has estimated the following probabilities for total hours usage over the next two years:

Total hours	Probabilities
1,000	0.05
2,000	0.10
3,000	0.20
4,000	0.40
5,000	0.20
6,000	0.05

Which machine should the company buy? Compute the expected savings for each of these two machines.

8. The Type-Right Secretarial Service obtains the greatest portion of its sales from relatively large service-oriented companies in the downtown New York City area.

The company charges customers $3.00 per hour for various secretarial services and pays employees $2.00 per hour. The company usually maintains a list of girls who work on a day-by-day basis; these girls are called as the requests from Type-Right customers are received. Once the company calls and reserves an employee for the day, however, it has to pay a full day's wage regardless of the hours that the employee actually works. If more requests for services are received than the company has employees in reserve to handle the work, it must honor the service request by utilizing emergency standby reserves, which cost $4.00 per hour, even though the customer charge remains at $3.00 per hour. Since management is uncertain about the exact number of employees that should be reserved, it computed the following probabilities based on the past 100 working days:

Number of employees requested per day	Number of days actually used
10	5
11	10
12	10
13	50
14	20
15	5

(a) Compute the optimum decision.
(b) Compute the amount of uncertainty.

9. The National Auto Supply Company is an automotive parts wholesaler. There has recently been an increasing demand from its customers for foreign car parts, a line of items that National does not currently stock. At the present time, the company obtains foreign parts for its customers from other wholesalers, but now the National is considering stocking its own line of foreign car parts.

In examining sales of foreign car parts for previous years, the following sales breakdown was made:

Sales per week, $	Occurrence, %
100	28.0
200	34.0
400	20.0
800	14.0
1,000	4.0

The company purchasing policy is to maintain an inventory level of at least twice the weekly sales level. On sales made from inventory, the gross margin is 40%. If sales are greater than stock carried in inventory (i.e., stock out), the parts must be procured on an emergency order basis, and only a 20% gross margin is made on these sales. However, there is also a penalty for carrying excess inventory; this is estimated to be 1% of the excess of the suggested average inventory level over the average sales level for the same period. What is the optimum inventory level that the company should maintain in foreign car parts?

10. John Appleton has been assigned to take over the duties of production manager at the Harbor Suit Manufacturing Company. John's duties will include improving the efficiency of both machines and personnel in order to minimize defective production. John has recently been informed that the $20 allowance per suit, currently given to retail clothiers to rework defective suits, is going to be charged to his department. With this in mind, John is looking for new or improved ways to decrease inefficiency in his department.

In his analysis, John is considering adding another inspector to the payroll at a cost of $500 per month. The average monthly production of the company is 5,000 suits with little deviation from this average figure. The production records for the past year are summarized in the following table:

Type of suit	Average monthly production	Percent defective
A	1,450	5
B	1,100	4
C	1,050	3
D	900	2
E	500	1

John Appleton is faced with the problem of deciding whether to add another inspector or to allow retailers to continue to rework the suits and have that cost charged to his department.

(a) Determine the optimum decision.

(b) What is the cost of uncertainty?

11. The Baxter Fertilizer Company is considering opening a new manufacturing plant in St. Louis to control all sales for the midwestern region. The major problem facing the company is determining the size of the plant to be built—whether it should build a large plant in anticipation of strong demand, or build a small one, assuming the large demand will not materialize. The size of a fertilizer plant is dependent mainly upon the prospective demand for its product over the next five years. The company marketing department, after a careful market survey of several potential retailers in this region, provided the following information concerning prospective demand: (a) The chances are 40% that demand will be strong and remain strong (A_1) in the region for the next five years; (b) there is a 10% chance that demand will initially be strong but will subsequently weaken due to competition (A_2); (c) there is a 30% probability that demand will initially be low and remain low (A_3); and (d) the chances are 20% that demand will initially be low, but will subsequently increase, owing to an organized advertising effort (A_4). Management is considering two options: (1) build a large plant (B_1), which costs $3,000,000 or (2) build a small plant (B_2), which costs $1,500,000, and later expand to a large plant (B_3) in case subsequent demand becomes strong. The B_3 option would cost the company an additional $2,000,000.

Given the projected demand size and the cost of building the different plants, the accounting department provided the following projected gross revenue matrix for the next five years.

149

Gross Revenue for the Baxter Company
($ millions)

Demand	Plant size		
size	B_1	B_2	B_3
A_1	$6.0	$3.5	$6.0
A_2	3.0	2.5	2.5
A_3	1.0	2.0	2.0
A_4	3.5	3.0	5.5

What is the optimum decision? Show your solution in a tree diagram.

12. The Sole Brother Shoe Manufacturing Company is currently losing $200,000 per year, mainly because of low prices of foreign shoes in the domestic market. The company actively campaigned in Congress to help pass the Mill's bill, which would limit the number of foreign-produced shoes that can be imported into the domestic market each year. If the bill is passed, the company expects to make at least a $1,200,000 profit annually. The company has an option to lease its total operation to the Holloway Leather Company of Italy for $700,000 per year.

Draw up the payoff table in this problem. What is the minimum probability that the bill will be passed, thus ensuring that the Sole Brother Company will remain in the shoe business rather than being forced to lease its plant to the Holloway Company?

13. Discuss both the similarities and the differences between the Wald minimax criterion and the Savage minimax criterion.

14. "The decision-making process using the Laplace criterion is essentially the same as the decision-making process under conditions of risk." Discuss this statement.

15. The manager of the Data Processing Department in a hospital is considering the acquisition of a multipurpose statistical package. There are currently three different packages available (A, B, and C), with each being developed by a leading consulting firm in New York. The costs of the packages are $800 for A, $900 for B, and $1,000 for C. There is also an additional charge for installing and debugging the computer product to fit the hospital's particular needs. These preparatory costs differ, depending upon whether the packages are poorly suited for a specific application (S_1), fairly well suited (S_2), or very well suited (S_3). The estimated additional costs associated with each of the above conditions are given in the table.

	Package costs, $		
Class	A	B	C
S_1	600	400	450
S_2	500	350	300
S_3	400	350	100

(a) What statistical package should the manager purchase if he is a confirmed optimist?

(b) Does the decision in (a) change if the manager is a confirmed pessimist?

(c) If the odds for each state of nature (i.e., S_1, S_2, and S_3) are estimated to be 0.25, 0.35, and 0.40, respectively, would this affect the manager's previous decision?

(d) If the odds change to 0.60 for S_1, 0.25 for S_2, and 0.15 for S_3, how would this affect the subsequent terminal decision?

(e) What is the optimum decision if the manager assigned the following probabilities to the states of nature in this problem?

	Packages		
	A	B	C
S_1	0.20	0.70	0.25
S_2	0.40	0.20	0.70
S_3	0.40	0.10	0.05

16. The Maxwell Tire Company is currently advertising that the company's best snow tire will be on sale during the summer for $30 each. The advertisement further indicates that the price of the same tire will be increased to $45 each in early winter.

Mr. Stapleton is considering whether he should buy two snow tires now to take advantage of the low cost or wait until winter or until a heavy snow actually comes. Since the area in which he lives has experienced few heavy snow days that would require the use of snow tires, he is very uncertain as to his decision, even though the past winter proved very troublesome as a result of unusually heavy snowfalls.

(a) If there is a 50:50 chance that this will be a heavy snowfall winter, should Mr. Stapleton buy the snow tires now?

(b) Repeat (a) above with the odds 2 to 1 in favor of snow. Does the previous decision change?

(c) Repeat (a) above with the odds now 3 to 1 in favor of snow.

(d) What decision should be made if Mr. Stapleton is a born optimist? A pessimist?

(e) What is the cost of perfect information if the odds are even?

SUGGESTED READINGS

References for Decision-Making Process with Prior Knowledge

Bierman, Harold, Jr.; Bonini, Charles P.; and Hauserman, Warren H. *Quantitative Analysis for Business Decisions* (3d ed.). Homewood, Ill.: Richard D. Irwin, 1973, chapter 4.

Chou, Ya-Lun. *Statistical Analysis with Business and Economic Application.* New York: Holt, Rinehart and Winston, 1969, chapter 21, sections 4 and 5.

Dyckman, T. R.; Smidt, S.; and McAdams, A. K. *Management Decision under Uncertainty.* New York: Macmillan, 1969, chapter 15, section 3, and chapter 17, section 2.

Halter, Albert N., and Dean, Gerald W. *Decisions under Uncertainty with Research Application.* Cincinnati: South-Western Publishing, 1971, chapters 4 and 10.

Hays, William L., and Winkler, Robert L. *Statistics: Probability, Inference, and Decision.* New York: Holt, Rinehart and Winston, 1971, chapter 9, sections 13 through 22.

Raiffa, Howard. *Decision Analysis: Introductory Lectures on Choices under Uncertainty.* Reading, Mass.: Addison-Wesley, 1968, chapter 2, sections 6 through 8.

Sasaki, K. *Statistics for Modern Business Decision Making.* Belmont, Calif.: Wadsworth Publishing, 1969, chapters 11 and 13.

Schlaifer, Robert. *Probability and Statistics for Business Decisions.* New York: McGraw-Hill, 1959, chapters 36 and 37.

Schlaifer, Robert. *Analysis of Decisions under Uncertainty.* New York: McGraw-Hill, 1969, chapters 10 and 11.

Schmitt, Samuel A. *Measuring Uncertainty: An Elementary Introduction to Bayesian Statistics.* Reading, Mass.: Addison-Wesley, 1969, chapter 3.

Spurr, William A., and Bonini, Charles P. *Statistical Analysis for Business Decisions.* Homewood, Ill.: Richard D. Irwin, 1973, chapters 7 and 8 (pp. 192–198).

Thompson, Gerald E. *Statistics for Decisions—An Elementary Introduction.* Boston: Little, Brown, 1972, chapter 10.

References for Decision-Making Process with No Prior Knowledge

Costis, Harry G. *Statistics for Business.* Columbus, Ohio: Charles E. Merrill, 1972, chapter 23.

Fellner, William. *Probability and Profit.* Homewood, Ill.: Richard D. Irwin, 1965, chapter 3.

Freund, John E., and Williams, Frank J. *Elementary Business Statistics* (2d ed.). Englewood Cliffs, N.J.: Prentice-Hall, 1972, chapter 6.

Halter, Albert N., and Dean, Gerald W. *Decisions under Uncertainty with Research Applications.* Cincinnati: South-Western Publishing, 1971, chapters 12 and 14.

Hamburg, Morris. *Statistical Analysis for Decision Making.* New York: Harcourt, Brace & World, 1970, chapter 17.

Newman, Joseph W. *Management Applications of Decision Theory.* New York: Harper & Row, 1973, chapter 11.

Raiffa, Howard. *Decision Analysis: Introductory Lectures on Choices under Uncertainty.* Reading, Mass.: Addison-Wesley, 1968, chapter 10.

Schlaifer, Robert, *Analysis of Decisions under Uncertainty.* New York: McGraw-Hill, 1969, chapters 1 and 2.

Specer, Milton H. *Managerial Economics.* Homewood, Ill.: Richard D. Irwin, 1968, chapter 1.

Thompson, Gerald E. *Statistics for Decisions.* Boston: Little, Brown, 1972, chapter 2.

6 / Decision-Making Process with Posterior Distribution and a Discrete Random Variable

6.1 / Purpose

The decision-making process under conditions of uncertainty, as previously discussed in Chapter 5, was based on the following assumptions:

(1) The problems under discussion contain a discrete random variable.
(2) The terminal decision is based solely on available or no prior information. In other words, the decision maker does not take into consideration the possibility of gaining additional information through further sampling and the subsequent benefits that could be derived from the use of these additional data.
(3) Decision makers behave strictly according to the results of expected monetary value.

Decision makers, however, seldom make a terminal decision based only on prior information. They are constantly striving to improve their problem-solving capabilities and subsequent expected monetary return. In general, any decision-making process that fails to recognize the value of additional information usually results in a suboptimum decision.

The main purpose of this chapter is to continue the study of the decision-making process under conditions of uncertainty, expanding the discussion of Chapter 5 by using additional sampling information. Accordingly, the prior assumption that the terminal decisions are made only on the basis of existing information will be relaxed in this chapter so as to accommodate a more realistic approach to the decision-making process. The other two assumptions (1) and (3) cited above are still presumed to be true; these two assumptions will be relaxed in subsequent chapters. Finally, the student is advised to review Section 2.4, concerning the explanation and use of the Bayes' theorem, before proceeding with the material contained in this chapter.

6.2 / Posterior Information and the Decision-Making Process

Two-way action case

Consider the following problem.

EXAMPLE 6.1 / The Wildwood Construction Company is considering submitting a bid for a construction contract to build a large shopping center in a New England suburb. The main concern of the company in the preparation of the bid proposal is the possible outcome of current labor contract negotiations. If the negotiations fail (i.e., labor does not receive all concessions it is seeking, and subsequently strikes), the projected profit on the shopping center venture is $500,000. If, on the other hand, labor accepts the company's current offer, the projected profit for the shopping-center construction job is $1,000,000. The management team that is negotiating for Wildwood has assigned an 80% probability that labor will accept the company offer. In the meantime, the company has been offered a construction contract that guarantees an $875,000 profit.

Since the union is affiliated with the National Union of Construction Workers, the Wildwood management decided to ask the National Association of Construction Companies in Washington, D.C., for further information regarding the possible results of contract negotiations. The Association has summarized the situation in the following manner:

Based upon nationwide experience, it is concluded that there is a close relationship between the degree of inflation and the subsequent incidences of nationwide strikes in construction unions. Specifically, the Association estimates that in seven of every ten incidences where the economy has experienced a major nationwide strike in the construction unions, a high rate of inflation has followed. On the other hand, in the past, there have been three of eight incidences where a major strike was not followed by a high rate of inflation in the economy. The Wildwood Construction Company decides to use this new information in conjunction with the prior information.

It has been reported that there will be a high rate of inflation for this year, compared to the previous year. Based on the information given in this case,

(1) compute the prior optimum decision.
(2) does sampling information alter this prior decision?
(3) compute the prior and posterior EVPI.

SOLUTION: Let

$P(S_i)$ = prior probability of labor negotiation results; $P(S_0)$ = probability of no strike; $P(S_1)$ = probability of a strike.

$P(HI|S_i) = $ conditional probability of a high inflation in the preceding year, given the negotiation results in the following year.

$P^*(S_i) = $ posterior probability of labor negotiation results.

From the given information,

$$P(S_0) = 0.8,$$
$$P(S_1) = 0.2,$$

$$P(HI|S_0) = \frac{3}{8} = 0.375,$$

$$P(HI|S_1) = \frac{7}{10} = 0.7.$$

Prior optimum decision: (a) Mathematical approach:

$$E(\text{prior}) = \sum_{i=1}^{n=2} P(S_i) \cdot X_i$$
$$= P(S_0) \cdot X_0 + P(S_1) \cdot X_1$$
$$= (0.8) \cdot (\$1,000,000) + (0.2) \times (\$500,000)$$
$$= \$900,000.$$

(b) Table approach: Based on the results in the table, the company should accept the shopping-center project because the expected profit for the project is $900,000, whereas the expected profit for the alternative project is only $875,000.

		Profit, $		Opportunity loss, $	
	Prior probability	Shopping center	Alternative	Shopping center	Alternative
No strike	0.8	1,000,000	875,000	0	125,000
Strike	0.2	500,000	875,000	375,000	0
	1.0	900,000*	875,000*	75,000[†]	100,000

* Expected profit.
[†] EVPI$_0$.

Posterior Decision: (a) Mathematical approach:

$$E(\text{posterior}) = \sum_{i=1}^{n=2} P^*(S_i) \cdot X_i$$
$$= P^*(S_0) \cdot X_0 + P^*(S_1) \cdot X_1,$$

where

$$P^*(S_0) = P^*(S_0|HI)$$

$$= \frac{P(S_0) \cdot P(HI|S_0)}{P^*(HI)}$$

$$= \frac{P(S_0) \cdot P(HI|S_0)}{P(S_0) \cdot P(HI|S_0) + P(S_1) \cdot P(HI|S_1)}$$

$$= \frac{0.8 \times 0.375}{0.8 \times 0.375 + 0.2 \times 0.7}$$

$$= \frac{0.3}{0.44} = 0.68.$$

Accordingly,

$$P^*(S_1) = 1 - P^*(S_0)$$
$$= 1 - 0.68$$
$$= 0.32.$$

Therefore, the posterior expected profit from the shopping center contract is

$$E(\text{posterior}) = P^*(S_0) \cdot X_0 + P^*(S_1) \cdot X_i$$
$$= 0.68 \times \$1,000,000 + 0.32 \times \$500,000$$
$$= \$840,000.$$

(b) Table approach: Tabulating our data, we find that the optimum posterior decision is to take the alternative job, which yields an \$875,000 profit. This is true because the posterior expected profit for the shopping center project is only \$840,000.

Events (S_i)	Prior probability $P(S_i)$	Conditional probability $P(HI|S_i)$	Joint probability $P(S_i)P(HI|S_i)$	Posterior probability $P(S_i|HI)$
No strike (S_0)	0.8	0.375	0.30	0.68
Strike (S_1)	0.2	0.700	0.14	0.32
	1.0		0.44*	1.00

* Marginal probability.

Profit, $		Opportunity loss, $	
Shopping center	Alternative	Shopping center	Alternative
1,000,000	875,000	0	125,000
500,000	875,000	375,000	0
840,000*	875,000*	120,000	85,000[†]

* Expected profit.
[†] $EVPI_1$.

Cost of uncertainty (EVPI). (a) Before additional information:

$$(\text{EVPI}_0) = \$75,000$$

(b) After additional information:

$$(\text{EVPI}_1) = \$85,000$$

Case with binomial assumptions

Let us reexamine the Nadar Safety Auto Parts Company originally analyzed in Chapter 5. The problem will be slightly revised in this chapter in order to reflect a more realistic business situation; i.e., the original inspection cost of $600 will be revised upward to $800. We assume that 2% defective parts are still produced, even after the machines have been inspected and adjusted, and the company still pays a $1.00 refund for each part discovered by a customer and subsequently returned as a defective item. Table 6.1 is reproduced from Table 5.8.

Table 6.1 / The Nadar Safety Auto
Parts Company

Rating	Proportion of defective items	Probability
A (Excellent)	0.02	0.35
B (Good)	0.05	0.30
C (Acceptable)	0.10	0.20
D (Fair)	0.15	0.10
E (Poor)	0.20	0.05
		1.00

The management of the Nadar Company feels that since the probability distribution has not been updated for an extended period of time, the validity of the probabilities contained in Table 6.1 are suspect. In an effort to determine a more reliable statistical base, the management of the company selects a random sample of 20 parts from several different lots.[1] The results of the survey showed that one of the samples is defective. This additional information (1 defective part in a sample of 20 parts) can be incorporated into the existing information base (prior probability) to estimate a posterior information. A posterior optimum decision can subsequently be made, based on this newly computed posterior information.

---

[1] We assume that this sample of 20 has been selected in an unbiased manner and is representative of the total population.

Optimum decision before sample information. Using the Nadar Company example, let us briefly review the decision-making process prior to using the additional sampling information.

Table 6.2 has been developed to illustrate the decision-making process based on a prior probability distribution. The logic used to compute the entries contained in columns (3), (4), (5), and (6) is the same as that used to construct the corresponding elements in Table 5.9. For column (3), the lump-sum fee for an inspection and corresponding machine adjustment is now $800, but there are still 2% defective parts produced per lot after the machines have been adjusted. The total cost for the inspection is therefore

$$\$800 + (10,000 \times 0.02 \times \$1.00) = \$1,000.$$

The entries in column (4) of Table 6.2 were computed in the following manner, similar to that used for Table 5.9:

$$
\begin{aligned}
\text{Lot A:} &\quad 10,000 \times 0.02 \times \$1.00 = \$200 \\
\text{B:} &\quad 10,000 \times 0.05 \times \$1.00 = \$500 \\
\text{C:} &\quad 10,000 \times 0.10 \times \$1.00 = \$1,000 \\
\text{D:} &\quad 10,000 \times 0.15 \times \$1.00 = \$1,500 \\
\text{E:} &\quad 10,000 \times 0.20 \times \$1.00 = \$2,000
\end{aligned}
$$

Table 6.2 / Decision Matrix for the Nadar Company (Prior to Sampling)

			Cost, $		Opportunity loss, $	
Rating	Defective rate (1)	Prior probability (2)	Inspection (3)	Refund[‡] (4)	Inspection (5)	Refund (6)
A	0.02	0.35	1,000	200	800	0
B	0.05	0.30	1,000	500	500	0
C	0.10	0.20	1,000	1,000	0	0
D	0.15	0.10	1,000	1,500	0	500
E	0.20	0.05	1,000	2,000	0	1,000
		1.00	1,000*	670*	430	100[†]

* Expected cost.
[†] EVPI.
[‡] An alternate method that can be used to derive the expected cost of the refund is to compute the expected proportion of defective parts produced and then multiply this result by the cost of the refund ($1.00), or

$$
E\,(\text{refund}) = \sum_{i=A}^{E} P(X_i) \cdot X_i \times 10,000 \times \$1.00
$$

$$
= (0.35 \times 0.02 + 0.30 \times 0.05 + 0.20 \times 0.10 \times 0.15 + 0.05 \times 0.20)
$$
$$
\times\ 10,000 \times \$1.00
$$
$$
= (0.0670) \times 10,000 \times \$1.00
$$
$$
= \$670.
$$

Using Table 6.2 as reference, the expected costs for both the inspection and the refund are therefore

$$E(\text{inspection}) = \sum_{i=A}^{E} P(X_i) \cdot X_i$$

$$= 0.35 \times \$1,000 + 0.30 \times \$1,000$$
$$+ 0.20 \times \$1,000 + 0.10 \times \$1,000$$
$$+ 0.05 \times \$1,000$$
$$= \$1,000.$$

$$E(\text{refund}) = \sum_{i=A}^{E} P(X_i) \cdot X_i$$

$$= 0.35 \times \$200.00 + 0.30 \times \$500$$
$$+ 0.20 \times \$1,000 + 0.10 \times \$1,500$$
$$+ 0.05 \times \$2,000$$
$$= \$670.$$

The entries contained in columns (5) and (6) were based on the following logic: If the machine produces lot A only, the optimum decision should be "refund," since this decision costs only $200. If, on the other hand, only lot A is produced and the decision maker, because of imperfect information, selected the alternative of "inspection," he would incur costs of $1,000; this is $800 more than the alternative optimum decision to refund. The opportunity losses for lots B, C, D, and E are similarly computed. The expected opportunity losses (EOL) for inspection and refund are therefore

$$\text{EOL (inspection)} = \sum_{i=A}^{B} P(X_i) \cdot X_i$$

$$= 0.35 \times \$800 + 0.30 \times \$500$$
$$= \$430.$$

$$\text{EOL (refund)} = \sum_{i=C}^{D} P(X_i) \cdot X_i$$

$$= 0.10 \times \$500 + 0.05 \times \$1,000$$
$$= \$100.$$

The optimum decision prior to obtaining the additional sampling information is to continue production as it is currently being performed; i.e., continue to let the machines produce the defective parts and refund $1.00 for each defective part that is returned. The expected optimum cost of this prior decision is $670, while the expected cost for the alternative decision to inspect is $1,000. The cost of uncertainty before the sample, or the prior expected value of perfect information (EVPI$_0$), is $100. This is, of course, identical with the lowest expected opportunity loss (EOL).

159

Optimum decision after sampling—conditional posterior optimum decision.
Let us see how the prior optimum decision is modified after additional
sampling information is added. The additional sampling information in the
Nadar Company case consisted of a random sample of 20 in which one sample was found to be defective. How does this additional information modify
and/or improve the ultimate decision that is made; i.e., does it reduce the
amount of uncertainty in the problem?

Table 6.3 illustrates the steps that are required to derive the posterior
probability distribution (column 5) and the corresponding conditional
posterior optimum decision. Columns (1) and (2) were reproduced from
columns (1) and (2) in Table 6.2. Column (3) shows the conditional probabilities that one defective part in the 20 random samples might have come
from each of the five different lots. For example, the conditional probability
of 0.273 in lot A was computed as follows: As indicated in column (1) in
Table 6.3, lot A contains 2% defective parts. According to Appendix Table
D.1, the probability of obtaining one defective part among 20 random
samples, when the population probability is 0.02, is 0.273, or,

$$P(X = 1|n = 20, P = 0.02) = 0.273.$$

The conditional probability of 0.273 means that the chances are 0.273 that
the one defective part among the 20 random samples came from lot A.[2]

The conditional probabilities for the other lots were similarly computed
as follows:

$$
\begin{array}{lll}
\text{Lot B:} & P(X = 1|n = 20, P = 0.05) & 0.377, \\
\text{C:} & P(X = 1|n = 20, P = 0.10) = 0.270, \\
\text{D:} & P(X = 1|n = 20, P = 0.15) = 0.137, \\
\text{E:} & P(X = 1|n = 20, P = 0.20) = 0.058.
\end{array}
$$

The conditional probabilities contained in column (3) only show the
chances that one defective part might have come from each of the different
lots. As column (2) in Table 6.3 indicates, there is variance concerning the
probabilities that each machine produces different types of lots. The probability that the machine produces a different type of lot and that one defective
part is found in the different lot is simply a joint probability; this is illustrated
in column (4). For example, the joint probability that a machine produces
lot A *and* one defective in 20 randomly selected samples is computed by using
Equation 2.8, or

$$P(X_A) \times P(X = 1|n = 20, P = 0.02)$$
$$= (0.35) \times (0.273) = 0.0955, \quad \text{or } 9.55\%.$$

In other words, the probability is 0.35 that a machine produces lot A (column 2), and the chances are 0.273 that one defective part in a sample of 20

---

[2] For a detailed explanation concerning the use of Appendix Table D.1, refer to Section 3.6.

Table 6.3 / Decision Matrix for the Nadar Parts Company
(After Sampling when $X = 1$, $n = 20$, $P = X_i$)

Lot	Defective rate (X_i) (1)	Prior probability $P(X_i)$ (2)	Conditional probability $P(X = 1\|n = 20, P = X_i)$ (3)	Joint probability: $P(X_i) \cdot P(X = 1\|n = 20, P = X_i)$ (4) = (2) · (3)	Posterior probability (5) = (4) ÷ $\sum$(4)
A	0.02	0.35	0.273	0.0955	0.3420
B	0.05	0.30	0.377	0.1131	0.4051
C	0.10	0.20	0.270	0.0540	0.1934
D	0.15	0.10	0.137	0.0137	0.0491
E	0.20	0.05	0.058*	0.0029	0.0104
		1.00		0.2792†	1.0000

* Students may be curious as to why the conditional probability of lot E is smaller than the other probabilities, even though the population probability (see column 1) for lot E is the highest among the five classifications. The reason for this apparent discrepancy is that although lot E contains the most defective items, it is highly unlikely to find only one defective item in the 20 samples that could have come from lot E; i.e., we would usually expect about four defective items from lot E ($\mu = np = 20 \times 0.2 = 4$), not just one. As shown in column 3 in Table 6.5, when four defective items appear among 20 samples, the conditional probability that those four items could have come from lot E is, indeed, higher than any of the conditional probabilities for the other lots.
† Marginal probability.

might have come from lot A (column 3). Therefore, the joint probability that machine A produces both lot A *and* the one defective part from a sample size of 20 is

$$0.35 \times 0.273 = 0.0955.$$

The joint probabilities for the other lots are similarly computed as follows:

Lot B:　　$0.30 \times 0.377 = 0.1131,$

　C:　　$0.20 \times 0.270 = 0.0540,$

　D:　　$0.10 \times 0.137 = 0.0137,$

　E:　　$0.05 \times 0.058 = 0.0029.$

The sum of the joint probabilities (0.279) contained in column (4) is simply the marginal probability. This marginal probability implies that there is a 27.92% chance of finding one defective part in a sample size of 20. It is important to realize that the marginal probability of 0.2792 is only one of 21 possible marginal probabilities that could arise in this problem. The marginal probability of 0.2792 is valid only when 1 of 20 samples is defective. If, for example, the sample results were to show three defective parts, the marginal probability would be different because the sampling results vary and therefore the conditional probabilities also differ; furthermore, the joint probabilities, which are based on these conditional probabilities, would also change. In other words, there are 21 possible marginal probabilities in a sample of 20, depending on whether the sampling results show no defective parts, 1 defective part, 2 defective parts $\cdots$ or 20 defective parts.[3]

The posterior probabilities contained in column (5) are computed by dividing each joint probability contained in column (4) by the marginal probability (0.2792). The logic behind this computation was discussed in detail under "Bayes' Theorem" in Section 2.4.

The significance of this posterior probability is obvious. The use of additional sampling information is assumed to reduce the degree of uncertainty in the problem. The prior probability contained in column (2) is, therefore, replaced by the posterior probability in column (5) as the primary decision-making variable or criterion. Table 6.4 shows an updated decision matrix for the Nadar Company.

As Table 6.4 illustrates, even with the additional sampling information the conditional posterior optimum decision is still to refund. The cost to the company, however, is now $558.80 (instead of the prior cost of $670), and the posterior expected value of perfect information ($EVPI_1$) is now $34.95 (in lieu of the prior value of $100). The computations of the data contained in columns (3), (4), (5), and (6) are similar to those performed to establish the corresponding elements in Table 6.2. Two factors should be

---

[3] For a complete list of all possible marginal probabilities applicable to this problem, see Table 6.7.

Table 6.4 / Posterior Optimum Decision for the Nadar Company
when $X = 1$, $n = 20$

Rating	Defective rate (1)	Posterior probability (2)	Cost, $ Inspection (3)	Cost, $ Refund (4)	Opportunity loss, $ Inspection (5)	Opportunity loss, $ Refund (6)
A	0.02	0.3420	1,000	200	800	0
B	0.05	0.4051	1,000	500	500	0
C	0.10	0.1934	1,000	1,000	0	0
D	0.15	0.0491	1,000	1,500	0	500
E	0.20	0.0104	1,000	2,000	0	1,000
		1.0000	1,000*	558.80*	476.15	34.95†

* Posterior expected cost.
† EVPI$_1$.

emphasized concerning the differences between Table 6.2 and 6.4. First, although the additional sampling information did not alter the prior decision to refund, management had previously felt (as shown in Table 6.2) that the expected proportion of defective parts produced prior to sampling was

$$E(\text{proportion})_0 = \sum_{i=A}^{E} P(X_i) \cdot X_i$$

$$= 0.35 \times 0.02 + 0.30 \times 0.05 + 0.20 \times 0.10$$
$$+ 0.10 \times 0.15 + 0.05 \times 0.20$$
$$= 0.067.$$

The additional information in the form of sampling results (i.e., 1 defective part in a sample size of 20) shows, however, that the expected proportion of defective parts decreased to

$$E(\text{proportion})_1 = \sum_{i=A}^{E} P(X_i) \cdot X_i$$

$$= 0.3420 \times 0.02 + 0.4051 \times 0.05 + 0.1934$$
$$\times 0.10 + 0.0491 \times 0.15 + 0.0104 \times 0.20$$
$$= 0.05589 \equiv 0.056.$$

Secondly, the posterior optimum decision derived in Table 6.4 is based on the assumption that only 1 defective part was found in a sample of 20. This decision might be further altered as additional sampling results are evaluated and the subsequent results vary; this situation is illustrated in Example 6.2 cited below.

Table 6.5 / Posterior Probability for the Nadar Company when $X = 4$, $N = 20$

Rating	Defective rate (X_i) (1)	Prior probability $P(X_i)$ (2)	Conditional probability $P(X = 4 \mid n = 20, P = X_i)$ (3)	Joint probability (4) = (2) · (3)	Posterior probability (5) = (4) ÷ $\sum$(4)
A	0.02	0.35	0.001	0.0004	0.0078
B	0.05	0.30	0.013	0.0039	0.0760
C	0.10	0.20	0.090	0.0180	0.3508
D	0.15	0.10	0.182	0.0182	0.3548
E	0.20	0.05	0.218	0.0108	0.2106
		1.00		0.0513*	1.0000

* Marginal probability.

164

EXAMPLE 6.2 / As discussed, when we take 20 samples, there are actually 21 possible sampling results. Let us consider one more possible result.

Suppose that the sampling results indicate that 4 parts among 20 samples are defective. What is the posterior optimum decision?

The optimum posterior decision and the posterior expected value of perfect information with the additional sampling information is shown in Table 6.6.

SOLUTION: The posterior optimum decision, when sampling results show 4 defective parts in a sample of 20, would be to inspect the machines; this is true because the inspection option costs only $1,000, whereas the alternative decision (i.e., to refund) costs the company $1,343.76. The corresponding posterior expected value of perfect information ($EVPI_1$) is now $44.24.

There might be some confusion as to the interpretation of the various optimum decisions derived in Tables 6.2, 6.4, and 6.6. The optimum decision in Tables 6.2 and 6.4 is to refund at a cost of $670 and $558.80, respectively, while the optimum decision reached when the additional sampling information in Table 6.6 was used is to shut down and inspect at a cost of $1,000. All these decisions are, however, based entirely on a conditional concept. For example, the prior optimum decision based on the data contained in Table 6.2 was derived from previously existing information only, while the posterior optimum decision derived in Table 6.4 was based upon the assumption that the results of sampling revealed only *1 defective part*. The posterior optimum decision illustrated in Table 6.6 was derived upon the assumption that *4 of the 20 sampled parts were*

Table 6.6 / Posterior Optimum Decision for the Nadar Company when $X = 4, n = 20$

Rating	Defective rate (1)	Posterior probability (2)	Cost, $ Inspection (3)	Cost, $ Refund (4)	Opportunity loss, $ Inspection (5)	Opportunity loss, $ Refund (6)
A	0.02	0.0078	1,000	200	800	0
B	0.05	0.0760	1,000	500	500	0
C	0.10	0.3508	1,000	1,000	0	0
D	0.15	0.3548	1,000	1,500	0	500
E	0.20	0.2106	1,000	2,000	0	1,000
		1.0000	1,000*	1,343.76*	44.24†‡	388.00‡

* Posterior expected cost.
† $EVPI_1$.
‡ Posterior EOL.

defective. Posterior optimum decisions vary, therefore, depending upon subsequent or additional sampling results. This is why the posterior optimum decision is sometimes called a *conditional posterior optimum decision.*

6.3 / Expected Posterior Optimum Decision

As indicated in Section 6.2, there is a maximum of 21 possible conditional posterior optimum decisions associated with a sample size of 20. Table 6.7 illustrates all 21 possible outcomes in the Nadar Company case. Since the individual marginal probability of obtaining more than six defective parts is extremely small, one cumulative marginal probability was used for any outcome that contained more than six defective parts.

In Table 6.7, column (1) shows the number of possible defective parts that can be obtained from the sample size of 20; the corresponding marginal probabilities for each of these possible outcomes are located in column (2). The marginal probabilities in this column indicate the probability of occurrence associated with each corresponding number of defective parts contained in column (1). For example, the chances are 0.3701 (or 37%) that no defective parts will be discovered from the samples; the chances are 0.2792 that only one defective part will be found, and so on. Since all the possibilities for obtaining defective parts from the total sample are exhausted, the sum of the marginal probabilities must be equal[4] to 1.0.

Table 6.7 / Expected Posterior Cost for the Nadar Company
for $X = i, n = 20$

Number of defective parts (1)	Marginal probability (2)	Conditional posterior optimum exp. cost, $ (3)	Conditional posterior alternative exp. cost, $ (4)	Posterior optimum decision (5)	Expected posterior optimum cost (EPOC), $ (6) = (2) × (3)
0	0.3701	356.49	1,000.00	Refund	131.94
1	0.2792	558.80	1,000.00	Refund	156.02
2	0.1619	846.16	1,000.00	Refund	136.99
3	0.0930	1,000.00	1,124.06	Inspect	93.00
4	0.0513	1,000.00	1,343.76	Inspect	51.30
5	0.0261	1,000.00	1,523.00	Inspect	26.10
6	0.0118	1,000.00	1,647.00	Inspect	11.80
≥7	0.0070	1,000.00	1,771.45	Inspect	7.00
	1.0004				614.15

[4] Note the discrepancy of 0.0004 due to a rounding error.

Column (3) in Table 6.7 indicates the conditional posterior optimum expected cost. For example, if the results of the sampling show no defective parts, the posterior optimum decision would be to refund at a cost of $356.49 because the cost of the alternative decision (i.e., to inspect) is $1,000 (column 4). If, on the other hand, the sampling results yield four defective parts, the optimum decision would be to inspect, with its associated cost of $1,000, because the cost for the alternative decision (to refund) is $1,343.76. From columns (3), (4), and (5) it can be seen that if the sampling results indicate more than 2 defective parts among 20 samples, the posterior optimum decision is to inspect and adjust the machines.

Column (6) in Table 6.7 shows the expected posterior optimum cost (EPOC). This *expected* posterior optimum cost should be differentiated from the *conditional* posterior optimum expected cost contained in column (3). This latter cost indicates that if the decision maker acts rationally (based on the additional sampling results), he will minimize his expected costs. For example, if the sampling results show no defective parts, the rational decision for a decision maker is to refund (column 5) at a cost of $356.41 (column 3); if the sampling results indicate five defective parts, the rational choice for a decision maker is to inspect and adjust the machines at a cost of $1,000; and so on. The likelihood that the decision makers in the Nadar Company will find no defective parts or as many as five defective parts among random samples of 20 are 0.3701 and 0.0261, respectively (see column 2). The expected posterior optimum cost (column 6) is the product of the marginal probability (column 2) and the conditional posterior optimum expected cost (column 3).

The sum of the values in column (6) is the expected posterior optimum cost. This means that management can expect to spend an average of $614.15 per decision period for either refunding or inspecting machines, depending upon the results of the posterior optimum decision (column 3). In other words, management should, at a minimum, allocate this amount ($614.15) for each decision as a projected expenditure for producing defective products during any given time period.

6.4 / Characteristics of the Posterior EVPI$_1$

It is interesting to examine the characteristics or behavior of the posterior EVPI$_1$. Table 6.8 shows the number of defective parts obtained from the sample size of 20 and the corresponding posterior[5] EVPI$_1$. The posterior EVPI$_1$ started with relatively small amounts ($6.95) when no defective parts were found and increased in size up to a maximum point when three defective parts were found ($117.39). After this point, the posterior EVPI$_1$ declined as the number of defective parts increased, becoming almost zero

---

[5] See Tables 6.4 and 6.6 for the computation of the posterior EVPI$_1$.

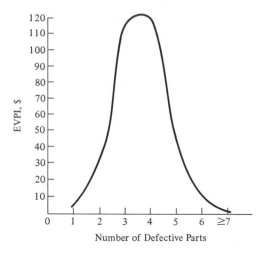

Figure 6.1 / Characteristics of the posterior
EVPI when $n = 20$.

when the sampling results yielded six or more defective parts. The behavior
is readily apparent when the amount of $EVPI_1$ is illustrated in a diagram
with the corresponding number of defective parts. In Figure 6.1 the Y axis
indicates the amount of posterior $EVPI_1$ and the X axis records the actual
sampling results. As shown, the posterior $EVPI_1$ is at its maximum point
when the sampling results yield three defective parts.

It is not too difficult to understand the type of behavior shown in
Figure 6.1. Let us take two extreme cases: (1) when the sampling results
indicate no defective parts ($X = 0|n = 20$) or (2) indicate six defective parts
($X = 6|n = 20$). In both cases, the required course of action for the decision
maker is clear. In case (1), the optimum decision is, of course, to refund at a
cost of $356.49 (column 3 in Table 6.7). This decision is obviously less

Table 6.8 / Posterior Expected Value of Perfect
Information when $X = i$, $n = 20$

No. of defective parts	Amount of posterior EVPI, $
0	6.95
1	34.95
2	112.70
3	117.39
4	44.24
5	11.50
6	0+
≥ 7	0+

expensive than the alternative decision to inspect ($1,000). Applying this same logic to case (2), it is relatively easy for the decision maker to select the optimum choice of "inspection," since the sampling results show a large ratio of defective parts (i.e., six). In either case, the degree of uncertainty and the subsequent cost of uncertainty should be smaller, owing to a conclusive sampling result.

If the sampling results indicate the number of defective parts as $0 < X < 6$, for example, where X represents a defective part, it is a great deal more difficult for the decision maker to arrive at an optimum decision. A good example of this condition is when the sampling results show three defective parts. As Table 6.7 indicates, the optimum decision under this condition is to inspect the machines at a cost of $1,000. The cost of the alternative decision (to refund at $1,124.09), however, is close to that of the optimum decision. This closeness of costs between the two alternative courses of action creates a great deal of uncertainty in the decision-making process. The selection of the wrong course of action by management in this case (due to inaccurate or incorrect sampling techniques, biased sampling techniques, inaccurate statistical analysis, etc.), while appearing to be only a minor statistical error, may result in heavy economic penalties. It follows, therefore, that even though the pure mathematics of the results indicate an optimum choice, in reality, both the degree of uncertainty and the subsequent or associated cost of perfect information in this case are relatively high.

6.5 / Expected Value of Sampling Information (EVSI)

Up to this point, we assumed that the decision maker chose to select additional samples that would increase the information base and help him derive an optimum decision. However, as every decision maker knows, sampling involves both increased cost and time. In this and subsequent sections, we will discuss whether it is worthwhile taking samples and, if it is, what the minimum, maximum, and optimum sample sizes should be.

The primary purpose of taking a sample or of obtaining additional information is to reduce uncertainty, since additional sampling information often improves the states of nature surrounding a problem and thus facilitates the choice of an optimum decision. Mathematically, since the degree of uncertainty is inversely related to the number and size of each sample, the larger the sample size, the smaller the degree of uncertainty.

Logically, EVSI is the difference between the expected monetary value of the prior sample and that of the posterior sample, or

$$\text{EVSI} = (\text{EMV prior to sample}) - (\text{EMV posterior to sample}). \quad (6.1)$$

In the Nadar Company case, for example, the expected cost prior to sampling was $670 (Table 6.2) and the expected posterior optimum cost was

$614.15 (Table 6.7). Therefore,

$$EVSI = \$670.00 - \$614.15 = \$55.85.$$

The logic for using Equation 6.1 to compute the EVSI in the Nadar problem can be explained as follows: Prior to taking the sample, management felt that the optimum decision was to refund at a cost of $670, since the alternative decision to inspect would cost $1,000. This optimum cost was, however, reduced to $614.15 after samples were taken. This expected reduction in cost (from $670 to $614.15) was made possible through sampling information only. The value of the sample is therefore $55.85.

6.6 / Expected Net Gain from the Sample (ENGS)

The EVSI is the value of sampling information without considering sampling cost. In other words, EVSI is the gross value of the sampling information. On the other hand, if the sampling information involves some costs, as almost all samples do, this sampling cost should be subtracted from the gross value of sampling information (EVSI); the net result is called the "expected net gain from sampling," or ENGS. Mathematically, ENGS is the difference between the EVSI and the cost of sampling, or

$$ENGS = EVSI - cost_n, \tag{6.2}$$

in which $cost_n = a + bn$, where

a = constant or fixed cost of taking the sample,

b = variable cost of taking one sample,

n = sample size.

Assume that the sampling cost for the Nadar Company has the following cost function:

$$Cost_n = \$10.00 + 0.25n.$$

Then the total cost of taking a sample size of 20 is

$$
\begin{aligned}
Cost\ (20) &= \$10.00 + \$0.25\ (20) \\
&= \$10.00 + \$5.00 \\
&= \$15.00.
\end{aligned}
$$

The net gain of sampling information in the preceding example is therefore

$$ENGS = \$55.85 - \$15.00 = \$40.85.$$

The positive value of the ENGS in this problem implies that it is worthwhile to take a sample size of 20. As a matter of fact, management may take as many additional samples as the EVSI allows; i.e., as long as the ENGS remains positive. In other words, the maximum number of samples (n_x)

within the EVSI limit is

$$\text{Cost}_{n_x} \leq \text{EVSI}_n$$

or

$$a + bn_x \leq \text{EVSI}_n,$$

$$n_x \leq \frac{\text{EVSI}_n - a}{b}, \tag{6.3}$$

where EVSI_n is the expected value of sampling information for a sample of size n.

In the Nadar Company case, using a cost function of $\text{cost}_n = 10 + 0.25n$, the maximum number of samples (n_x) would be

$$a + bn_x \leq \text{EVSI}_n,$$

$$10 + 0.25n_x \leq \text{EVSI}_n,$$

$$n_x \leq \frac{\text{EVSI}_n - 10.00}{0.25}.$$

In other words, the management of the company can take as large a sample as n_x in an effort to reduce uncertainty. Any sample size beyond n_x would, however, cost more than the additional information derived from the sample could contribute toward reducing uncertainty.

The maximum number of samples within the EVSI limit (i.e., n_x) should not be confused with the maximum number of samples within the prior EVPI_0 limit. As explained in Section 5.3, the practical implication of the EVPI_0 is that management is allowed to spend up to the amount indicated by the EVPI_0 to reduce uncertainty. EVPI_0 does not, however, tell management how and where it should spend company funds to achieve these maximum results. In other words, investing in additional sampling information is only one of many ways that management can choose to spend funds to achieve a more optimal solution. But, unlike other expenses in business, the expenses associated with the EVPI cannot be eliminated or even minimized; this is due to the fact that the EVPI is, by definition, already a minimum expense. Any attempt to economize on the EVPI, therefore, only widens the area of uncertainty, and this, of course, eventually reduces the ultimate expected gain of the optimum decision.

If management so desires, the entire expense associated with the EVPI_0 can be spent on taking additional samples, even if this figure exceeds the maximum sample size indicated by the EVSI. Under these circumstances, the maximum number of samples within the EVPI_0 is

$$\text{Cost}_n \leq \text{EVPI}_0,$$

$$a + bn_x \leq \text{EVPI}_0,$$

$$n_x \leq \frac{\text{EVPI}_0 - a}{b}. \tag{6.4}$$

In the Nadar Company case, where the $EVPI_0 = \$100$ (refer to Table 6.2), the maximum number of samples is

$$10 + 0.25n_x \leq 100,$$

$$n_x \leq \frac{100 - 10}{0.25},$$

$$n_x \leq 360.$$

In other words, if management feels that a very large sample size will facilitate and ultimately optimize the decision-making process (in terms of reducing uncertainty), they may take a sample with a size as large as 360 and still not exceed the $EVPI_0$.

6.7 / Optimum Sample Sizes

As an aid to reducing uncertainty, large sample sizes are normally preferred to small ones.[6] Large sample sizes, however, by their very nature have several disadvantages when compared with a relatively small sample. Since the cost of the sample naturally plays an important role in the decision-making process, it is obvious that a large sample costs more than a small sample. If the sampling cost is higher than the expected gains to be received from the results of the sample, it is certainly not desirable to take a large sample.

An optimum sample size is defined as that size which produces the maximum net gain. Figure 6.2 illustrates a method for determining the optimum sample size and also shows the relationship between the optimum sample size, the minimum acceptable sample size, the maximum sample size *within the EVSI limit*, and the maximum sample size *within the EVPI_0 limit*. In Figure 6.2 the size of the sample is located on the X axis, while the $EVPI_0$, EVSI, ENGS, and costs of the corresponding sample sizes are all located on the Y axis. It is further assumed that the principle of marginal diminishing returns is considered in computing and analyzing the results of the expected value of sampling information. In other words, the first several samples provide the most valuable information; as more samples are added, the value, in terms of the utility that is derived from these additional samples, decreases.

From Figure 6.2 it can be seen that sample size n_1 is the minimum sample size. At sample size n_1, the total cost of the sample is C_1; accordingly, the total gain derived from this sample (EVSI) is S_1, where $C_1 = S_1$. Any sample size smaller than n_1, such as n_0, is both unrealistic and undesirable. This is

---

[6] This statement assumes, of course, that any samples taken, whether large or small, are selected random unbiased, representative, well compiled, and thoroughly analyzed.

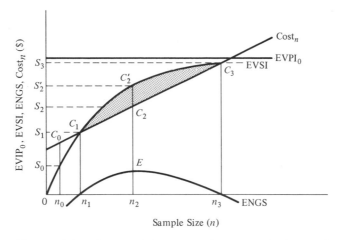

Figure 6.2 / Optimum, minimum, maximum sample sizes, and the cost of sampling.

true because the total sampling cost at n_0 is C_0, whereas the total gain obtained from the sample is S_0 and, in this case, $S_0 < C_0$.

According to Figure 6.2, any sample size greater than n_1 should yield a net positive gain (ENGS) until a sample size of n_3 is reached. From n_1 to n_2, the slope of the EVSI is higher than the slope of the cost line. This indicates that the marginal gain (MG) derived from the sample is greater than the marginal cost (MC) of taking the sample.[7] Therefore, since management gains more than it spends by taking the extra sample from n_1 to n_2, it is desirable to take additional samples up to the n_2 limit. In other words, it is advantageous to take additional samples up to the point where the marginal gain derived from the additional sampling information equals the marginal cost of obtaining the information; at this point, $MG = MC$, and we have reached the optimum sample size. As Figure 6.2 illustrates, the slope of the cost line at C_2 is identical to that of the EVSI at C_2'. Furthermore, at this optimum sample size, the total cost of the sampling information is C_2, while the total gain is S_2', where $C_2 < S_2'$; as a result the total net gain is E.

At all points subsequent to n_2, however, the slope of the cost line is higher than that of the EVSI. This indicates that the marginal gain is smaller than the marginal cost of the sampling information. As a result, the margin of net gain becomes smaller until point n_3 is reached; beyond this point, the net gain becomes negative. It is important to note that although the marginal cost is higher than the marginal gain in the region (from n_2 to n_3), the total cost is still smaller than the total gain. Sample sizes greater than n_3 will, however, not only result in a marginal cost that exceeds the marginal gain, but will also

---

[7] It should be emphasized that marginal gain (cost) refers only to the extra or incremental (not the total) gain (cost) that results from taking the extra sample.

Table 6.9 / Optimum Sample Sizes for the Nadar Parts Company*

Sample size (1)	Prior optimum cost, \$ (2)	EPOC, \$ (3)	EVSI, \$ (4) = (2) − (3)	Sampling cost, \$ \$10 + \$0.25n (5)	ENGS, \$ (6) = (4) − (5)
15	670	622.93	47.07	13.75	33.32
20	670	614.15	55.85	15.00	40.85
30	670	602.15	67.85	17.50	50.35
40	670	597.33	72.67	20.00	52.67
50	670	589.11	80.89	22.50	58.38
55	670	587.24	82.76	23.75	59.01
60	670	585.27	84.73	25.00	59.73
65	670	584.30	85.70	26.25	59.45
70	670	583.45	86.55	27.50	59.05
80	670	581.28	88.72	30.00	58.72
90	670	579.35	90.65	32.50	58.15
100	670	576.97	93.03	35.00	58.03

* The computation of the ENGS for many different sample sizes can rapidly become very time consuming. In most practical applications where a series of ENGS values are required, the computations are performed with the aid of a computer.

result in the *total* cost exceeding the total gain. The maximum sample size is therefore n_3; at this point, the total cost is equal to the total gain, or $C_3 = S_3$, thus leaving ENGS equal to zero.

Theoretically, it is possible that the value of $EVPI_0$ may intersect EVSI. This would imply that the additional sampling information has completely eliminated uncertainty. In reality, however, the EVSI never meets $EVPI_0$, for (as mentioned before) regardless of the sample size, complete elimination of uncertainty in forecasting future events is not possible, even under a very tightly controlled environment.

In the Nadar Company case, the optimum sample size is approximately 60. This conclusion is shown in Table 6.9.

There are some instances where additional sampling information is not worth using. As illustrated in Figure 6.3, if the sampling cost is higher than the expected gains to be derived from the sample, the best decision is *not* to take *any* sample. Since the total cost of sampling in the chart is greater than the total gain throughout the entire sampling zone, there is no minimum, maximum, or optimum sample size.

EXAMPLE 6.3 / The Hill-Trail Electronic Company recently completed preliminary manufacturing tests of a new electronic desk calculator. The new calculator is more efficient, more compact, and less expensive than a current competitor's well-known brand. The market potential of the Hill-Trail product is based mainly upon the results of a market survey of the 250

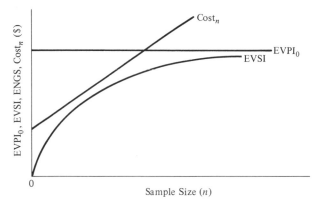

Figure 6.3 / Sample sizes, sampling cost, and the expected value of sampling when $cost_n$ > EVSI for all sample sizes.

largest wholesale distributors of electronic desk calculators. Management is uncertain, however, as to the total number of wholesalers who will accept the new Hill-Trail product for sale to their retail outlets. For analytical purposes, and prior to performing actual survey, the company has assigned probabilities concerning the distribution potential of their product, as listed in Table 6.10.

The company management has decided that if it can reasonably expect that 50% or more of the 250 largest customers will be willing to market its product, the company will manufacture and market the electronic desk calculator. An initial sample of 25 wholesalers reveals that 13 customers are interested in marketing the product. What is the prior and posterior optimum decision?

SOLUTION: Table 6.11 illustrates a complete solution of this problem. Column (1) shows the proportion of wholesalers that the company estimates

Table 6.10 / Prior Probability
for the Hill-Trail
Electronic Company

Number of wholesalers who will market the product	Probability
50	0.1
100	0.3
150	0.5
200	0.1
	1.0

Table 6.11 / Posterior Probability Distribution for the Hill-Trail Electronic Company

Proportion of wholesalers who would market (X_i) (1)	Prior probability $P(X_i)$ (2)	Conditional probability $P(X = 13\|n = 25, P = X_i)$ (3)	Joint probability $P(X_i) \times P(X = 13\|n = 25, P = X_i)$ $(4) = (2) \cdot (3)$	Posterior probability $(5) = (4) \div \sum(4)$
0.2 (50/250)	0.1	0+	0+	0+
0.4 (100/250)	0.3	0.076	0.0228	0.2853
0.6 (150/250)	0.5	0.114	0.0570	0.7134
0.8 (200/250)	0.1	0.001	0.0001	0.0013
			0.0799*	1.0000

* Marginal probability.

would be interested in marketing its product. Column (2) contains the prior probabilities, while column (3)—computed using Appendix Table D.1—contains the conditional probabilities. As shown in column (3), there is practically no chance that the 13 customers, of the 25 who expressed a willingness to market the product, came from this first group; there is a 0.076 chance that the 13 potential customers came from the second group, and so on. The marginal probability of 0.0799 implies that the chance is almost 8% that the 13 potential customers, in a sample of 25, will actually be willing to market the product.

The current sampling information seems to substantiate the company's prior decision that it should market the product, since the expected posterior sample proportion does not deviate significantly from the original estimate. This fact is illustrated by

$$E(\text{prior}) = \sum_{i=1}^{4} P(X_i) \cdot X_i \qquad (\text{column } 2 \times \text{column } 1)$$

$$= 0.1 \times 0.2 + 0.3 \times 0.4 + 0.5 \times 0.6 + 0.1 \times 0.8$$

$$= 0.52.$$

$$E(\text{posterior}) = \sum_{i=1}^{4} P(X_i) \cdot X_i \qquad (\text{column } 1 \times \text{column } 5)$$

$$= (0+) \times 0.2 + 0.2853 \times 0.4 + 0.7134 \times 0.6$$

$$+ 0.0013 \times 0.8$$

$$= 0.5431.$$

In other words, prior to the sample, the company projected that the average proportion of customers who would market its product would be 0.52. The sampling results of 0.5431 more or less confirmed that this projection would be validated. The prior and posterior optimum decisions are to market the products, since both prior and posterior expected proportions are higher than the expectation (0.5).

6.8 / Summary

The main purpose of this chapter has been to show the various methods and techniques that decision makers need and use to make an optimum decision under conditions of uncertainty when additional sampling information is provided. Unlike the decision-making process explained in Chapter 5, where the optimum decision was based only on given or existing information, the concepts discussed in this chapter have employed new information (posterior probability) as well as existing information (prior probability). The following summary should be helpful to the student in understanding the contents of this chapter:

(1) The posterior optimum decision always has the largest posterior expected monetary value.

(2) The posterior $EVPI_1$ is the lowest posterior expected opportunity loss among at least two alternative courses of actions.

(3) The expected posterior optimum decision is the one based on all possible marginal probabilities in any given sample size.

(4) The EVSI is the difference of value between the prior $EVPI_0$ and the optimum posterior expected value per decision period.

(5) The ENGS is the net value of sampling information when sampling costs are also considered. Mathematically, $ENGS = EVSI - Cost_n$.

(6) If the sampling cost is higher than the EVSI, ENGS becomes negative, which implies that samples should not be taken.

(7) The optimum sample size occurs when the marginal gain is the same as the marginal cost, at which point ENGS is at its maximum. The minimum sample size occurs when the EVSI line cuts the cost line from below, and the maximum sample size occurs when the EVSI line cuts the cost line from above.

(8) In reality, the EVPI line never meets the EVSI line. This implies that additional information obtained from a sample never completely eliminates uncertainty.

EXERCISES

1. Discuss the differences between a prior decision-making process and the posterior decision-making process.

2. Differentiate between the concept of a posterior expected optimum value and the expected posterior optimum value.

3. Explain the role of sampling in decision theory.

4. Discuss the difference between the maximum sample size within the EVSI limit and the maximum sample size within the $EVPI_0$ limit.

5. Discuss the following statement: "A decision maker, if he so desires, could take samples beyond the optimum sample size range, but not more than the maximum sample size within the EVSI limit."

6. Explain the implication of the statement that the $EVSI_0$, in reality, never exceeds the EVPI.

7. Using Tables 6.7 and 6.8 pertaining to the Nadar Auto Safety Company case in this chapter, confirm that when there are 2 defective parts among 20 samples, (a) the marginal probability is 0.1619, (b) the conditional posterior optimum expected cost is $846.16, and (c) the $EVPI_1$ is $112.70. (*Hint*: construct tables similar to Tables 6.5 and 6.6.)

8. The results of a survey (*M*) of 500 business executives across the country indicate the following distribution concerning the projected state of the economy in the next five years:

State of the economy, X_i	$P(X_i)$	$P(M\|X_i)$	$P(X_i)\cdot P(M\|X_i)$	$P(X_i\|M)$
Expansion (X_1)	0.3	0.7		
Status quo (X_2)	0.5	0.1		
Recession (X_3)		0.2		

(a) Fill in the missing numbers.

(b) Interpret the results both prior and posterior to the survey.

9. The management of the Columbo Soap Company is considering a design change of its label to attract consumer attention. The management has assigned an even chance for success (S) or failure (F) for the new design. The results of the initial test (X) in a limited market reveal that $P(X|S) = 0.2$ and $P(X|F) = 0.9$. What is the posterior optimum decision?

10. The following table gives partial information regarding the opportunity loss of accepting a lot of color television sets offered to a company for retail sale.

Proportion of defective parts	Prior probability	Opportunity loss, $
0.15	0.4	0
0.20	0.3	100,000
0.25	0.2	150,000
0.30	0.1	200,000

A preliminary sample of 20 television sets indicates that five sets are defective. Compute the EOL before and after sampling.

11. Mr. John Coke currently has $10,000 worth of Colorado Uranium Ltd. common stock. By selling the stock now, he would realize a 10% capital gain. The Wallstreet Report, however, has carried a story that there is a 40% chance that his stock will appreciate more than 10% in the next year; on the other hand, there is a 60% chance that it will appreciate less than 10% during the same time period.

Mr. Coke performed his own independent study and found that, historically, there has been a close relationship between the behavior of prices for uranium stocks in general and the prices quoted for oil stocks. Specifically, his investigation revealed that in 8 of 10 incidences in the past year, price increases for oil stocks were preceded by a sharp decline in the prices of uranium stock in general. On the other hand, in 2 of 4 incidences where uranium stock prices increased sharply, the prices of oil stocks also increased sharply.

A leading Wall Street analyst has determined that, owing mainly to an anticipated rapid economic recovery, there will be a strong demand for oil in the next year, with a corresponding increase in the price of oil stock. Should Mr. Coke sell his stock now?

12. An appliance wholesaler is considering handling a new brand of television sets. The order size is based on 1,000 sets each month, allocated so that 25% will be Class A; 25%, Class B; and 50%, Class C. In case the supplier cannot make the delivery of the specific order (i.e., 25%, 25%, and 50% for Class A, B, and C, respectively), the wholesaler will take any combination of television sets based on a maximum of

1,000 sets per month. The corresponding cost of a television set is $100 for Class A, $200 for B, and $300 for C. The dealer plans to set a retail price of $120 for Class A, $250 for B, and $400 for C. The supplier has informed him, however, that because of a shortage of certain materials, he cannot guarantee a complete shipment of his original order. He estimates that the chances are 0.8, 0.8, and 0.6, respectively, for full shipment of the three different types of sets. In order for the appliance dealer to stay in business, he must make at least a 25% net return per month on his cost. What are his prior and posterior optimum decisions?

13. The Fashion Shoe Manufacturing Company has designed a new style of sandal for women. Because the design requires a different manufacturing process, extensive tooling costs are required to produce the sandal. The production manager estimates these initial one-time tooling costs to be $1,000,000. The break-even sales point, excluding tooling costs, is estimated to be $10,000,000. The sales department forecasts that the women's sandal market for the coming year will be as follows:

Age groups, years	Size of market ($ million)	Prior probability
8–14	40	0.10
15–21	50	0.60
22–35	20	0.25
36	10	0.05

The company's share of the market has been 25%. The sales department assigned the tabulated prior probabilities as an estimate of the true states of nature. Since the company has to spend $1,000,000 for tooling costs, it performed a survey of major department stores which carry the company's brands. The results were as follows: There is a 50% chance that the market estimate for the 8–14 year-age group may materialize; a 40% chance for the 15–21 age-group market estimate; and a 10% chance for the 22–35 age-group market estimate. According to this survey, there is no basis for the estimate concerning the 36-year-and-over age group. What are the prior and posterior probabilities concerning the actual market potential for the new sandals? Should the company undertake the venture?

14. Each February, the Great Western Electronics Company offers to sell 2,000 shares of common stock to employees at market price. You have an opportunity to buy 50 shares of the stock at $50 per share. Based on the past experience of the stock, you are aware that its value generally follows the Dow Jones trend (DJ). If the DJ rises continuously for the next year, the stock may be worth $60 per share; however, on the other hand, if the DJ drops, the stock may be worth only $40 per share. Due to present economic trends, you assign a probability of 0.3 that the DJ would increase continuously for the next 12 months and 0.7 that it would drop for the next 12 months. Upon further investigation, you discover that for the last ten decision periods when the machine orders were up, the DJ index also had been up on nine occasions.

On the other hand, based on the same information (ten randomly selected decision periods when the machine orders were increasing), the overall DJ average actually decreased on two occasions. The latest report indicates that the machine orders for last January were up. Should you buy the stock? Assume that you have to borrow $2,500 at a 6% annual rate in order to make the purchase. Would you still be interested in buying the stocks? Would new information change the prior optimum decision? If it costs you $50 for better information, will it be worthwhile for you to spend that sum?

SUGGESTED READING

Bierman, Harold, Jr.; Bonini, Charles P.; and Hauserman, Warren H. *Quantitative Analysis for Business Decisions* (3d ed.). Homewood, Ill.: Richard D. Irwin, 1973, chapter 5.

Costis, Harry G. *Statistics for Business.* Columbus, Ohio: Charles E. Merrill, 1972, chapter 26.

Dyckman, T. R.; Smidt, S.; and McAdams, A. K. *Management Decision Making under Uncertainty.* New York: Macmillan, 1969, chapter 15, section 3; chapter 17, section 2.

Halter, Albert N., and Dean, Gerald W. *Decisions under Uncertainty with Research Application.* Cincinnati, Ohio: South-Western Publishing, 1971, chapters 4 and 10.

Hays, William L., and Winkler, Robert L. *Statistics: Probability, Inference, and Decision.* New York: Holt, Rinehart and Winston, 1971, chapter 9, sections 13 through 22.

Raiffa, Howard. *Decision Analysis: Introductory Lectures on Choices under Uncertainty.* Reading, Mass.: Addison-Wesley, 1968, chapter 2, sections 6 through 8.

Sasaki, Kyohei. *Statistics for Modern Business Decision Making.* Belmont, Calif.: Wadsworth Publishing, 1969, chapters 11 and 13.

Schlaifer, Robert. *Probability and Statistics for Business Decisions.* New York: McGraw-Hill, 1959, chapters 36 and 37.

Schlaifer, Robert. *Analysis of Decisions under Uncertainty.* New York: McGraw-Hill, 1969, chapters 10 and 11.

Schmitt, Samuel A. *Measuring Uncertainty: An Elementary Introduction to Bayesian Statistics.* Reading, Mass.: Addison-Wesley, 1969, chapter 3.

Spurr, William A., and Bonini, Charles P. *Statistical Analysis for Business Decision.* Homewood, Ill.: Richard D. Irwin, 1973, chapter 13.

Thompson, Gerald E. *Statistics for Decisions—An Elementary Introduction.* Boston: Little, Brown, 1972, chapter 10.

Winkler, Robert L. *Introduction to Bayesian Inference and Decision.* New York: Holt, Rinehart and Winston, 1972, chapter 3.

7 / Decision-Making Process with Prior Information and a Continuous Random Variable—Normal Distribution

7.1 / Purpose

The decision-making process discussed in Chapters 5 and 6 was based on the assumption that the states of nature comprised discrete random variables. There are many cases, however, in which the states of nature can be more realistically expressed in terms of a continuous distribution rather than as discrete distribution. For example, suppose that the potential yearly demand for a particular product is classified as high, average, or low; further, assume that the decision maker has assigned corresponding probabilities to the three classes of demand as 0.3, 0.5, and 0.2, respectively. In this case it is quite reasonable (and to some extent more realistic) to assume that the real probabilities associated with these states of nature may actually vary anywhere from zero to 1, and not precisely occupy the discrete points of 0.3, 0.5, or 0.2.

When the states of nature surrounding a problem are assumed to constitute a continuous distribution, the construction of a decision matrix becomes a great deal more complicated (if not impossible), since there is an almost infinite range of decision criteria that could be included in the matrix. Under these circumstances, the states of nature should be expressed in terms of a continuous function. The purpose of this chapter is to survey several statistical methods that can be used to derive an optimum decision under conditions where a continuous probability distribution exists. Among the several decision models that could be discussed, however, only the decision model with the normal probability function (with prior information) will be discussed in this chapter and in the following chapter (with posterior information).

7.2 / Assumptions

Several assumptions are inherent in the remaining contents of this chapter. These assumptions include the following:

(1) The probability distribution under consideration is normally dispersed around the mean value; i.e., a normal distribution.

(2) The problems under consideration are expressed in terms of a linear function with respect to the states of nature; e.g., $y = a + b\mu$, where a and b are constant, μ is a random variable, and y is the value to be optimized.

(3) Only a two-action case will be considered; i.e., there are only two choices available to the decision maker.

(4) The terminal decision made is based upon prior information only.

The complexity of problem solution and the amount of work required to derive an optimum decision is greatly reduced when the variables in a problem can be stated in terms of a linear function under the normal probability conditions. This is true because the expected value of a linear equation requires only the computation of the expected value of the random variable being analyzed (the concept was discussed in Chapter 3). Under conditions where a normal probability distribution is applicable, we can express the linear profit function as

$$y = a + b\mu. \tag{7.1}$$

In this case, the expected value of y is related directly to the expected value of μ, or

$$E(y) = E(a + b\mu)$$
$$= E(a) + E(b\mu)$$
$$= a + bE(\mu) \quad \text{(from Equations 3.5 and 3.6)}.$$

Let us examine the profit function (Equation 7.1) in terms of the following case.

A company is considering two courses of action, A_1 and A_2. The corresponding payoffs for these two alternatives are

$$A_1 = a_1 + b_1\mu, \tag{7.2}$$

and

$$A_2 = a_2 + b_2\mu, \tag{7.3}$$

where $b_1 > b_2$, and $a_1 < a_2$.

Two different methods will be considered here to determine the optimum decision between choices A_1 and A_2. First, the expected value of each course of action is computed and the course of action that provides the largest expected monetary value is selected as an optimum decision. If we take the expected value of Equation 7.2 and Equation 7.3,

$$E(A_1) = a_1 + b_1E(\mu), \tag{7.4}$$

and

$$E(A_2) = a_2 + b_2E(\mu). \tag{7.5}$$

If $E(A_1) > E(A_2)$, the choice A_1 should be selected; if, on the other hand, $E(A_1) < E(A_2)$, then alternative action A_2 should be taken.

Secondly, a break-even point (μ_k) is computed and compared with the random variable μ. The break-even point is defined as the amount of random variable μ that makes the expected values of A_1 and A_2 identical. In other words, at the break-even point,

$$E(A_1) = E(A_2).$$

From Equations 7.4 and 7.5,

$$a_1 + b_1 E(\mu) = a_2 + b_2 E(\mu).$$

Subtracting a_1 and $b_2 E(\mu)$ from both sides, we obtain

$$a_1 - a_1 + b_1 E(\mu) - b_2 E(\mu) = a_2 - a_1 + b_2 E(\mu) - b_2 E(\mu),$$
$$b_1 E(\mu) - b_2 E(\mu) = a_2 - a_1,$$
$$E(\mu)(b_1 - b_2) = a_2 - a_1.$$

Solving for $E(\mu)$,

$$E(\mu) = \frac{a_2}{b_1} \frac{a_1}{- b_2}. \tag{7.6}$$

To differentiate the expected value of μ at the break-even point from the usual expected value of μ (random variable) Equation 7.6 is further modified to read

$$\mu_k = \frac{a_2 - a_1}{b_1 - b_2}, \tag{7.7}$$

where the subscript k indicates the value at the break-even point.

The optimum decision, using Equation 7.7, depends upon the value of $E(\mu)$ in relation to μ_k. If $E(\mu) > \mu_k$, the optimum decision is to select alternative A_1; this is shown in Figure 7.1. On the other hand, if $E(\mu) < \mu_k$,

Figure 7.1 / Optimum decision when $E(\mu) > E(\mu)_k$ and $b_1 > b_2$.

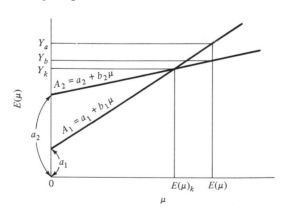

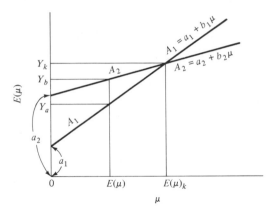

Figure 7.2 / Optimum decision when
$E(\mu) < E(\mu)_k$ and $b_1 > b_2$.

the optimum decision is to select choice A_2; this decision is illustrated in Figure 7.2.

In Figure 7.1, when $E(\mu) > \mu_k$, the expected value of y, using choice A_1, is Y_a, and the expected value of y, using choice A_2, is Y_b, where $Y_a > Y_b$. The optimum decision in this case is therefore to select alternative A_1. If, on the other hand, $E(\mu) < \mu_k$, as illustrated in Figure 7.2, the expected value of alternative A_1 is Y_a, and that of choice A_2 is Y_b, where $Y_b > Y_a$. The optimum decision in this case is therefore to select alternative A_2.

If the problem being considered is concerned with a cost function instead of a profit or revenue function (such as Example 7.7), the Y axis takes a negative sign. Therefore, if A_1 and A_2 are cost functions and $b_1 > b_2$, the optimum decision, when $\mu_k < E(\mu)$, is action A_2, since

$$-E(y)_b > -E(y)_a.$$

If, on the other hand, $\mu_k > E(\mu)$, the optimum decision is A_1, since

$$-E(y)_a > -E(y)_b.$$

It should be noted that if $b_1 > b_2$, then the value of a_1 should be greater than that of a_2; if this were not the case, there would be no intersection of the two linear equations to form a break-even point; this is shown in Figure 7.3. In Figure 7.3, since $b_1 > b_2$ and $a_1 > a_2$, the two decision lines, rather than reach a point of intersection, diverge more widely as the value of μ increases. Consequently, there is no break-even point for these two decision lines, and no subsequent decision-making problem, since course of action A_1 is always superior to alternative A_2.

Practical applications of the above decision-making process are contained in the following examples.

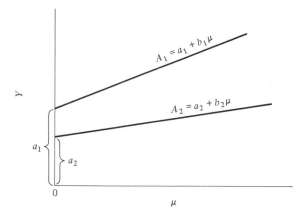

Figure 7.3 / Linear equations where $b_1 > b_2$
and $a_1 > a_2$.

EXAMPLE 7.1 / The Sintax Shorts Company is considering spending an additional $2,000,000 for television advertising to promote the sale of its products. The management of the company believes that the increased advertising expense will result in a more than commensurate increase in sales. If each pair of boxer shorts that the company sells results in a $0.50 net contribution (selling price minus variable cost), how many extra pairs of boxer shorts must be sold to recover the additional company advertising advertising expenses?

SOLUTION: If the company does not spend any additional money on advertising, it is assumed that no extra revenue will be generated. This means that the profit functions for the decision to advertise (W_1) or not to advertise (W_0) are, respectively,

$$W_1 = -2,000,000 + 0.5q \quad \text{and} \quad W_0 = 0,$$

where q = the number of boxer shorts to be sold. The break-even point, q_k, is accordingly computed as

$$q_k = \frac{a_2 - a_1}{b_1 - b_2}$$

$$= \frac{0 - (-2,000,000)}{0.5 - 0} = 4,000,000.$$

In order to recover the advertising expenses, the company must sell at least 4,000,000 extra pairs of boxer shorts.

EXAMPLE 7.2 / The Lions Club of Oak Farm, Maine, is considering bringing a circus into the community to raise funds for charity. A brief consultation with the circus agent revealed that the contract price required to obtain the service of the circus for a three-night performance is $5,000. The sponsor believes that if the price per ticket is set at $0.50, a total of 20,000 people will come to see the circus performance. Furthermore, he projects that there is a 50/50 chance that the range of attendance will be from 16,650 to 23,350 people. Assuming that there is a linear relationship between the total revenue received and the number of people who purchase tickets, what is the break-even value? Should the Lions Club contract to bring the circus to town?

SOLUTION: Let

C_1 = the decision to bring in the circus.

C_0 = the decision not to bring in the circus.

Then, according to Equation 7.2 and Equation 7.3, the profit functions for these two decisions are

$$C_1 = a_1 + b_1\mu = -5,000 + 0.5\mu$$
$$C_0 = a_2 + b_2\mu = 0.$$

Break-even value. According to Equation 7.7, the break-even value μ_k is computed as follows:

$$\mu_k = \frac{a_2 - a_1}{b_1 - b_2}$$

$$= \frac{0 - (-5,000)}{0.5 - 0} = 10,000,$$

or

$$\mu_k = 10,000 \text{ people.}$$

Optimum decision. According to the solution (1), the break-even value in this venture is 10,000 people; this attendance is a great deal less than the sponsor's expectation ($\mu = 20,000$). Accordingly, the optimum decision is to contract to bring in the circus, since $\mu > \mu_k$. Based on these results, this decision would produce an expected profit of

$$C_1 = -\$5,000 + \$0.5\mu$$
$$= -\$5,000 + \$0.5(20,0000)$$
$$= \$5,000.$$

The expected profit for the alternative decision is, of course, zero.

From the information given above, Figure 7.4 illustrates the relationship between the expected profit and the break-even value in this case. Any number of customers greater than 10,000 would generate a net profit, and any number less than 10,000 would contribute toward a loss.

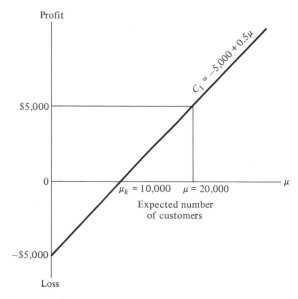

Figure 7.4 / Break-even value and expected number of circus customers.

7.3 / Opportunity-Loss Function

As previously discussed, the opportunity loss is defined as the amount of loss that could have been avoided if the optimum decision had been selected under conditions of perfect information. In our present discussion, if the payoff function in a two-action problem is linear with respect to the random variables, the loss function is also assumed to be linear with respect to the appropriate states of nature.

In general, if there are two courses of action, A_1 and A_2, both of which exhibit linear functions with respect to their states of nature (μ), or

$$A_1 = a_1 + b_1\mu, \qquad A_2 = a_2 + b_2u,$$

where $b_1 > b_2$, then the opportunity-loss functions (L) for the two different courses of action are as follows:

For A_1,

$$
\begin{aligned}
L &= 0 \qquad (\text{if } \mu > \mu_k), \\
L &= E(A_2) - E(A_1) \\
&= E(a_2 + b_2\mu) - E(a_1 + b_1\mu) \\
&= (a_2 - a_1) - (b_1 - b_2)E(\mu) \qquad (\text{if } \mu < \mu_k).
\end{aligned}
\tag{7.8}
$$

189

For A_2,

$$L = 0 \qquad (\text{if } \mu < \mu_k),$$

$$
\begin{aligned}
L &= E(A_1) - E(A_2) \\
&= E(a_1 + b_1\mu) - E(a_2 + b_2\mu) \\
&= E(a_1 - a_2) + (b_1 - b_2)E(\mu) \\
&= (a_1 - a_2) - (b_2 - b_1)E(\mu) \qquad (\text{if } \mu > \mu_k).
\end{aligned}
\qquad (7.9)
$$

EXAMPLE 7.3 / Let us reexamine the Lions Club case. As illustrated in Example 7.2, since the total profit for alternative C_1 (to bring in the circus) is greater than the profit for the course of action C_0 (not to bring in the circus), the optimum decision was to bring in the circus. Given that the Lions Club actually decided to bring in the circus, the amount of opportunity loss is zero (i.e., $L = 0$), since the projected number of customers is greater than the break-even value; i.e., 20,000 people versus 10,000 people.

On the other hand, how much would the Lions Club lose if $\mu < \mu_k$ and the club decided to bring in the circus (C_1)?

SOLUTION: As previously discussed, if $\mu < \mu_k$, the optimum decision should be not to bring in the circus (C_0). The amount of opportunity loss when $\mu < \mu_k$ is simply the difference in the expected values between the optimum decision (C_0) and the suboptimum decision (C_1), or

$$
\begin{aligned}
L &= E(C_0) - E(C_1) \\
&= E(0 - 0) - E(-5{,}000 + 0.5\mu) \qquad (\text{Equation 7.8}) \\
&= 5{,}000 - 0.5E(\mu) \\
&= 0.5[10{,}000 - E(\mu)] \qquad (\text{if } \mu < \mu_k).
\end{aligned}
$$

Table 7.1 / Amounts of Opportunity Loss when $\mu < \mu_k$ and the Lions Club Decides to Bring in the Circus (C_1)

Random variable (μ), number of customers	Opportunity loss (L), $\mu < \mu_k \vert C_1$, \$
10,000	0
8,000	1,000
6,000	2,000
4,000	3,000
2,000	4,000

In other words, if the Lions Club decided to bring in the circus (C_1), the opportunity loss for C_1 would be

$L = 0$ (if $\mu > \mu_k$),
$L = 0.5[10,000 - E(\mu)]$ (if $\mu < \mu_k$).

The actual amounts of opportunity loss for the various random values of μ are shown in Table 7.1 and Figure 7.5.

If, on the other hand, the projected number of customers is determined to be less than 10,000 people ($\mu < \mu_k$), the optimum decision would be not to bring in the circus (C_0). Accordingly, if the Lions Club actually decided not to bring in the circus, the opportunity loss would be zero. If, however, $\mu > \mu_k$ and the Lions Club still decided not to bring in the circus, the opportunity loss would be the difference in the expected value between the optimum decision (C_1) and the suboptimum decision (C_0), or

$$L = E(C_1) - E(C_0)$$
$$= E(-5,000 + 0.5\mu) - E(0 - 0) \quad \text{(Equation 7.9)}$$
$$= -5,000 + 0.5E(\mu)$$
$$= 0.5[E(\mu) - 10,000] \quad \text{(if } \mu > \mu_k\text{)}.$$

The opportunity losses when the Lions Club decided not to bring in the circus (C_0) in this case are therefore

$L = 0$ (if $\mu < \mu_k$),
$L = 0.5[E(\mu) - 10,000]$ (if $\mu > \mu_k$).

Figure 7.5 / Opportunity loss function when alternative C_1 is taken and $\mu < \mu_k$.

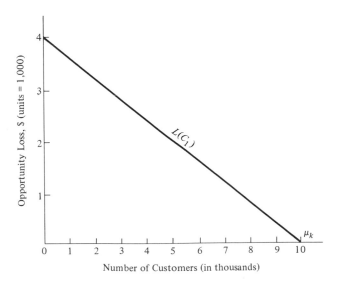

Table 7.2 / Amount of Opportunity Loss when $\mu > \mu_k$ and the Lions Club Decided Not to Bring in the Circus (C_0)

Customers, random variable (μ)	Opportunity loss (L), \$
10,000	0
12,000	1,000
14,000	2,000
16,000	3,000
18,000	4,000
$\vdots$	$\vdots$

The amounts of opportunity loss for the random variable μ when $\mu > \mu_k$, given that the Lions Club actually decided not to bring in the circus, are shown in Table 7.2 and Figure 7.6.

Figure 7.7 shows a combined opportunity-loss function that was compiled from the data contained in Figures 7.5 and 7.6.

EXAMPLE 7.4 / Express the opportunity-loss functions for the Sintax Shorts Manufacturing Company (Example 7.1).

SOLUTION: From the given payoff functions of

$$W_1(\text{advertise}) = -2{,}000{,}000 + 0.5q$$

and

$$W_0(\text{not advertise}) = 0,$$

Figure 7.6 / Opportunity loss function when alternative C_0 is taken and $\mu > \mu_k$.

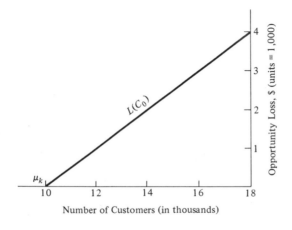

Number of Customers (in thousands)

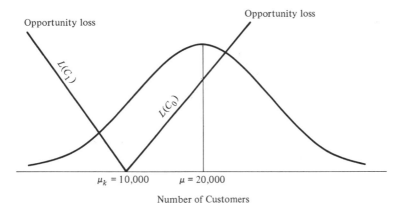

Figure 7.7 / Opportunity loss for two different courses of action.

the opportunity-loss functions for the decisions to advertise (L_{w1}) and not to advertise (L_{w0}) are

(1) To advertise (W_1):

$$L_{w1} = 0 \quad \text{(if } q > q_k\text{),}$$
$$L_{w1} = E(W_0) - E(W_1)$$
$$= 0 - (-2{,}000{,}000) - (0.5 - 0)E(q)$$
$$= 2{,}000{,}000 - 0.5E(q)$$
$$= 0.5[4{,}000{,}000 - E(q)] \quad \text{(if } q < q_k\text{).}$$

(2) Not to advertise (W_0):

$$L_{w0} = 0 \quad \text{(if } q < q_k\text{),}$$
$$L_{w0} = E(W_1) - E(W_0)$$
$$- (-2{,}000{,}000 - 0) - (0 - 0.5)E(q)$$
$$= -2{,}000{,}000 + 0.5E(q)$$
$$= 0.5[E(q) - 4{,}000{,}000] \quad \text{(if } q > q_k\text{).}$$

If the company decided to advertise (W_1) and $q > q_k$, then there would be no opportunity loss, since the company made an optimum decision. If, on the other hand, $q < q_k$, then the optimum decision would be not to advertise (W_0). Accordingly, the opportunity loss is the difference in expected profit between the optimum decision (W_0) and the alternative decision (W_1), given in decision (1).

If the company decided not to advertise (W_0) and $q < q_k$, then the company would have made a correct decision and the subsequent opportunity loss would have been zero. However, if $q > q_k$, then the opportunity loss would have been the difference in the expected values between the optimum decision (W_1) and the alternative decision (W_0), given in decision (2).

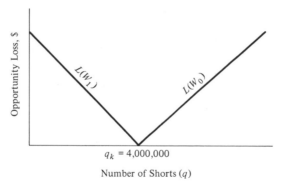

Figure 7.8 / Opportunity loss for the Sintax Shorts Company.

Figure 7.8 illustrates the opportunity-loss functions for the two courses of action in this example.

7.4 / Cost of Uncertainty with a Linear Payoff Function (EVPI)

The cost of uncertainty, or expected opportunity loss, of the optimum decision (which by definition, is the same as the EVPI) under the terms of a normal probability distribution is also assumed to be linear with respect to the existing states of nature. This assumption of linearity greatly simplifies the computation of the EVPI; this is shown in the following equation[1]:

$$\text{EVPI} = d\sigma L_n(D), \tag{7.10}$$

where

$d = |b_1 - b_2| =$ per unit opportunity loss, or slope of loss function,

$\sigma =$ standard deviation of the mean,

$L_n(D) =$ unit normal-loss function,

where

$$D = \left| \frac{\mu_k - \mu}{\sigma} \right|. \tag{7.11}$$

Occasionally the value of D is called a "unit normal-loss deviate"; in this case, it represents the standard normal unit deviations between the mean

---

[1] For a mathematical proof, see Note A at the end of this chapter.

and the break-even value (μ_k). The corresponding probability of the D value can be determined from Appendix Table I. A more detailed explanation of Equation 7.11 is contained in Section 7.5.

EXAMPLE 7.5 / Compute the EVPI for the Lions Club circus problem (Example 7.2).

SOLUTION: According to Equation 7.10, before we attempt to determine the EVPI, the standard deviation (σ) must first be computed. In the Lions Club circus problem, there is a 50/50 chance that the actual number of customers will range from 16,650 to 23,350 people. In other words, and as shown in the normal probability curve in Figure 7.9, the area under the normal curve is 50%; i.e., 25% on each side of the mean value of 20,000 potential customers. Appendix Table G indicates that the standard normal deviate (Z) that corresponds to this 25 percentile is ± 0.67.

The distance from the mean value to each standard deviate boundary is 3,350 people (20,000 − 16,650 and 23,350 − 20,000). This distance must now be recomputed so that it is stated in terms of, or is equivalent to, one standard normal deviate ($\pm 1Z$); this computation is performed as follows: From Equation 4.12,

$$Z = \frac{x - \mu}{\sigma},$$

$$\sigma = \frac{x - \mu}{Z}$$

$$= \frac{23,350 - 20,000}{0.67} = 5,000 \text{ customers.}$$

Figure 7.9 / Normal curve for the Lions Club.

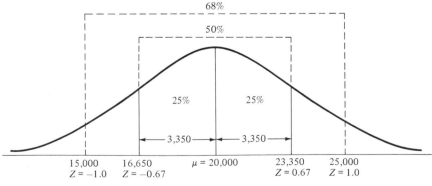

Number of Customers

Therefore, from Equation 7.10, $\text{EVPI} = d\sigma L_n(D)$, where

$$d = |b_1 - b_2| = |0.5 - 0| = \$0.5,$$
$$\sigma = 5{,}000 \text{ customers,}$$

$$D = \left| \frac{\mu_k - \mu}{\sigma} \right|$$

$$= \left| \frac{10{,}000 - 20{,}000}{5{,}000} \right| = 2.0,$$

$$L_n(D = 2.0) = 0.008491 \qquad \text{(from Appendix Table I).}$$

Then,

$$\text{EVPI} = \$0.5 \times 5{,}000 \times 0.008491$$
$$= \$21.2275 \equiv \$21.23.$$

This amount of uncertainty (\$21.23) seems rather small when compared with the expected profit of \$5,000, and therefore it implies that a terminal decision could be made without taking additional sampling or obtaining further information.

7.5 / Characteristics of $D, L_n(D)$, and EVPI

There is an interesting and very important relationship between the values of D, $L_n(D)$, and EVPI, which should be explained in order to understand the full implication of the decision-making process under conditions of a normal probability distribution.

Table 7.3 shows three arbitrary values of D and their corresponding values of $L_n(D)$ and EVPI. Suppose that there are three events (A, B, and C)

Table 7.3 / Relationships among D, $L_n(D)$, and EVPI

Event (1)	D (2)	$L_n(D)$ (3)	EVPI* (4)
A	0.5	0.1978	Small
B	1.0	0.08332	Smaller
C	3.0	0.0003822	Smallest

* The values of d and σ are assumed to be constant for the three cases.

which have corresponding D values of 0.5, 1.0, and 3.0, respectively. These are shown in column (2) of Table 7.3. The corresponding unit-opportunity losses (from Appendix Table I) are contained in column (3). It is clear from the data in columns (2) and (3) that as the D value increases, the value of $L_n(D)$ decreases, and the subsequent value of the EVPI also declines. In summary, as the value of D increases, the value of EVPI becomes less, and accordingly the cost of uncertainty becomes less expensive.

It is not too difficult to understand this inverse relationship between the D value and the EVPI if the mathematical properties inherent in D are fully analyzed. Equation 7.11 gives the D value, and from that equation it can be seen that the D value simply shows the relative position of μ_k in terms of μ. If we assume that the standard deviation (σ) is constant, the D value becomes larger as μ moves farther away from the break-even point (μ_k), and vice versa; this is shown in Figure 7.10.

In Figure 7.10, the two different break-even points (μ_{k1} and μ_{k2}) are arbitrarily depicted with one mean value (μ). In this case, since the break-even point (μ_{k1}) is obviously located farther from the mean (μ) than from the other break-even value (μ_{k2}), it becomes relatively easy for a decision maker to choose the optimum decision.

For example, assume that we are analyzing a retail business. The break-even point (μ_{k1}) suggests that the owner needs to sell up to that point, whereas the mean (μ) shows the amount that management actually believes that it can sell; in this latter case, $\mu > \mu_{k1}$. In other words, the distance between μ_{k1} and μ provides a clear-cut profit margin.

In the μ_{k2} case, however, the difference between μ_{k2} and μ is rather small; this means that a minor error in computing the value of either μ_{k2} or μ may result in either the decision maker's realizing a profit (if μ_{k2} is indeed smaller than μ) or suffering a loss (if μ_{k2} is greater than μ). In this case, the decision maker may end up spending more time and resources on

Figure 7.10 / Hypothetical opportunity-loss lines.

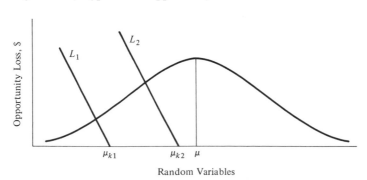

Random Variables

the μ_{k2} case than on μ_{k1} in order to derive an optimum decision. Accordingly, the EVPI for μ_{k2} is higher than that for the μ_{k1} condition; this also partially explains the inverse relationship between the value of D and the EVPI.[2]

7.6 / Decision Making with Discrete Information

Example 7.6 is a unique case, which differs to some degree from the previous discussions and examples in this chapter; i.e., this case assumes discrete prior information rather than continuous information. Nevertheless, it can be shown that the basic assumptions contained in the chapter (e.g., linearity of relationship, two-actions case) can still be applied to problems other than those with normal prior information.[3]

EXAMPLE 7.6 / The Oriental Carpet Import Company of San Francisco buys the majority of the carpets it sells from the Far Eastern Carpet Corporation of Bombay, India. Each shipment contains 500 carpets at an average wholesale price of $986.80 per carpet; this price includes both shipping and insurance costs. If the carpets are received in San Francisco in excellent condition, each carpet is dusted and cleaned at a cost of $64 and subsequently sold for home use on the retail market for $2,500 each. However, if the carpets are not received in an excellent condition, each carpet is dusted and cleaned at a cost of $16 and subsequently sold for office use at a retail price of $800 each. Based on past experience with the Far Eastern Corporation, the management of Oriental Carpet has assigned the following probabilities concerning the proportion of carpets that they expect will be received in excellent condition:

Proportion of carpets rated as excellent condition (q)	Probability, $p(q)$
0.05	0.4
0.10	0.3
0.15	0.2
0.20	0.1

[2] A common example of this relationship is often experienced by college professors at the end of a school term when an analysis is made of the grade that should be assigned to each student. If, for example, a course average of 70 is the dividing line in determining whether a student passes or fails the course, then it is relatively easy for the teacher to assign a passing grade to a student who has an average of 75; it is also relatively easy to assign a failing grade to those students who accumulate an average of 60. However, it is extremely difficult for the teacher to assign a proper grade to those students who have an average in the range of 68–69 or 70–71. In this case, any mistake by the instructor could have a serious adverse impact on the student who receives the grade.

[3] Exercises 12 and 13 at the end of this chapter are similar to this illustration.

(1) Construct the profit functions.
(2) What is the break-even value?
(3) Express the opportunity-loss functions for the two alternatives.
(4) What is the optimum decision?

SOLUTION: Let

N_1 = decision to buy the carpets,
N_0 = decision not to buy the carpets.

Profit functions.

$$E(N_1) = E(\text{cost} - \text{revenue})$$
$$= E[-(500 \times \$986.80) + 500(\$2,500 - \$65)q$$
$$+ 500(\$800 - \$16)(1 - q)]$$
$$= -\$101,400 + \$826,000E(q).$$
$$E(N_0) = 0.$$

Break-even value (q_k). According to Equation 7.7,

$$q_k = \frac{a_2 - a_1}{b_1 - b_2} = \frac{0 - (-101,400)}{826,000 - 0}$$
$$= 0.12276.$$

Accordingly, if the proportion of carpets received in excellent condition, when compared to the total number of carpets purchased, is greater than 0.12276, the optimum decision is N_1, and vice versa.

Opportunity-loss functions. (1) Decision to buy the carpets (N_1):

$$L = 0 \quad (\text{if } q > q_k),$$
$$L = E(N_0) - E(N_1) \quad (\text{if } q < q_k)$$
$$= 0 - E(-101,400 + 826,000q)$$
$$= 826,000[0.12276 - E(q)].$$

(2) Decision not to buy the carpets (N_0):

$$L = 0 \quad (\text{if } q < q_k),$$
$$L = E(N_1) - E(N_0)$$
$$= [-101,400 + 826,000E(q)] - 0$$
$$= 826,000[E(q) - 0.12276] \quad (\text{if } q > q_k).$$

Optimum decision. From the given information we can compute the expected value of carpets that are received in excellent condition (q) as follows:

$$E(q) = \sum P(q_i) \cdot (q_i)$$
$$= 0.4 \times 0.05 + 0.3 \times 0.10 + 0.2 \times 0.15 + 0.1 \times 0.2$$
$$= 0.10.$$

Since the break-even value ($q_k = 0.12276$) is greater than the expected value ($q = 0.1$), the optimum decision is not to buy the carpets (N_0). If the Oriental Company bought the carpets, the expected profit would be negative (i.e., a loss), or

$$E(N_1) = -\$101,400 + \$826,000E(q)$$
$$= -\$101,400 + \$826,000(0.1)$$
$$= -\$18,800.$$

EXAMPLE 7.7 / The Fox Creek Soft Drink Company usually replaces its bottling machines every five years. Currently, there are two machines available for purchase on the market, the PX205 and the QL200. Both machines have a life expectancy of five years. When purchased new, the PX205 costs $20,000 and the QL200 costs $10,000. The variable costs of the PX205 are, however, $4.00 per hour, whereas the variable costs to operate the QL200 are $6.00 per hour.

The major concern of the firm is the number of hours that the company will actually use the machine. A brief study of past history reveals that the company used similar machines for approximately 6,000 hours during the last five-year period, with a standard deviation of 1,000 hours. Assuming that the total costs of the machines are represented by a linear function in relation to the number of hours used and that the number of hours of machine use is normally distributed,

(1) compute the break-even point.
(2) state the optimum decision.
(3) express the opportunity-loss functions.
(4) compute the EVPI.

SOLUTION: Let q = hours of machine use and TC = total cost. Then, according to Equations 7.2 and 7.3,

$$TC(\text{QL200}) = 10,000 + 6q,$$
$$TC(\text{PX205}) = 20,000 + 4q.$$

Break-even value. According to Equation 7.7, the break-even point μ_k is computed as follows:

$$\mu_k = \frac{a_2 - a_1}{b_1 - b_2}$$
$$= \frac{10,000 - 20,000}{4 - 6} = 5,000 \text{ hours.}$$

Optimum decision. The total expected costs for the PX205 and QL200 machines are, respectively,

$$E(\text{PX205}) = -E(\$20,000 + \$4q)$$
$$= -E(\$20,000 + \$4 \times 6,000)$$
$$= -E(\$44,000)$$
$$= -\$44,000,$$

and

$$E(\text{QL200}) = -E(\$10,000 + \$6q)$$
$$= -E(\$10,000 + \$6 \times 6,000)$$
$$= -E(\$46,000)$$
$$= -\$46,000.$$

Since $E(\text{PX205}) > E(\text{QL200})$, the optimum decision is to buy machine PX205; this is true because the total annual cost of machine PX205 is $44,000, whereas the total cost of the QL200 machine is $46,000.

Opportunity-loss[4] functions (L). (1) Decision to buy PX205:

$$L = 0 \quad (\text{if } q > q_k),$$
$$L = E[\text{TC(QL200)} - \text{TC(PX205)}]$$
$$= -E[(10,000 + 6q) - (20,000 + 4q)]$$
$$= -[-10,000 + 2E(q)]$$
$$= 2[5,000 - E(q)] \quad (\text{if } q < q_k).$$

(2) Decision to buy QL200:

$$L = 0 \quad (\text{if } q < q_k),$$
$$L = E[\text{TC(PX205)} - \text{TC(QL200)}]$$
$$= -E[(20,000 + 4q) - (10,000 + 6q)]$$
$$= -[10,000 - 2E(q)]$$
$$= 2[E(q) - 5,000] \quad (\text{if } q > q_k).$$

In the opportunity-loss function cited in paragraph (1), if $q > q_k$ the management made the right decision by purchasing the machine PX205. If $q < q_k$, however, the optimum decision would be to purchase the machine QL200. Since the management decided to buy the machine PX205, the opportunity loss would be the difference in expected cost between alternative choices PX205 (suboptimum choice) and QL200 (optimum choice).

This same reasoning applies to the opportunity-loss function shown in paragraph (2) above. If $q < q_k$, the optimum decision would be to purchase

---

[4] Since this problem has a cost function, the minus sign is attached to the opportunity-loss function (E).

Table 7.4 / Opportunity-Loss Tables for Two
Different Decisions

Opportunity loss (PX205), $2[5{,}000 - E(q)]$		Opportunity loss (QL200), $2[E(q) - 5{,}000]$	
q	L	q	L
5,000	$ 0	5,000	$ 0
4,000	2,000	6,000	2,000
3,000	4,000	7,000	4,000
2,000	6,000	8,000	6,000
1,000	8,000	9,000	8,000
0	10,000	10,000	10,000

the machine QL200, and the subsequent opportunity loss would be zero
if the management actually decided to buy the machine QL200. If, on
the other hand, $q > q_k$, and the management still decided to buy the
machine QL200, the opportunity loss in this case is the difference in the
expected cost between courses of action PX205 (optimum decision) and
action QL200 (suboptimum decision).

The results of both actions are further illustrated in Table 7.4 and
Figure 7.11.

EVPI. According to Equation 7.10,

$$\text{EVPI} = d\sigma L_n(D),$$

Figure 7.11 / Opportunity loss for the Fox Creek
Manufacturing Company.

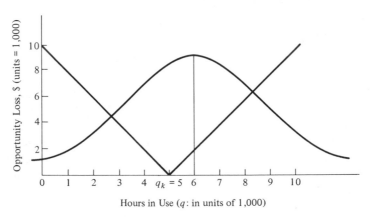

Hours in Use (q: in units of 1,000)

where

$$d = |b_1 - b_2| = |6 - 4| = \$2.00$$
$$\sigma = 1,000 \text{ hours}$$
$$q_k = 5,000 \text{ hours}$$
$$q = 6,000 \text{ hours}$$

According to Equation 7.11,

$$D = \left| \frac{5,000 - 6,000}{1,000} \right| = 1.0.$$

From Appendix Table I,

$$L_n(D = 1.0) = 0.08332.$$

Therefore,

$$\text{EVPI} = \$2 \times 1,000 \times 0.08332$$
$$= \$166.64.$$

The expected opportunity loss for the optimum decision, or EVPI, is therefore $166.64. This may also be interpreted as the maximum dollar amount that the company should be willing to spend to reduce uncertainty.

7.7 / Summary

In this chapter we have discussed the decision-making process under the assumptions of a normal probability distribution without the benefit of additional sampling information. The following assumptions were used in analyzing the problems contained in the chapter: (1) The probability distribution was assumed to be normally distributed; (2) the payoff functions were linear with respect to the existing states of nature; (3) only two-action problems were considered; and (4) the terminal decision was derived from prior information only. The following summary may help the readers to understand the contents of this chapter:

(1) Payoff functions are usually expressed in terms of a linear relationship; e.g., $y = a + b\mu$.

(2) The expected value of a linear payoff function is expressed as

$$E(y) = E(a + b\mu)$$
$$= E(a) + E(b\mu)$$
$$= a + bE(\mu).$$

(3) The opportunity-loss function is dependent upon the break-even value (μ_k). In general, the opportunity-loss functions for two alternative

decisions of the type $A_1 = a_1 + b_1\mu$ and $A_2 = a_2 + b_2\mu$, where $b_1 > b_2$, are expressed for A_1 as

$$L = 0 \quad \text{(if } \mu > \mu_k),$$
$$L = d(\mu_k - \mu) \quad \text{(if } \mu < \mu_k);$$

and for A_2 as

$$L = 0 \quad \text{(if } \mu < \mu_k),$$
$$L = d(\mu - \mu_k) \quad \text{(if } \mu > \mu_k);$$

where $d = |b_1 - b_2|$.

(4) The amount of expected opportunity loss of the optimum decision (EVPI) is also given in linear terms and expressed as

$$\text{EVPI} = d\sigma L_n(D),$$

where

$$d = |b_1 - b_2|,$$
$$\sigma = \text{standard deviation},$$
$$L_n(D) = \text{probability of } D.$$

In $L_n(D)$,

$$D = \left| \frac{\mu_k - \mu}{\sigma} \right|$$

EXERCISES

1. Discuss both the difference and similarities between the decision-making process using a discrete random variable and a continuous random variable.

2. Compare the implication of normal standard deviate (Z) and unit normal loss deviate (D) in decision theory.

3. Explain the importance of determining the prior probability distribution in the decision-making process.

4. A manufacturer is considering major modification of its primary product; this change will require a capital investment of $500,000. Currently, the product contributes an incremental profit of $10 per unit. The product modification should result in approximately 5,000 additional retail stores that would not ordinarily handle the product. The sales manager estimates that the mean sales per new outlet will be about 10 units, with a 60% chance that the mean sales could be less than 8 or more than 12 units.

   What is the optimum decision for the sales manager, and what is the expected value of prior information?

5. The Quick Approval Loan Company is considering an additional branch office in St. Louis. The total annual operating cost of the branch is estimated at $40,000. The

company usually makes loans to high-risk clients and reaps higher rates of return. For this type of customer, the company charges 20% per year on the money loaned to its customers. Based on past experience with branch offices in other cities, the company estimates that the total volume of loans would be approximately $150,000 per year. Pressed for more information, the manager admitted that there are two chances in three that the actual amount loaned could range from $100,000 to $200,000. If the revenue of the company is a linear function of the loans demanded, should the company open the branch office in St. Louis? Compute the corresponding loss functions. What is the EVPI?

6. The Coldlager Brewing Company is considering the purchase of a new machine that will automatically detect and replace defective bottles of beer as they come through the bottling line. The machine will cost $10,000 and have a life expectancy of ten years. The company uses a straight-line depreciation method. The new machine would reduce the amount of annual inspection process and thereby save a considerable amount of labor hours. The company inspectors are presently paid $5.00 per hour. The management of the production department estimates that, with the new machine, approximately 2,100 manual inspection hours can be saved during the ten years, with an even chance that the actual hours saved will be less than 1,800 or more than 2,400 hours.
   (a) Express the saving functions for the decision.
   (b) Compute the opportunity-loss functions.
   (c) Compute the EVPI.
   (d) What is the optimum decision?

7. A manufacturer of a leading food product is considering a change in the package of one of its less profitable products in hopes of attracting more customers. The total annual costs, including such items as design, machine adjustment, and advertising, has been estimated at $100,000. The product, which costs $0.60, is currently selling for $0.75 per item. The present level of sales is 1,000,000 units, but the management estimates that the new design would double sales, with corresponding standard deviation of 150,000 units.
   (a) Should the company change the design?
   (b) Compute the opportunity-loss functions.
   (c) Calculate the EVPI.

8. The General National Auto Supply Company (GNAS) is an automotive parts jobber in Missouri. As part of an expansion movement, the company is examining the possible purchase of a retail outlet in a small midwestern town. Similar experiences with retail stores in small cities indicate an average annual sales of $200,000. An initial investment of $75,000 is required to purchase the new branch store. A standard company policy followed in the acquisition of a new retail outlet requires that the initial investment must be recovered within the first five years, with a minimum return of 6% of gross sales per year. The marketing manager at GNAS, however, feels that with effective management and an aggressive advertising campaign, annual sales could be as high as $260,000, with an even chance that the sales would be outside the range of $220,000 to $300,000 per year.
   (a) What is the optimum decision?
   (b) If the prior mean is correct, what is the minimum return required for the company to break even?
   (c) What is the cost of uncertainty?

9. The management of a large discount department store is considering the introduction of free delivery service to customers when the total purchase at any one time exceeds $50. The annual cost of the service is estimated as $120,000. The company management believes that such a service would attract 4,000 additional $50-plus purchase customers each year, with each customer making an average of four such purchases per year. There is an even chance, however, that the number of purchases could be less than two or more than six per year. Historically, the store is making a 20% profit on sales of this size. What is the optimum decision? Compute the loss functions. What is the EVPI? (Hint: Use $50 × 0.2 = $10 as a minimum contribution.)

10. A company is considering whether or not to hire an additional part-time student salesman on a campus. The annual compensation of the salesman is estimated as $4,500 salary plus an 8% commission on gross sales. Based on past history, an aggressive salesman can usually contact 650 students, with the average purchase being $800 per student if he persuades the students to buy the products. The company is making an average of 35% profit on gross sales, considering everything except the salesman's salary and commission. Assume that w represents the proportion of customers with whom the student salesman talks, which eventually results in a sale:
    (a) Compute the profit functions.
    (b) Compute the opportunity-loss functions.

11. A commercial bank is appraising its service charge for checking-account customers. At present, the average checking-account balance is $40, and the bank charges $1.00 for service every three months. If the bank were to offer free checking for accounts with an average balance over $100, the bank feels that it could significantly increase the average balance. In fact, management has estimated that the average balance would jump to $120, with a standard deviation of $10, and the bank would make 6% return on increased deposits. Currently, there are 20,000 checking accounts. Assuming that the number of accounts remains the same,
    (a) should the bank offer free checking?
    (b) what is the break-even deposit level?
    (c) what is the EVPI?

12. The School Supply Company of Atlanta manufactures approximately 1,000,000 fine ball-point pens each year. Each pen costs 50¢ and is sold to retailers for 70¢. If, however, the pen is defective in any way, the wholesale price is reduced by 60¢ to 10¢ per pen. Because of the dramatic drop in price that can be charged for defective pens, the company management feels that it must revise the manufacturing process in order to reduce the rate of defective pens produced. Under the proposed new process, the ratio of nondefective pens produced would vary as shown in the following table:

Ratio of nondefective pens	Probability
0.85	0.20
0.90	0.40
0.95	0.30
1.00	0.10

Assuming that q represents the proportion of defective ball pens,
(a) express the profit functions.
(b) express the loss functions.
(c) compute the optimum decision.

13. The Triple Farm Market owns and operates a chain of high-quality specialty meat markets. Triple Farm, which normally buys meat in large quantities, has been offered 1,000 head of beef at a price of $250 per head. If the beef is of prime quality, the cost of specialty meat cutting is $150 per head, and the cuts will normally be sold for an average of $1.00 per pound. The average net weight after cutting is 750 pounds per head. If, on the other hand, the beef is of "average" grade, the cutting cost is $100 per head, and the cuts are sold for an average of $0.50 per pound. The average net weight of this grade after cutting is 500 pounds per head. Based on previous experience with the producer of this herd of cattle, the owner assigns the following probabilities to the proportion of beef that will be graded as prime quality:

Proportion of prime quality	Probability
0.15	0.20
0.20	0.40
0.25	0.30
0.40	0.10

(a) Let q represent the proportion of prime beef in the herd. Write the expression of profit as a function of q.
(b) Construct the opportunity-loss functions.
(c) What is the optimum decision?

SUGGESTED READING

Bierman, Harold, Jr.; Bonini, Charles P.; and Hauserman, Warren H. *Quantitative Analysis for Business Decisions* (3d ed.). Homewood, Ill.: Richard D. Irwin, 1973, chapter 7.

Costis, Harry G. *Statistics for Business*. Columbus, Ohio: Charles G. Merrill, 1972, chapter 25, section 25.4.

Dyckman, T. R.; Smidt, S.; and McAdams, A. K. *Management Decision Making Under Uncertainty*. New York: Macmillan, 1969, chapter 12.

Hays, William L., and Winkler, Robert L. *Statistics: Probability, Inference, and Decision*. New York: Holt, Rinehart and Winston, 1971, chapter 9, section 12.

Sasaki, Kyohei. *Statistics for Modern Decision Making*. Belmont, Calif.: Wadsworth Publishing, 1969, chapter 17.

Schlaifer, Robert. *Probability and Statistics for Business Decision*. New York: McGraw-Hill, 1959, chapters 33 through 35.

Spurr, William A., and Bonini, Charles P. *Statistical Analysis for Business Decisions*. Homewood, Ill.: Richard D. Irwin, 1973, chapter 14.

Thompson, Gerald E. *Statistics for Decisions—An Elementary Introduction*. Boston: Little, Brown, 1972, chapter 15.

NOTE A

DERIVATION OF EVPI

The total opportunity-loss difference between the optimum decision W_1 and an alternative decision W_2 is

$$W_2 - W_1 = |b_2 - b_1|(\mu - \mu_k).$$

Since u, expressed as $F(\mu)$, is assumed to be normally distributed, then the expected opportunity loss for the optimum decision (EVPI) is

$$\text{EVPI} = \int_{-\infty}^{\mu_k} (W_2 - W_1)f(\mu)\, d\mu$$

$$= (|b_2 - b_1|)\sigma \left[\frac{e^{-Z_b^2/2}}{\sqrt{2\pi}} - Z_b P(\mu > Z_b) \right],$$

where $Z_b = (\mu_k - \mu)/\sigma$. The portion of the formula in brackets represents a normal loss function (formula answers are given in Appendix Table I). It follows, therefore, that if we let the loss function $= L_n(D)$, then

$$\text{EVPI} = |b_2 - b_1| L_n(D)$$
$$= d\sigma L_n(D).$$

8 / Decision-Making Process with Posterior Information and a Continuous Random Variable —Normal Distribution

8.1 / Purpose

In Chapter 7 we discussed the decision-making process without the benefit of additional sampling information. As briefly mentioned in Chapter 6, however, the use of additional sampling information normally improves both the quality and applicability of current information, and therefore usually enhances the reliability of the subsequent statistical inference that is derived from the data. The value of additional information is especially important in a society where technological change is constantly occurring and where available information is exchanged quite rapidly through continuously improving communications systems. In other words, in the dynamic environment in which we live, the states of nature are constantly changing; and it is this constantly changing environment that causes the greatest amount of uncertainty in the decision-making process.

The main purpose of this chapter is to study the decision-making process when new information is added to an existing data base of states of nature. In other words, the terminal decision is made only after additional information is received and evaluated. In conjunction with this topic, other related subjects, such as the expected opportunity loss (EOL), the expected value of perfect information (EVPI), the expected value of sampling information (EVSI) and the expected net gain from sampling information (ENGS) will also be discussed.

The discussions in this chapter are based on the following assumptions:

(1) The problems under consideration exhibit the properties of a normal probability distribution.
(2) The data being analyzed can be expressed in terms of a linear function with respect to the existing states of nature.
(3) The problems are linear with a two-action case; i.e., there are only two alternative choices available.
(4) The decision maker has the choice of taking additional samples.

8.2 / Optimum Decision without Sampling Information: Review

Let us briefly review the decision-making process under normal probability conditions without sampling information. The case illustrated below will also be used in Section 8.3 to show how a prior optimum decision can be modified by additional sampling information.

EXAMPLE 8.1 / The Sure-Fix Auto Parts Manufacturing Company of New York is considering whether or not it should open a new wholesale branch office in Chicago to handle the company's business in the Midwest region. It is estimated that the new branch will cost the company approximately $1,000,000 in annual operating expenses; these costs include such expenses as wages, rent, and maintenance. The management of the company believes that by opening the new branch, an additional 5,000 retail outlets through-out the region could be serviced, with each outlet purchasing approximately 120 parts annually. This estimated sale of 120 parts per outlet is, however, somewhat uncertain. The management of Sure-Fix believes that there are two chances in three that actual sales may range between 80 and 160 parts per outlet. The average wholesale price of parts sold to retailers is $5.00 per part; based on this price, the company makes a profit of 40%. Suppose that the frequency of potential customers, in terms of the number of parts purchased annually, is normally distributed and that the revenue received from the sale of these parts is a linear function of sales volume. Based on these conditions,

(1) express the profit functions for the two decisions; i.e., open the branch and not open the branch.
(2) compute the break-even value.
(3) what is the optimum decision?
(4) express the appropriate loss functions.
(5) compute the EVPI.

SOLUTION: Let

G_1 = decision to open the branch,

G_0 = decision not to open the branch,

μ_0 = prior mean or estimated sales per outlet,

S_0 = standard deviation of the prior mean.

Profit functions. From a linear function of $Y = a + bX$,

$$G_1 = -\$1,000,000 + (\$5.00 \times 0.4\mu_0 \times 5,000)$$
$$= -1,000,000 + 2 \times 5000\mu_0$$
$$= -1,000,000 + 10,000_{\mu_0},$$
$$G_0 = 0.$$

Break-even value (μ_k). At the break-even value, $G_1 = G_0$, or $-1,000,000 + 10,000\mu_0 = 0$, where $\mu_0 = 100$ or $\mu_k = 100$ parts per outlet. In other words, the company needs to sell an average 100 parts per retail outlet to break even.

Prior optimum decision. Since the prior mean $(\mu_0 = 120)$ is greater than the break-even value $(\mu_k = 100)$, the prior optimum decision is to open the branch in Chicago (G_1). The expected profit for this prior optimum decision is

$$E(G_1) = -1,000,000 + 2(120) \times 5,000$$
$$= \$200,000,$$

while the expected profit for the prior alternative decision (G_0) is $E(G_0) = 0$.

Opportunity-loss functions (L). (1) The opportunity-loss function for the decision to open the Chicago branch is

$$L_1 = 0 \quad (\text{if } \mu_0 > \mu_k),$$
$$L_1 = E(G_0 - G_1)$$
$$= 0 - E(-1,000,000 + 2\mu_0 \times 5,000)$$
$$= 2[100 - E(\mu_0)] \times 5,000 \quad \text{if } \mu_0 < \mu_k).$$

(2) The opportunity-loss function, if the company decides not to open the branch, is

$$L_0 = 0 \quad (\text{if } \mu_0 < \mu_k),$$
$$L_0 = E(G_1 - G_0)$$
$$= E(-1,000,000 + 2\mu_0 \times 5,000) - 0$$
$$- 2[E(\mu_0) - 100] \times 5,000 \quad (\text{if } \mu_0 > \mu_k).$$

These opportunity-loss functions are shown in Figure 8.1.

Figure 8.1 / Opportunity-loss function for the Sure-Fix Auto Parts Company.

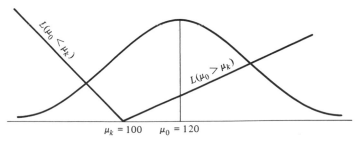

$\mu_k = 100 \qquad \mu_0 = 120$

Expected value of perfect information $(EVPI_0)$. Using Equation 7.10,

$$EVPI_0 = dS_0L_n(D_0),$$

where

$$d = \text{unit loss } or$$

$$d = |b_1 - b_2| = |2 - 0| = \$2.00,$$

$S_0 = $ standard deviation of the prior mean, or 40 parts per outlet,

$D = $ unit normal loss deviate *or*

$$D = \left| \frac{100 - 120}{40} \right| = 0.5,$$

$$L_n(D_0 = 0.5) = 0.1978 \qquad \text{(from Appendix Table I)}.$$

Therefore,

$$\begin{aligned} \text{EVPI}_0 &= 2 \times 40 \times 0.1978 \times 5{,}000 \\ &= \$79{,}120. \end{aligned}$$

The value of perfect information is therefore \$79,120.

8.3 / Optimum Decision with Sampling Information: Posterior Data Analysis

The prior optimum decision discussed in Example 8.1 was derived from the given or previously available information in the problem; i.e., $\mu_0 = 120$. This prior information was estimated by the decision maker before any sampling information was obtained. It is very important, however, for a decision maker to determine the reliability of the original information ($\mu_0 = 120$), particularly in the case where the break-even value and the prior mean (estimate) are not significantly different.

One way to determine the reliability of the original estimate in the Sure-Fix Company case is to obtain additional information by taking samples. For example, the company could take a survey of potential retail outlets in order to forecast more objectively the number of parts that each outlet would actually order during a given time period. This course of action is illustrated in Example 8.2.

EXAMPLE 8.2 / Suppose that the management of the Sure-Fix Auto Parts Company decided to sample 50 outlets as a test of the reliability of the prior estimated mean ($\mu_0 = 120$).[1] Assume that the sampling results show an actual sample mean ($\bar{X}$) of 105 parts per outlet, with a standard deviation

---

[1] We assume that the sample size of 50 outlets is large enough so that the distribution of the sample mean ($\bar{X}$) is approximately normal and that the standard error of the sample mean ($S_{\bar{x}}$) has a small sampling error. For further discussion of $S_{\bar{x}}$, see definitions for Equation 8.1.

(S) of 38 parts.[2] How does this additional information affect the prior optimum decision derived in Example 8.1?

SOLUTION: The classical approach to this problem stresses only the importance and use of the most recent sampling results ($\bar{X} = 105$ and $S = 38$) in the problem solution and ignores the value of the prior information ($\mu_0 = 120$ and $S_0 = 40$).[3] Although new information is important in the decision-making process under conditions of uncertainty, it is also important to use prior information; this is necessary for two reasons: First, when a decision maker assigns a prior probability (information) to the states of nature, he usually bases it on either knowledgeable subjective judgment or objective observation. It follows, therefore, that some knowledge (confidence) of the particular subject being examined is assumed to have existed at the time that the decision was made. Secondly, by using both prior and sampling information, it is possible (and indeed desirable) to assign different weights to the information available for use in the problem.

Bayesian decision theory takes the position that prior information is as important as additional or newly acquired sampling information in the decision-making process. Consequently, the former (prior information) should be incorporated into the latter (current information) to generate posterior information; these data can then be used to measure the new states of nature.[4] Since this posterior information is the joint product of both prior experience and current sampling information, it is assumed that the use of posterior information will yield a better approximation of the true states of nature than if only prior or new sampling information were used alone; this, in turn, should allow for a better decision to be reached.

However, as shown in Equation 8.4, the posterior information is greatly influenced by the size of samples that the decision maker is considering using. If he starts with very little knowledge on the state of nature (diffuse prior knowledge), the sample information will have a large weight in deriving the posterior information. On the other hand, the sample information will exert little influence, if any, on posterior information if it (sample information) is based on small sample size and subsequently on large sampling error.

---

[2] The sample mean ($\bar{X}$) and the standard deviation of the sample mean (S) are derived as follows:

$$\bar{X} = \frac{\sum X_i}{n}, \qquad S = \sqrt{\frac{\sum (X_i - X)^2}{n - 1}}$$

[3] Some students may be curious as to the difference between the concepts of the standard deviation of the prior mean (S_0) and the standard deviation of the sample mean (S). The former indicates the amount of dispersion from the *prior* mean (M_0), whereas the latter measures the amount of dispersion from the *sample* mean ($\bar{X}$). Accordingly, we can mathematically compute S, using the procedures shown in footnote 2 of this chapter, whereas there is no precise mathematical method to compute S_0. In essence, S_0 is determined by a subjective calculation (guesswork) of the decision maker.

[4] For an explanation of the mathematical logic of the basic Bayes' theorem, see Section 2.4.

To apply the preceding logic to the Sure-Fix Auto Parts Company case, we need to determine the posterior mean (μ_1) and to compare this value with the break-even value (μ_k). If the posterior mean is greater than the break-even value, the optimum decision is still to open the Chicago branch. On the other hand, if the posterior mean is less than the break-even value, the optimum decision is to discard the Chicago branch plan.

Posterior mean (μ_1) and the posterior standard deviation (S_1)

Since the standard deviation is itself a measurement of the degree of uncertainty, every piece of new information added to the existing information base is assumed to decrease the amount of uncertainty in a problem; this is subsequently shown in a reduced standard deviation. In other words, as the uncertainty over the states of nature lessens, the measures of information is assumed to be the reciprocal of the variance. Accordingly, if

I_0 = information existing prior to sampling,

I_x = information resulting from the sample,

I_1 = improved information resulting from the posterior analysis,

then

$$I_1 = I_0 + I_{\bar{x}} = 1/S_1{}^2, \tag{8.1}$$

where

$I_0 = 1/S_0{}^2,$

$I_{\bar{x}} = 1/S_{\bar{x}}{}^2$

$S_{\bar{x}}$ = standard error of the sample mean, which is computed by $S/\sqrt{n}$. The standard error of the sample mean ($S_{\bar{x}}$) measures the extent to which sample means vary owing to chance.[5]

From Equation 8.1, we can compute a prior information adjustment factor (I_{P_0}). This factor merely shows the ratio of the prior information (I_0) to the improved information (I_1), or

$$I_{P_0} = \frac{I_0}{I_0 + I_x} = \frac{I_0}{I_1} \tag{8.2}$$

On the other hand, the sampling information adjustment factor (I_{P_1}) is determined as

$$I_{P_1} = \frac{I_{\bar{x}}}{I_0 + I_{\bar{x}}} = \frac{I_{\bar{x}}}{I_1} \tag{8.3}$$

[5] For detailed illustrations of S_x, see William A. Spurr and Charles P. Bonini, *Statistical Analysis for Business Decisions*. Homewood, Ill.: Richard D. Irwin, 1973, pp. 231–235.

Computation of posterior mean (μ_1). The posterior mean is the sum of the prior mean, adjusted by the prior information adjustment factor (Equation 8.2), and the sample mean, adjusted by the sampling information adjustment factor (Equation 8.3), or

$$u_1 = \frac{I_0}{I_0 + I_{\bar{x}}}(\mu_0) + \frac{I_{\bar{x}}}{I_0 + I_{\bar{x}}}(\bar{X})$$

$$= \frac{I_0}{I_1}(\mu_0) + \frac{I_{\bar{x}}}{I_1}(\bar{X})$$

$$= \frac{1}{I_1}(I_0\mu_0 + I_{\bar{x}}\bar{X}). \tag{8.4}$$

Standard deviation of the posterior mean (S_1). Using Equation 8.1, we can compute the standard deviation of the posterior mean in the following manner[6]:

$$I_1 = I_0 + I_{\bar{x}},$$

$$\frac{1}{S_1{}^2} = \frac{1}{S_0{}^2} + \frac{1}{S_{\bar{x}}{}^2},$$

$$S_1 = \sqrt{\frac{(S_0{}^2) \times (S_{\bar{x}}{}^2)}{S_{\bar{x}}{}^2 + S_0{}^2}} = \sqrt{\frac{S^2}{w + n}}, \tag{8.5}$$

where $w = S^2/S_0{}^2$.

In the Sure-Fix Auto Parts Company case, $\mu_0 = 120$, $S_0 = 40$, $n = 50$, $\bar{X} = 105$, $S = 38$; therefore,

$$I_0 = \frac{1}{S_0{}^2} = \frac{1}{(40)^2} = 0.000625,$$

$$I_{\bar{x}} = \frac{1}{S_{\bar{x}}{}^2} = \frac{n}{S^2} = \frac{50}{(38)^2} = 0.034626,$$

$$I_1 = I_0 + I_{\bar{x}} = 0.000625 + 0.034626$$
$$= 0.035251.$$

Then μ_1, the posterior mean (from Equation 8.4), is equal to

$$\mu_1 = \frac{1}{0.035251}(0.000625 \times 120 + 0.034626 \times 105)$$

$$= \frac{0.075000 + 3.635730}{0.035251}$$

$$= 105.26 \text{ parts per outlet,}$$

[6] For a mathematical derivation of this model, see Note A at the end of this chapter.

and S_1, the posterior standard deviation (from Equation 8.5), is

$$S_1 = \sqrt{\frac{S^1}{w + n}} = \sqrt{\frac{(38)^2}{[(38)^2/(40)^2] + 50}}$$

$$= \sqrt{28.37} = 5.33.$$

Optimum posterior decision. Since the posterior mean ($\mu_1 = 105.26$) is greater than the break-even value ($\mu_k = 100$), the optimum posterior decision is still to open a branch office in Chicago. The posterior expected profits under the optimum decision (G_1) and the alternative decision (G_0) are, respectively,

$$E(G_1) = -\$1,000,000 + 2 \times 105.26 \times 5,000$$

$$= \$52,600,$$

$$E(G_0) = 0.$$

8.4 / Expected Value of Perfect Information Using Posterior Information (EVPI$_1$)

An optimum strategy in business usually requires reliable and relatively constant information bases or states of nature. But, as previously discussed, we live in a dynamic environment where no amount of additional information can entirely eliminate uncertainty. Therefore, in the real world, while additional sampling information may reduce the current degree of uncertainty, it is unlikely that a relatively small sample can completely eliminate uncertainty concerning future events. It is imperative, therefore, for a decision maker to properly assess the accuracy and adequacy of a sample and the resulting degree of uncertainty that still remains (after the sampling), before a final decision is made. Only after a thorough analysis is performed of the degree of uncertainty that still exists after the sample has been taken can a decision maker select an appropriate course of action.

In the Sure-Fix Auto Parts Company case, the posterior cost of perfect information (EVPI$_1$) is

$$EVPI_1 = dS_1 L_n(D_1), \tag{8.6}$$

where

$$d = |b_1 - b_2| = |2 - 0| = \$2.00,$$

$$S_1 = 5.33 \text{ parts per outlet,}$$

$$D_1 = \left| \frac{u_k - \mu_1}{S_1} \right| = \left| \frac{100 - 105.27}{5.33} \right| = 0.99.$$

From Appendix Table I,

$$L_n(D_1 = 0.99) = 0.08491.$$

Therefore,

$$\text{EVPI}_1 = \$2.00 \times 5.33 \times 0.08491 \times 5,000$$
$$= \$4,525.70.$$

From the preceding computation it is obvious that there is a considerable reduction in the degree of uncertainty after sampling. In this case, the cost of perfect information before sampling was $79,120; with the additional sampling information, this cost was reduced to $4,525.70.

8.5 / Expected Value of Sampling Information (EVSI)

How much does a decision maker gain from taking an additional sample? For our purpose, it will be shown that the amount of gain that results from the additional sampling information is linearly related to its states of nature, or

$$\text{EVSI} = dS_* L_n(D^*), \tag{8.7}$$

where

$$d = |b_1 - b_2|,$$
$$S_* = \text{amount of uncertainty reduced by the sample} = \sqrt{S_0^2 - S_1^2}$$
$$D^* = \text{unit normal loss deviate} = |(\mu_k - \mu_0)/S_*|.$$

In Equation 8.7 it is important to note that S_* is the amount of uncertainty that is *reduced* by the additional sampling information. Since S_0^2 represents the amount of uncertainty that existed prior to sampling, and S_1^2 represents the amount of uncertainty remaining after sampling, the difference in uncertainty between S_0^2 and S_1^2 gives the degree of uncertainty reduced by the sampling. This reduced uncertainty (S_*) can be reformulated and reduced to

$$S_* = \sqrt{S_0^2 - S_1^2}.$$

Then, from Equation 8.5,

$$S_* = \sqrt{S_0^2 - \frac{S^2}{w + n}}. \tag{8.8}$$

In the Sure-Fix Auto Parts Company case,

$$\text{EVSI} = dS_* L_n(D^*),$$

where

$$d = |2 - 0| = \$2.00,$$

$$S_* = \sqrt{(40)^2 - \frac{(38)^2}{[(38)^2/(40)^2] + 50}}$$

$$= \sqrt{1571.63}$$

$$= 39.6 \text{ parts per outlet,}$$

$$D^* = \left| \frac{100 - 120}{39.6} \right|$$

$$= 0.50.$$

From Appendix Table I,

$$L_n(D^* = 0.50) = 0.1978.$$

Therefore,

$$\text{EVSI} = dS_* L_n(D^*)$$

$$= \$2.00 \times 39.6 \times 0.1978 \times 5{,}000$$

$$= \$78{,}328.80.$$

In other words, by taking an additional sample size of 50, the amount of uncertainty was reduced by almost 99% (i.e., $78{,}328.80 \div 79{,}120$).

In general, the size of the EVSI increases as the sample size increases. This occurs because, as the sample size increases, there is a corresponding

Table 8.1 / Sample Sizes, S_1, S_*, D^*, and EVSI for the Sure-Fix Auto Parts Company

n (1)	S_1 (2)	S_* (3)	D^* (4)	EVSI, \$ (5)	Marginal increase, \$ (6)
0	—	—	—	—	7,342.11
10	11.4	38.30	0.52	73,421.11	
					270.66
20	8.3	39.10	0.51	76,027.70	
					107.35
30	6.8	39.40	0.51	77,101.20	
					102.98
40	5.9	39.50	0.50	78,131.00	
					19.78
50	5.3	39.60	0.50	78,328.80	
					9.89
60	4.8	39.65	0.50	78,427.70	
					7.91
70	4.4	39.69	0.50	78,506.82	

Note: Column (2) was computed by using Equation 8.5; Column (3) was derived from Equation 8.8, and the data for column (5) were obtained by using Equation 8.7. We assume here that $d = 2$ and $S = 38$. n = sample size.

reduction in the variance of the posterior mean (see Equation 8.5), which in turn increases the size of S_*. This effect is illustrated in Table 8.1.

Table 8.1 shows the characteristics of EVSI for several different sample sizes (n). By definition, it is evident that the EVSI will never be greater than the $EVPI_0$ ($79,120), since no amount of additional sampling information can ever reduce uncertainty to zero. Therefore, although the EVSI increases as the sample size increases, the incremental rate increases at a decreasing rate; this is indicated in column (6). The marginal EVSI from sample sizes ranging from zero to 10 is $7,342.11; from 10 to 20 this value is reduced to $270.66; and finally, the value decreases dramatically to a marginal EVSI of only $7.91 with sample sizes ranging from 60 to 70. In other words, the value of additional sampling information (i.e., the marginal gain) decreases rather rapidly as more sample data is added to the decision information base.

8.6 / Expected Net Gain from a Sample (ENGS)

As discussed in Chapter 6, the EVSI represents the gross value of sampling information.[7] It is advantageous for the decision maker to obtain as much additional information as possible, providing the cost of obtaining this additional data does not exceed the benefits derived from the additional information; the desirable sample sizes are limited, therefore, by the costs required to take the sample.

The ENGS measures the net contribution (i.e., costs versus benefits) of sampling information in terms of reducing uncertainty. As previously discussed, the ENGS is the difference of monetary value between the EVSI and the cost of sampling, or

$$ENGS = EVSI - Cost_n. \tag{8.9}$$

Here

$$Cost_n = a + b(n),$$

where

a = fixed cost of taking the sample,

b = variable cost of taking one sample,

n = sample size(s).

In the Sure-Fix Auto Parts Company case, if we assume that the sampling cost exhibits the following linear cost function,

$$Cost_n = \$10,000 + \$50n,$$

[7] See Section 6.6.

then the ENGS for the company becomes

$$\begin{aligned} \text{ENGS} &= \text{EVSI} - \text{cost}_n \\ &= \$78{,}328.80 - (\$10{,}000 + \$50 \times 50) \\ &= \$78{,}328.80 - 12{,}500 \\ &= \$65{,}828.80. \end{aligned}$$

There is therefore a net contribution of \$65,828.80 obtained from taking an additional sample size of 50. The positive value of ENGS suggests that it is worthwhile for the company to proceed with taking the sample; as a matter of fact, if the company desires, it may continue to take samples as long as the value of the ENGS remains positive.

The maximum sample sizes within the EVSI limit (n_x) is determined by and restricted to

$$\text{Cost}_n \leq \text{EVSI}_n,$$
$$a - bn_x \leq \text{EVSI}_n,$$
$$n_x \leq \frac{\text{EVSI}_n - a}{b}, \tag{8.10}$$

where EVSI_n is the expected value of sampling information for the nth sample size. In the Sure-Fix Auto Parts Company case, the maximum sample size is

$$a + bn_x \leq \text{EVSI}_n,$$
$$10{,}000 + 50n_x \leq \text{EVSI}_n,$$
$$50n_x \leq \text{EVSI}_n - 10{,}000,$$
$$n_x \leq \frac{\text{EVSI}_n - 10{,}000}{50}.$$

The maximum sample size (n_x) within the EVSI should not be confused with the maximum sample size within the EVPI prior to sampling (EVPI_0). As explained in Section 6.6, the maximum sample size within the EVPI_0 is given as

$$\text{Cost}_n \leq \text{EVPI}_0$$
$$a - bn_x \leq \text{EVPI}_0$$
$$n_x \leq \frac{\text{EVPI}_0 - a}{b} \tag{8.11}$$

In the Sure-Fix Auto Parts Company case, where $\text{EVPI}_0 = \$79{,}120$ and $10{,}000 + 50n_x \leq 79{,}120$, the maximum sample size is

$$n_x \leq \frac{79{,}120 - 10{,}000}{50}$$
$$= 1{,}382.4$$
$$\equiv 1{,}382.$$

In other words, the receipt of additional information from a sample size up to 1,382 is preferable to the alternative of obtaining no sample data at all.

8.7 / Optimum Sample Size

The optimum sample size is defined as the number or size of the sample which produces the maximum expected gain per sample. Mathematically, the optimum sample size is determined at the point where the marginal gain (MG) obtained from the sample is equal to the marginal cost (MC) required to take and analyze the sample data. Any sample size that is greater than this equilibrium point (MG = MC) causes the marginal cost to exceed the marginal gain. On the other hand, any sample size that is less than the equilibrium point is undesirable, since management would gain more by taking additional samples than it would spend on obtaining the extra information; this relationship[8] is illustrated in Figure 8.2.

Figure 8.2 is based on Figure 6.2. As explained in Chapter 6, the minimum sample size is n_1, the optimum sample size is n_2, and the maximum sample size within the $EVSI_n$ is n_3. As shown in Figure 8.2, the ENGS is maximum at the point where the marginal gain is equal to the marginal cost. Sample sizes of less than n_1, therefore, result in obtaining a total gain less than the total cost; conversely, any sample size greater than n_3 results in obtaining a total cost in excess of the total gain.

Figure 8.2 / Relationship between $EVPI_0$, EVSI, ENGS, cost of sampling and sample size.

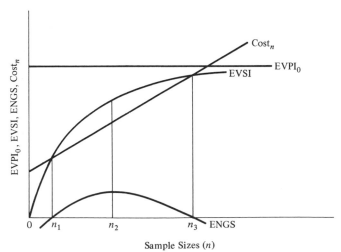

Sample Sizes (n)

[8] For a detailed discussion of this subject, see section 6.7 in chapter 6.

Table 8.2 / Optimum Sample Sizes for the
Sure-Fix Auto Parts Company

N (1)	EVSI, \$ (2)	Cost of sample, \$ (3)	ENGS, \$ (4) = (2) − (3)
10	73,421.11	10,500	62,921.11
20	76,027.70	11,000	65,027.70
30	77,101.20	11,500	65,601.20
40	78,131.00	12,000	66,131.00
50	78,328.80	12,500	65,828.80
60	78,427.70	13,000	65,427.70
70	78,506.82	13,500	65,006.82

In the Sure-Fix Auto Parts Company case, the optimum sample size is a sample size between 40 and 50; this is shown in Table 8.2.

8.8 / Market Projection Based on Sampling:
Further illustrations of typical market projections follow.

EXAMPLE 8.3 / The Metropolitan Catering Service Company is considering leasing a turkey-cooking machine with an automatic timer. This machine costs $1,400 more per month to operate than the present cooker, but it is estimated that the new machine can increase cooking productivity by 80 pounds per 100 birds; there is, however, considerable uncertainty concerning this projected increased yield. The management of the company estimates that there is a 50/50 chance that the actual increased yield may range from 60 to 100 pounds per 100 birds. Currently, the company charges 50¢ per pound for turkey, and the present monthly consumption of birds is approximately 4,000 turkeys. After an initial trial of the new machine, management took a random sample of 10 batches of 100 birds each. The results yielded a sample mean of 70 pounds per 100 birds with a 20-pound deviation. The company that analyzed the sample forwarded a bill of $100 for its services. Should the company continue to lease the new machine?

SOLUTION: From the information given above, the following statistics will be used in the solution of the problem:

$$d = 50\text{¢ per pound,}$$
$$M_0 = 80 \text{ pounds per 100 birds (prior mean),}$$
$$S_0 = 30 \text{ pounds per 100 birds,}[9]$$
$$n = 10 \text{ batches,}$$

[9] For the derivation of this number, see Note B at the end of the chapter.

$$\bar{X} = 70 \text{ pounds per 100 birds,}$$
$$S = 20 \text{ pounds per 100 birds,}$$
$$\text{Leasing cost} = \$1,400 \text{ per month.}$$

Prior optimum decision

(1) Profit functions:

$$W_1 \text{ (lease)} = -1,400 + 0.5M_0 \, (4,000/100)$$
$$= -1,400 + 20M_0,$$
$$W_0 \text{ (not lease)} = 0.$$

(2) Break-even value: From Equation 7.7,

$$M_k = \frac{a_2 - a_1}{b_1 - b_2}$$

$$= \frac{0 - (-1,400)}{20 - 0}$$

$$= 70 \text{ pounds per 100 birds.}$$

(3) Optimum decision: Since the prior mean ($M_0 = 80$ pounds) is greater than the break-even value ($M_k = 70$ pounds), the prior optimum decision is to lease the new machine. The expected net profit of this decision is

$$W_1 = -1,400 + 0.5\left(80 \times \frac{4,000}{100}\right)$$

$$= -1,400 + 1,600$$
$$= \$200 \text{ per month.}$$

(4) Opportunity-loss functions (L):

    (a) Opportunity-loss function under W_0:

$$L_0 = 0 \qquad \text{(if } M_0 < M_k),$$
$$L_0 = E(W_1 - W_0)$$
$$= E(-1,400 + 20M_0) - 0$$
$$= 0.5[E(M_0) - 70] \times 40 \qquad \text{(if } M_0 > M_k).$$

    (b) Opportunity-loss function under W_1:

$$L_1 = 0 \qquad \text{(if } M_0 > M_k),$$
$$L_1 = E(W_0 - W_1)$$
$$= 0 - E(-1,400 + 20M_0)$$
$$= 1,400 - 20E(M_0)$$
$$= 0.5[70 - E(M_0)] \times 40 \qquad \text{(if } M_0 < M_k).$$

Figure 8.3 illustrates the opportunity loss functions for this problem.

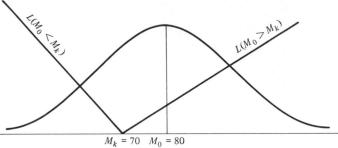

$M_k = 70$ $M_0 = 80$

Figure 8.3 / Opportunity losses for the Metropolitan Catering Service Company.

(5) Cost of perfect information (EVPI$_0$). According to Equation 7.10 and Equation 7.11,

$$EVPI_0 = dS_0 L_n(D_0),$$

where

$$D_0 = \left| \frac{M_k - M_0}{S_0} \right| = \left| \frac{70 - 80}{30} \right|$$

$$= 0.33.$$

From Appendix Table I, $L_n(D_0 = 0.33) = 0.2555$. Therefore,

$$EVPI_0 = 0.5 \times 30 \times 0.2555 \times \left(\frac{4{,}000}{100} \right)$$

$$= \$153.33 \text{ per month.}$$

Posterior optimum decision

(1) Posterior Mean and Standard Deviation: The determination of the posterior optimum decision requires the computation of the posterior mean (M_1) and standard deviation (S_1). According to Equation 8.4,

$$M_1 = \frac{1}{I_1} (I_0 M_0 + I_{\bar{x}} \bar{X}),$$

where

$$I_0 = \frac{1}{S_0{}^2} = \frac{1}{(30)^2} = 0.0011,$$

$$I_{\bar{x}} = \frac{1}{S_{\bar{x}}{}^2} = \frac{1}{(S/\sqrt{n})^2} = \frac{n}{S^2} = \frac{10}{(20)^2} = 0.025,$$

$$I_1 = I_0 + I_{\bar{x}} = 0.00111 + 0.025 = 0.02611,$$

$$M_1 = \frac{1}{0.026111} (0.00111 \times 80 + 0.025 \times 70)$$

$$= 70.42 \text{ pounds per 100 birds.}$$

Since the posterior mean ($M_1 = 70.42$) is still greater than the break-even value ($M_k = 70$), the posterior optimum decision is to continue leasing the new machine.

(2) Posterior Cost of Perfect Information (EVPI$_1$): According to Equation 8.6,

$$\text{EVPI}_1 = dS_1 L_n(D_1),$$

where, from Equation 8.5,

$$S_1 = \sqrt{\frac{S^2}{w + n}} = \sqrt{\frac{(20)^2}{[(20)^2/(30)^2] + 10}}$$

$$= \sqrt{38.3} = 6.16 \text{ pounds per 100 birds.}$$

$$D_1 = \left|\frac{M_k - M_1}{S_1}\right| = \left|\frac{70 - 70.42}{6.16}\right|$$

$$= 0.07.$$

From Appendix Table I, $L_n(D_1 = 0.07) = 0.3649$. Therefore,

$$\text{EVPI}_1 = 0.5 \times 6.16 \times 0.3649 \times \left(\frac{4,000}{100}\right)$$

$$= \$44.96 \text{ per month.}$$

The analysis of the sample of ten batches, therefore, allowed a sizable reduction in uncertainty; i.e., from \$153.33 to \$44.96 per month.

(3) Expected Value of Sampling Information (EVSI): According to Equation 8.7,

$$\text{EVSI} = dS_* L_n(D^*),$$

where

$$S_* = \sqrt{S_0^2 - S_1^2} = \sqrt{(30)^2 - (6.16)^2} = \sqrt{862.05}$$

$$= 29.4 \text{ pounds per 100 birds,}$$

$$D^* = \left|\frac{M_k - M_0}{S_0}\right| = \left|\frac{70 - 80}{29.4}\right|$$

$$= 0.34.$$

From Appendix Table I, $L_n(D^* = 0.34) = 0.2518$. Therefore,

$$\text{EVSI} = 0.5 \times 29.4 \times 0.2518 \times \frac{4,000}{100}$$

$$= \$148.06 \text{ per month.}$$

(4) Expected Net Gain from the Sample (ENGS): According to Equation 8.8,

$$\text{ENGS} = \text{EVSI} - \text{cost}_n = \$148.06 - \$100$$

$$= \$48.06.$$

It follows, therefore, that it is worthwhile for the company to take a sample size of 10; the net gain in this case is $48.06.

EXAMPLE 8.4 / Let us reexamine the Oriental Carpet Import Company case (Example 7.6).[10] Suppose that the management of the company took a sample of 100 carpets to determine the quality of the carpets. An analysis of the sample data revealed that 20 carpets were in excellent condition. Using a normal approximation to the binomial probability distribution, compute the posterior optimum decision.

SOLUTION: In order to use the normal approximation of a binomial distribution, it is necessary to estimate the mean (μ) and standard deviation (σ). Table 8.3 illustrates the steps that are required to obtain the appropriate statistics and posterior probabilities in this case. Columns (1) and (2) were taken from the table in Exercise 7.6. The mean and standard deviation for Rating A, for example, contained in columns (3) and (4) in Table 8.3 were computed as follows:

$$\mu = nq = 100 \times 0.05$$
$$= 5 \text{ excellent carpets,}$$
$$\sigma = \sqrt{nqp} = \sqrt{100 \times 0.05 \times 0.95}$$
$$= 2.18 \text{ excellent carpets.}$$

The conditional probabilities contained in column (5) were computed in accordance with the steps outlined in Section 4.4. Therefore, for Rating A, the conditional probability is zero, since

$$P\left(Z = \frac{20.5 - 5}{2.18}\right) - P\left(Z = \frac{19.5 - 5}{2.18}\right)$$
$$= P(Z = 7.11) = P(Z = 6.65)$$
$$= 0.5000 - 0.5000 = 0.$$

The conditional probabilities for the remaining ratings were similarly computed. The marginal probability of 0.0189 in column (6) indicates that there is almost a 2% chance that the company management will find 20 carpets in excellent condition among the 100 samples selected. Finally, the posterior probabilities in column (7) were computed using the method discussed in Section 6.2.

The prior optimum decision determined in Chapter 7 was not to buy (N_0) the carpets; this was true because the expected prior proportion of excellent quality (q) carpets of 0.10 was less than the break-even value of 0.12267. In order to compute the posterior optimum decision, it is necessary to estimate the posterior optimum proportion of carpets received

---

[10] As explained in Section 7.6, this problem also deviates to some extent from the main stream of this chapter; i.e., this problem assumes discrete prior information. Exercises 11 and 12 at the end of this chapter are also similar to this illustration.

Table 8.3 / Derivation of the Posterior Probability for the Oriental Carpet Company

Rating	Excellent condition, (q) (1)	Probability, $p_0(q)$ (2)	Mean, $\mu = nq$ (3)	Standard deviation, $\sigma = \sqrt{nqp}$ (4)	Conditional probability, $P(X = 20\|n = 100, p = q)$ (5)	Joint probability, $(2) \times (5)$ (6)	Posterior probability, $p_1(q)$ (7)
A	0.05	0.4	5	2.18	0	0	0
B	0.10	0.3	10	3.00	0.001	0.0003	0.0159
C	0.15	0.2	15	3.60	0.043	0.0086	0.4550
D	0.20	0.1	20	4.00	0.100	0.0100	0.5291
		$\overline{1.0}$				$\overline{0.0189^*}$	$\overline{1.0000}$

* Marginal probability.

in an excellent condition; this is computed as follows:

$$E_1(q) = 0.05 \times 0.1 + 0.1 \times 0.0159 + 0.15 \times 0.455$$
$$+ 0.2 \times 0.5291$$
$$= 0.17566.$$

Since the posterior expected proportion of carpets received in an excellent condition (0.17566) is greater than the break-even value (0.12267), the company should buy the carpets (N_1). The expected posterior profit under these conditions is

$$E_1(N_1) = -\$101{,}400 + \$826{,}000E(q)$$
$$= -\$101{,}400 + \$826{,}000(0.17566)$$
$$= \$74{,}260.$$

In this case, the use of additional information reversed the prior optimum decision.

8.9 / Summary

This chapter is essentially a continuation of Chapter 7 to the extent that all the assumptions contained there were also used in this chapter. However, the approach is different because we now base the terminal decision on posterior information. The highlights of this chapter are listed below.

(1) The following symbols denoting the mean and standard deviation are most commonly used in a posterior analysis.

Mean

μ_0 = prior mean or estimate,

$\bar{X}$ = sample mean,

μ_1 = posterior mean.

Standard Deviation

S_0 = deviation of the prior mean,

S = deviation of the sample mean,

S_1 = deviation of the posterior mean,[11]

$S_{\bar{x}}$ = standard error of the sample mean.

Computational Equalities

$$\bar{X} = \frac{\sum X_i}{n},$$

---

[11] For derivation of the posterior standard derivation S_1, see Note A at end of this chapter.

$$\mu_1 = \frac{1}{I_1}(I_0\mu_0 + I_{\bar{x}}\bar{X}),$$

$$S = \sqrt{\frac{\sum(X_i - \bar{X})^2}{n-1}},$$

$$S_1 = \sqrt{\frac{S^2}{w+n}} \qquad \text{(where } w = S^2/S_0{}^2\text{)},$$

$$S_{\bar{x}} = \frac{S}{\sqrt{n}}.$$

(2) The posterior optimum decision is the one that yields the largest posterior expected value.

(3) The posterior cost of perfect information (EVPI$_1$) is determined by

$$\text{EVPI}_1 = dS_1 L_n(D_1),$$

where

$$D_1 = \left|\frac{\mu_k - \mu_1}{S_1}\right|.$$

(4) The expected value of sampling information (EVSI) is computed by

$$\text{EVSI} = dS_* L_n(D^*),$$

where

$$S_* = \sqrt{S_0{}^2 - S_1{}^2},$$

$$D^* = \left|\frac{\mu_k - \mu_0}{S_*}\right|.$$

(5) The expected net gain (ENGS) derived from sampling information is determined by

$$\text{ENGS} = \text{EVSI} - \text{cost}_n,$$

where

Cost$_n$ = cost of taking a sample of size n.

(6) The optimum sample size is determined at the point where the marginal gain obtained from the sample is equal to the marginal cost of taking the sample.

(7) The maximum sample size within the EVSI is determined by

$$n_x \leq \frac{\text{EVSI}_n - a}{b},$$

where

a = fixed cost of the sample,

b = variable cost of taking one sample.

(8) The maximum sample size within the EVPI_0 is determined by

$$n_x \leq \frac{\text{EVPI}_0 - a}{b}.$$

EXERCISES

1. What benefits are derived from the use of sampling information in the decision-making process?

2. Discuss the difference between the prior information adjustment factor (I_{p_0}) and the sampling information adjustment factor (I_{p_1}) as they relate to the decision-making process under conditions of uncertainty.

3. Explain any differences in the meaning and use between the following family of means and standard deviations:

Mean	Standard deviation
μ_0	σ_0
$\bar{X}$	S
μ_1	σ_1

4. Discuss the differences of a decision-making process under the assumption of discrete random variable and continuous random variable.

5. A research study indicates that if course of action A_1 is followed, the expected profit would be $10,000 per year; this is compared to an expected profit of zero for the alternative decision (A_2). The same study reveals that the cost of uncertainty would be $15,000. Is it possible that the cost of uncertainty is higher than the expected profit? Is it worthwhile for the company to pay the cost of uncertainty?

6. The following data pertain to the leading product for a large household manufacturing company:

Prior mean $(M_0) = 50$,
Prior standard deviation $(S_0) = 10$,
Profit function $(A_1) = -120 + 2X$,
Profit function $(A_2) = 0$,
Sample size $(n) = 100$,
Sample mean $(\bar{X}) = 45$,
Standard deviation of sample mean $(S) = 9$,
Cost of sample $(C_n) = \$1.00 + \$0.5n$.

From these data,
(a) compute the break-even value.
(b) compute the prior optimum decision.
(c) express loss functions.
(d) compute the posterior optimum decision.
(e) compute both the prior and posterior EVPI.
(f) decide whether it is worthwhile for the company to take a sample as large as 100.

7. The U.S. Chemical Corporation is considering the manufacture and subsequent marketing of an outdoor carpet for use as a playing surface for various types of sporting events. Since there is a limited market for this type of product, the company management wants to determine the advisability of opening several retail specialty outlets to handle the new product. Since the market for this type of product is very competitive, advertising and other promotional costs are very high; these costs are currently estimated to be $2,000,000 per year. The long-range company plan is to eventually establish as many as 500 retail stores, with each store selling an average of 5,000 yards annually and having a standard deviation of 2,000 yards. The wholesale price at which the company plans to sell the product to retailers is $2.50 per square yard; at this figure, the company would realize a profit of 40% of the wholesale price. Assuming that the resulting revenue is a linear function of sales volume,

(a) should the company open the retail outlets?

(b) what is the cost of uncertainty?

(c) suppose the company decided to take samples to determine the reliability of the prior information and that the results of sampling were as follows:

$n = 25, \bar{X} = 4,500$ square yards;
$S = 1,000$ square yards;
Cost of sample $= \$5,000$ per outlet.

What is the posterior decision? Is it worthwhile for the company to take a sample of size 25?

8. The Boyte-Shine Wax Company has developed a new acrylic floor wax that appears to have a considerable market potential. Based on information supplied by the company's market research department, it has been determined that there is a 68% chance that the company can market between 230 and 310 units per year at each of their 2,000 retail outlets. The net contribution to the new project is $1.00 per unit. The production department estimates that the costs for the new product will be approximately $500,000 per year, including such costs as advertising and public relations activity. Since the initial cost for the new product is considerable, the company decided to hire a consultant at a cost of $10,000 to perform a feasibility study concerning the potential worth of the venture. Of the 100 retail outlets selected by the consultant, the sampling results showed that $\bar{X} = 280$ outlets and $S = 35$ outlets.

(a) Determine both the prior and posterior optimum decision.

(b) Is it worthwhile for the company to use the consulting company?

9. The Midwest Drug Manufacturing Company of Kansas City has just received Federal Drug Administration approval to market a new prescription drug for control of a wide range of allergies. The company has sent a sample unit of ten doses of drug to 10,000 physicians (primarily allergists) in the Midwest to be used for test-acceptance purposes. The cost of introducing the product is estimated at $150,000 per year. Present estimates indicate that the company makes $0.75 profit per unit. Since the drug is both new and unique, however, it is difficult to accurately estimate the number of actual units that the company will sell. The current consensus is that annual sales will be 150,000 units, with a standard deviation of 100,000 units. However, the result of pretesting the drug by 10,000 physicians revealed that the standard error of the sample mean ($S_{\bar{x}}$) was actually 15,000 units. If the cost function of the test is given as $6,000 + \$0.5n$, where n is the sample size,

(a) what is the expected profit, based on the prior information? What is the cost of uncertainty?

(b) what are the EVSI and ENGS?

10. A consumer finance company is considering a direct-mail advertising campaign to 1,000,000 households. The costs of the mail campaign are broken down as follows: $50,000 for materials, $30,000 for handling and preparation, and 10¢ per unit mailed. Because of this relatively high cost of the direct-mail campaign, management is undecided about the potential revenue benefits to be derived from this method of advertising. While there is usually a high degree of uncertainty concerning the value of direct-mail campaigns, the company estimates that 0.1% of those individuals solicited will eventually apply for, and subsequently be given approval for a loan from the company, with a 2/3 chance that the number of new loans would be between $800 and $1,200. Previous experience indicates that the initial loan to new customers normally averages $1,000, with a corresponding profit to the company of 15% of the principal. In order to reduce the amount of uncertainty, the company decided to take a sample of 3,000 families at a cost of $5,250. The results yielded a sampling error ($S_{\bar{x}}$) of 40 loans and a sample mean of 1,200 loans.

(a) Based only on prior information, what is the expected profit? What is the cost of uncertainty?

(b) What is the posterior optimum decision? Is it worthwhile for the company to undertake the sampling?

11. In Exercise 10, Chapter 7, assume that the following additional information was provided:

W	$P_0(W)$	P(he tries hard/W)
0.02	1/3	0.2
0.03	1/3	0.3
0.04	1/3	0.9

The third column in the tabulation indicates the conditional probabilities of making a sale, given that the salesman puts extra effort into making the sale.

(a) Based on the prior information, should the company hire the part-time salesman?

(b) Does the posterior information alter the prior decision?

12. A National League baseball team uses 1,000 new baseballs during each month of the season. Currently, the team buys baseballs from two suppliers in Haiti, supplier A and supplier B. Supplier A charges $0.70 per ball, whereas B charges $0.85 per ball. However, supplier A charges $0.50 to replace a defective baseball, but supplier B replaces defective balls free of charge. From past experience with these two suppliers, the team estimates that they will receive the following proportion of defective balls (p):

P	$P_0(p)$
0.05	0.20
0.10	0.30
0.15	0.40
0.20	0.10

(a) What break-even rate of defective balls will make the team indifferent as to the choice between the two suppliers?

(b) Based on the information given in (a), from which supplier should the team buy the balls?

(c) The manager of the team examines ten random balls from a fresh shipment and discovers that two balls are defective. Does this new information alter the original decision in (b)?

SUGGESTED READING

Bierman, Harold, Jr.; Bonini, Charles P.; and Hauserman, Warren H. *Quantitative Analysis for Business Decisions* (3d ed.). Homewood, Ill.: Richard D. Irwin, 1973, chapter 8.

Braverman, Jerome D., and Stewart, William C. *Statistics for Business and Economics.* New York: Ronald Press, 1973, chapter 13.

Costis, Harry G. *Statistics for Business.* Columbus, Ohio: Charles E. Merrill, 1972, chapter 27.

Dyckman, T.R.; Smidt, S.; and McAdams, A.K. *Management Decision Making under Uncertainty.* New York: Macmillian, 1969, chapter 15, section 4.

Hays, William L., and Winkler, Robert L. *Statistics: Probability, Inference, and Decision.* New York: Holt, Rinehart and Winston, 1971, chapter 9, sections 23 through 27.

Jedamus, Paul, and Frame, Robert. *Business Decision Theory.* McGraw-Hill, 1969, chapter 12.

Sasaki, Kyohei. *Statistics for Modern Decision Making.* Belmont, Calif.: Wadsworth Publishing, 1969, chapter 18.

Schlaifer, Robert. *Probability and Statistics for Business Decisions.* New York: McGraw-Hill, 1959, chapters 33 through 35.

Spurr, William A., and Bonini, Charles P. *Statistical Analysis for Business Decisions.* Homewood, Ill.: Richard D. Irwin, 1973, chapter 14.

Thompson, Gerald E. *Statistics for Decisions—An Elementary Introduction.* Boston: Little, Brown, 1972, chapter 15.

Winkler, Robert L. *Introduction to Bayesian Inference and Decision.* New York: Holt, Rinehart and Winston, 1972, chapter 6, section 6.

NOTES

A / Derivation of the Posterior Standard Deviation (S_1)

$$S_1 = \sqrt{\frac{(S_0^2)(S_{\bar{x}})^2}{S_{\bar{x}}^2 + S_0^2}} = \sqrt{\frac{S_0^2[(S/\sqrt{n})^2]}{[(S/\sqrt{n})^2] + S_0^2}}$$

$$= \sqrt{\frac{[(S_0^2)(S^2)]/n}{(S^2/n) + S_0^2}}.$$

By multiplying both the numerator and denominator inside the squared root by $n/S_0{}^2$, we obtain

$$S_1 = \sqrt{\frac{[((S_0{}^2)(S^2))/n](n/S_0{}^2)}{(S^2/n) + S_0{}^2(n/S_0{}^2)}}$$

$$= \sqrt{\frac{S^2}{(S^2/S_0{}^2) + n}}.$$

If we let $w = S^2/S_0{}^2$, then

$$S_1 = \sqrt{\frac{S^2}{w + n}}. \tag{A–1}$$

B / Standard Deviation Equivalent

According to Appendix Table G, the Z value that corresponds to a 50% probability is 0.67. Therefore, the standard deviation equivalent to $1Z$ is approximately 30 pounds (20 pounds/0.67) per 100 birds; this is illustrated in the following graph.

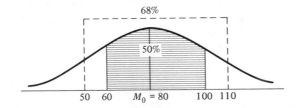

9 / Utility Theory as a Decision-Making Tool

9.1 / Purpose

The decision-making process under conditions of uncertainty, previously discussed in Chapter 5 through Chapter 8, was based on the following two important assumptions: (1) The utility of money to the decision maker was a linear function of the amount of money received, and (2) there were no predetermined constraints (tolerance level) imposed on the decision maker when he evaluated all possible alternative courses of action.

The first assumption indicates that the decision maker will always select as an optimum decision the course of action that generates the largest expected monetary value (EMV) among several alternative courses of action. It is not always true, however, that the utility of money to a decision maker is linear. Therefore, it is not necessarily true that the optimum decision is the one that yields the largest EMV.

The second assumption is probably more unrealistic than the first assumption because, in the real world, the decision maker is often forced to select a suboptimum choice, owing to external constraints. Examples of these constraints include a limited market size in relation to the optimum production level, the maximum amount of economic loss in a given venture that a company can absorb and still remain solvent, and the nonavailability of the outside capital required to finance an optimum venture.

9.2 / Limitations of EMV Theory as a Decision-Making Tool

Examples that illustrate the limitations of EMV theory in the decision-making process under conditions of uncertainty are given in Table 9.1, where the optimum choices, according to EMV theory, are the sets under choice B. This is true because the expected returns for choice B exceed the corresponding returns for choice A. The following equations show that the expected returns for choice B are, respectively,

$$E(\text{B-1}) = \$21,000 \times 0.5 + \$0 \times 0.5 = \$10,500,$$
$$E(\text{B-2}) = \$100,000(0.2 \times 0.6 - 0.1 \times 0.4) = \$8,000,$$
$$E(\text{B-3}) = \$5,000,000 \times 0.5 + (-\$1,000,000) \times 0.5$$
$$= \$2,000,000.$$

Table 9.1 / Choice of Actions under Various Conditions of Uncertainty

Choice A	Choice B
1. A $10,000 tax-free gift with certainty.	1. A $21,000 tax-free gift with a 50% chance of success and failure, with the penalty for failure being no gift.
2. A 7.5% tax-free guaranteed income from municipal bonds on a $100,000 investment.	2. A 60% chance of a 20% annual after-tax appreciation or a 40% chance of a 10% depreciation on a $100,000 investment converted into common stock.
3. A 50% chance of a $1,000,000 after-tax profit from success or a 50% chance of a $500,000 loss if the project fails, with the stipulation that the maximum loss that the company can take without going bankrupt is $800,000.	3. A 50% chance of a $5,000,000 after-tax profit if gold is discovered or a 50% chance of a $1,000,000 loss if gold is not discovered. Again the maximum loss that the company can absorb and still remain solvent is $800,000.

The corresponding expected monetary values for the alternatives under choice A are, respectively,

$$E(A\text{-}1) = \$10,000,$$
$$E(A\text{-}2) = \$100,000 \times 0.075 = \$7,500,$$
$$E(A\text{-}3) = \$1,000,000 \times 0.5 + (-\$500,000) \times 0.5$$
$$= \$250,000.$$

In most practical business situations, however, individuals or management would prefer the choices under A, even though the expected returns for the set of alternatives under choice B were higher. The reason for this choice, is that, although the $E(B\text{-}1)$ is higher than the $E(A\text{-}1)$, the former choice involves uncertainty; i.e., in the alternative under choice B, there is some chance (50%) that no appreciative return on investment will be realized. This is contrasted against the smaller yet certain return that results from the selection of choice A. The same is true in choice B-2. In this case, the expected return after taxes for choice A-2 is only $7,500, whereas the choice of B-2 results in $8,000. Choice A-2, however, has no associated uncertainty, whereas choice B-2 contains a 40% chance that the original investment may depreciate by 10%. Both choices A-3 and B-3 contain some elements of uncertainty, and both projects involve a contingency that stipulates $800,000 as the maximum loss the company can absorb and still remain solvent. As Table 9.1 indicates, the maximum loss that can result from the selection of choice A-3 is $500,000, whereas the selection of choice B-3 may yield a maximum loss of $1,000,000. Therefore, although the expected monetary return from choice B-3 is much higher than that from choice A-3, and even though both projects contain a certain degree of uncertainty, choice A-3 is preferred because of the critical nature of the loss constraint (tolerance level) stipulated by management. In other words, the potential loss that may result from choice B is above the maximum limit set by management.

The conditions set forth in Table 9.1, especially case 3, create a serious problem in the decision-making process when some of the conditions implicitly assumed in EMV theory are either not applicable or simply do not exist. In the examples given above, there are clear contradictions between what EMV theory suggests as the desired course of action and the actual decisions that are, or should be, made. This type of conflict in the decision-making process under conditions of uncertainty is known as the "St. Petersburg paradox."

Daniel Bernoulli first suggested that when such contradictions exist, a utility scale (utility theory) should be used rather than a monetary scale to maximize the expected return. For our purpose, utility is defined as a measurement of value (not necessarily only monetary value) that is used to assess conditions which involve risk, and a utility index is a scale that contains the real numbers used to measure utility. Although many different approaches can be used to measure value in addition to utility theory, only the von Neumann–Morgenstern utility theory will be discussed here.

The main purpose of this chapter is to explain the mechanics necessary to construct an expected utility index; this index can then be used as an optimum decision criterion under conditions of uncertainty. The optimum decision in utility theory is the one which yields the largest expected utility index among at least two alternative courses of action. The case below was designed to illustrate this principle and also to show the problems that may be encountered by using only EMV theory in the solution of a problem.

EXAMPLE 9.1 / The Shelton Construction Company wants to submit bids for the construction of two shopping centers, one in New York (NY) and the other in Los Angeles (LA). Because of the financial condition of the company, only one of the shopping center projects can be undertaken at this time. The main concern of the company is the status of labor unions in both cities. Company records show that contract negotiations with the unions in both NY and LA will be held during the proposed construction period. The profit from each construction job will vary, depending upon the results of these upcoming labor negotiations. The management of the company has assigned the probabilities and corresponding estimated profits to each of three possible construction events. These data are shown in Table 9.2.

Using the EMV approach (e.g., assuming the linear utility of money), what is the optimum decision for the Shelton Construction Company?

SOLUTION: According to EMV theory, as mentioned previously, the optimum decision is the one that yields the largest expected monetary return.

Table 9.2 / Probability and Estimated Profits for Two Construction Jobs

| | New York | | Los Angeles | |
Events	Probability	Profit, $	Probability	Profit, $
A_1: No strike	0.40	90,000	0.40	70,000
A_2: Strike with compromise	0.35	50,000	0.50	10,000
A_3: Strike with no compromise	0.25	– 80,000	0.10	– 50,000

The expected profit from the NY and LA construction jobs are, respectively,

$$E(NY) = \sum_{i=1}^{3} P(X_i) \cdot X_i$$

$$= 0.40 \times \$90,000 + 0.35 \times \$50,000 + 0.25 \times (-\$80,000)$$

$$= \$33,500.$$

$$E(LA) = \sum_{i=1}^{3} P(X_i) \cdot X_i$$

$$= 0.40 \times \$70,000 + 0.50 \times \$10,000 + 0.10 \times (-\$50,000)$$

$$= \$28,000.$$

The optimum decision using EMV theory is, of course, to accept the NY job; this choice yields a $33,500 expected profit, whereas the expected profit from the LA contract is only $28,000.

Suppose, however, that the utility of profit or loss to the decision maker is not linear with respect to the amount of profit or loss generated by a specific venture. For example, if the utility of the first $500 profit to the decision maker is, say, 10 units, the utility of the second $500 profit could be less than 10 (if he has a diminishing marginal return of utility of money) or more than 10 (if he has an increasing marginal return of utility of money).

When the utility of money is not a linear function of the amount of money received, and there are predetermined constraints in the decision-making process, a utility theory approach is often used to solve problems that have the conditions of the St. Petersburg paradox. In other words, under certain conditions, an expected utility index may be used in lieu of using EMV as a decision criterion.

9.3 / Utility Theory as a Substitute Decision Criterion

Underlying assumptions

There are several important assumptions used in constructing a utility index; they include preference, transitivity, critical probability, preference of higher probability, and subjectivity.

Preference. In any situation, regardless of the size of the monetary reward or penalty, a decision maker is either indifferent to two or more projects or he prefers one project over the other(s). For example, if a decision maker is considering two jobs, A and B, he may be indifferent to jobs A and B (if all conditions are the same) or may prefer job A to B if job A has less uncertainty.

Transitivity. If a decision maker prefers job A to job B, and also prefers job B to job C, then it follows that he prefers job A to job C. Also, if he is indifferent to jobs A and B, and also jobs B and C, then he is indifferent to jobs A and C as well. For example, if job A yields a $100,000 profit, job B a $50,000 profit, and job C a $20,000 profit, an individual would certainly prefer job A to B, B to C, and A to C.

Critical probability (indifference probability). If a decision maker prefers job A to job B, and also prefers job B to job C, then there should be a critical probability that can be constructed so that a person is indifferent to jobs A and C with uncertainty, on the one hand, and between jobs A and B with certainty on the other. For example, in a gambling contest, three outcomes are possible: (1) winning $1,000, A; (2) neither winning nor losing, B; and (3) losing $500, C. In this case, there should be a critical probability, P, which makes an individual indifferent between choices A and C with uncertainty, on the one hand, and indifferent between choices A and B with certainty, on the other. Let

$$P = \text{critical probability of winning \$1,000,}$$
$$1 - P = \text{critical probability of losing \$500.}$$

Then

$$1,000P + (1 - P)(-500) = 0,$$
$$P = \tfrac{1}{3}.$$

Preference of higher probability. If a decision maker has two or more jobs that pay identical monetary returns but have different probabilities of success, then he would always prefer the job that yields the highest probability of success. For example, suppose that a decision maker has two jobs, A and B, both of which yield a $10,000 expected return. If the probability of making a profit from job A is 0.9 and from B is 0.5, then he would prefer job A to job B even though the expected monetary return from the two jobs is identical.

Subjectivity. As mentioned previously, utility is strictly a subjective concept. Accordingly, the utility index of event A for one individual might be quite different from the utility that is derived from the same event for another individual. This is simply because each individual's personal and business values differ in many ways, including the attitude toward the utility of money.

Construction of a utility index

Keeping the preceding five assumptions in mind, the expected utility index is computed in the following manner:

$$E(U) = P(X_1) \cdot U_1 + P(X_2) \cdot U_2 + \cdots + P(X_n) \cdot U_n$$

$$= \sum_{i=1}^{n} P(X_i) \cdot U_i, \tag{9.1}$$

where

$$E(U) = \text{expected utility,}$$
$$P(X_i) = \text{probability for event } i,$$
$$U_i = \text{utility index for event } i,$$
$$i = 1,2,3,\ldots n.$$

Given this set of circumstances, the optimum decision is the one that yields the highest expected utility index among at least two alternative decisions.

The following steps are necessary to derive a utility index for any event i:

Step 1. Assign utility-index values to the two extreme events in a given problem. The two extreme cases for the Shelton Construction Company, for example, are either a $90,000 profit (the largest projected profit) or an $80,000 loss (the largest potential loss). In order to make the utility-index theory more practical and easy to manipulate, the range of the utility scale in this case will be expanded to $+\$100,000$ and $-\$100,000$, respectively. This not only makes subsequent computations easier but also means that the index can accommodate any value within these expanded extremes in the Shelton Company.

Two arbitrary utility indexes will be assigned to these two extreme values:

$$U(+\$100,000) = 1.0$$
$$U(-\$100,000) = -1.0.$$

These values are interpreted as follows: The utility index for a $100,000 profit to the Shelton Company is assumed to be 1.0, and the utility index for a $100,000 loss to the company is assumed to be -1.0. While there is no specific rule that dictates the sizes of the utility indexes that are used to depict these extreme values, it is normally assumed that the utility index for profit should be greater than the utility index that represents a loss. Therefore, if the utility index for a $100,000 profit is 1.0, then the utility index for a $100,000 loss should be $U(-\$100,000) < 1.0$.

Step 2. Determine utility indexes for all possible values $U(X_i)$ that fall within the two extreme values established in Step 1, or

$$-\$100,000 < U(X_i) < +\$100,000.$$

The best way to determine the utility indexes for all possible events that fall within the extreme ranges is to draw a continuous utility-index curve based on several random utility indexes as follows:

Event (a). The management of the company has asked: "What is the amount at which the company is indifferent if the chances are 50/50 that a proposed project will yield either a $100,000 profit or a $100,000 loss?" If management considers that it can absorb up to a $30,000 loss for the project, the indifference amount to the company is $-\$30,000$; thus the utility index for $-\$30,000$ is

$$U(-\$30,000) = U(+\$100,000) \times 0.5 + U(-\$100,000) \times 0.5$$
$$= 1.0 \times 0.5 + (-1.0) \times 0.5$$
$$= 0.0,$$

since

$$U(+\$100,000) = 1.0 \text{ by assumption,}$$
$$U(-\$100,000) = -1.0 \text{ by assumption.}$$

This is interpreted to mean that management is aware of the fact that it may earn a $100,000 profit, but it is also aware that it faces a possible loss of $100,000. Since the loss of $100,000 may lead to bankruptcy for the company, it may be willing to pay some amount (in the preceding case, $-\$30,000$) to assure that the company does not face this potentially undesirable outcome. In other words, the company is willing to pay $30,000 for insurance to cover a potential $100,000 loss.

Event (b). One may continue to ask similar questions for other alternative courses of action. For example: "What is the indifference amount if the project has a 50/50 chance of yielding a $100,000 profit or a $30,000 loss?" If the indifference amount to the company is $+\$20,000$, the utility index for $+\$20,000$ is

$$U(+\$20,000) = U(+\$100,000) \times 0.5$$
$$+ U(-\$30,000) \times 0.5$$
$$= 1.0 \times 0.5 + 0.0 \times 0.5 \quad \text{(from event (a))},$$
$$= 0.5.$$

Unlike event (a), however, the company now wants to "sell" the project for $20,000, since the loss of $30,000 to the company is relatively less painful than the case in which the potential profit of $100,000 was accompanied by a potential loss of $100,000, with each result having the same chance of success and failure.

Event (c). If the indifference amount is $-\$40,000$ when the project may produce either a $20,000 profit or a $100,000 loss, with a 50/50 chance the utility index for this "insurance" of $40,000 is

$$U(-\$40,000) = U(+\$20,000) \times 0.5$$
$$+ U(-\$100,000) \times 0.5$$
$$= 0.5 \times 0.5 + (-1.0) \times 0.5$$
$$= -0.25.$$

Event (d). If the difference amount is $+\$40,000$ for certainty when the project has a 50/50 chance of yielding either a \$100,000 profit or a \$20,000 profit, the utility of $+\$45,000$ is

$$U(+\$45,000) = U(+\$100,000) \times 0.5$$
$$+ U(+\$20,000) \times 0.5$$
$$= 1.0 \times 0.5 + 0.5 \times 0.5$$
$$= 0.75.$$

Event (e). Finally, if the indifference amount is $-\$60,000$ for certainty when the project yields either a \$100,000 loss or a \$40,000 loss with a 50/50 chance, the utility index for $-\$60,000$ is

$$U(-\$60,000) = U(-\$100,000) \times 0.5$$
$$+ U(-\$40,000) \times 0.5$$
$$= (-1.0) \times 0.5 + (-0.25) \times 0.5$$
$$= -0.625.$$

By using the approach outlined in (a) through (e), as many utility indexes as desired may be computed within the two extreme ranges. Table 9.3 summarizes the utility indexes for the random values computed for events (a) through (e) for the Shelton Construction Company.

Step 3. Draw a continuous utility-index curve based on the utility index developed in Table 9.3. This curve is illustrated in Figure 9.1.

Table 9.3 / Utility Indexes for $-\$100,000 \leq U(X_i) \leq +\$100,000$

	Monetary Return, $		Probability		Indifference	Utility
	Success	Failure	Success	Failure	Amounts, $	Index
Events	(1)	(2)	(3)	(4)	(5)	(6)
(a)	+ 100,000	− 100,000	0.5	0.5	− 30,000	0.0
(b)	+ 100,000	− 30,000	0.5	0.5	+ 20,000	0.5
(c)	+ 20,000	− 100,000	0.5	0.5	− 40,000	−0.25
(d)	+ 100,000	+ 20,000	0.5	0.5	+ 45,000	0.75
(e)	− 100,000	− 40,000	0.5	0.5	− 60,000	−0.625

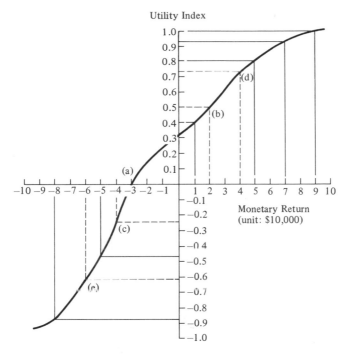

Figure 9.1 / Utility-index curve for the Shelton
Construction Company.

In Figure 9.1, the X axis shows the indifference amounts contained in
column (5) of Table 9.3. If the indifference amounts are positive, they are
located on the right side of the origin; if they are negative, they are shown
on the left side of the origin. The Y axis indicates the utility indexes contained
in column (6) of Table 9.3 for each corresponding monetary value. Points (a),
(b), (c), (d), and (e) correspond to each index point calculated in Table 9.3.
Each point in Figure 9.1 is then connected to form the continuous utility-
index curve for the Shelton Construction Company.[1]

*Step 4. Finally, by using Figure 9.1, determine the utility indexes for
each monetary value for the Shelton Company whose data are contained in
Table 9.2.*

For a $90,000 profit, draw a straight line from $+90,000$ on the X axis
(right side of the origin) toward the utility-index curve. Then draw another
straight line from the utility-index curve toward the Y axis; the resulting
point of intersection is approximately 0.98 and represents the utility index

[1] It should be noted that, unlike the EMV theory, the utility-index curve (Figure 9.1) for the
Shelton Company would probably be quite different for other companies, even if another
company estimated the same monetary returns and probabilities. This is likely because each
company would have a different capital structure and set of external constraints; therefore, the
utility values associated with each alternative choice would differ.

Table 9.4 / Utility Index for the Shelton Construction Company

	NY Project			LA Project		
Events	Probability (1)	Profit, $ (2)	Utility (3)	Probability (4)	Profit, $ (5)	Utility (6)
A_1	0.40	90,000	0.98	0.40	70,000	0.92
A_2	0.35	50,000	0.81	0.50	10,000	0.42
A_3	0.25	−80,000	−0.85	0.10	−50,000	−0.46

for a $90,000 profit, or

$$U(+90,000) = 0.98.$$

For a $50,000 loss, draw a straight line from −50,000 on the X axis (left side of the origin) toward the utility-index curve (downward). Then draw another straight line from the utility curve toward the Y axis; this point of intersection, which is approximately −0.46, represents the utility index for a $50,000 loss, or

$$U(-50,000) = -0.46.$$

The utility indexes for the remaining events are similarly computed; the results of these computations (which are illustrated graphically in Figure 9.1) are shown in columns (3) and (4) in Table 9.4.

Optimum decision under utility theory

As discussed briefly in Section 9.2, the optimum decision using utility-index theory is the one that produces the largest expected utility index among at least two alternative projects. Using Equation 9.1 and the data provided in Table 9.4, the expected utilities for the NY and LA projects are, respectively,

$$E(U)_{NY} = \sum_{i=1}^{3} P(X_i) \cdot U_i$$

$$= 0.40 \times 0.98 + 0.35 \times 0.81$$
$$+ 0.25 \times (-0.85)$$
$$= 0.463.$$

$$E(U)_{LA} = \sum_{i=1}^{3} P(X_i) \cdot U_i$$

$$= 0.40 \times 0.92 + 0.50 \times 0.42$$
$$+ 0.10 \times (-0.46)$$
$$= 0.532.$$

The optimum decision, therefore, is to select the LA project, since this alternative yields an expected utility index of 0.532, whereas the expected utility index for the alternative course of action (NY project) is only 0.463.

9.4 / Measuring Utility

A common question that is often asked when discussing utility theory is: "Can utility actually be measured?" The three measuring scales generally used to measure utility are the nominal, ordinal, and cardinal.

Nominal measurement. A nominal scale assigns descriptive terms to a set of elements. It may be a number, or it may be an adjective such as "large" or "small." This type of measurement is probably most commonly used to describe the measurement of utility. For example, a decision may be divided into "success" or "failure," "acceptable" or "unacceptable." In terms of utility, a nominal scale may assign either "negative utility" or "positive utility."

Ordinal measurement. This type of measurement assigns *relative* terms to a set of elements. For example, the monetary return from project A may be *higher* than that from project B; or the chance of success in an investment in oil is *less* than that of an investment in mining. Furthermore, an ordinal measurement refers to a riskless decision, such as an indifference-curve analysis in economic theory.

Cardinal measurement. This type of measurement assigns a *number* to a set of elements, such as the number of pounds of weight or the number of cubic inches in volume. With a cardinal measurement, one can measure such diverse items as distance, weight, light, sound, or time.

9.5 / Characteristics of the Utility-Index Curve

There are several different kinds of utility-index curves; in most cases, the development and subsequent shape of the curves vary, depending upon the individual's attitude toward risk. Three of these curves are shown in Figure 9.2 and are explained in subsequent paragraphs.

Linear utility-index curve. When the utility index varies in the same proportion as the change in monetary return, utility is measured in terms of a linear utility-index curve. The utility-index curve A in Figure 9.2 is a linear utility-index curve; this means that the slope of the curve is identical for the entire utility line.

When an individual's preferences exhibit the form of a linear utility-index curve, the individual is indifferent as to whether either the expected monetary value or the expected utility index is maximized. In this case,

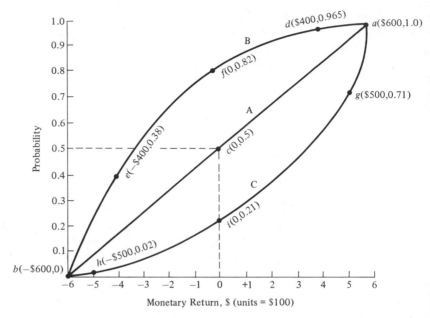

Figure 9.2 / Different utility-index curves.

EMV theory can be used to determine the optimum decision. In Figure 9.2, the utility index for maximum profit (+ $600) is 1.0 (point a); for maximum loss (− $600), the index is 0.0 (point b); therefore, the utility index for no reward is 0.5 (point c). Furthermore, the utility-index curve varies proportionally with the monetary reward.

Concave utility-index curve. When the utility curve is increasing at a decreasing rate, such as utility curve B in Figure 9.2, utility is plotted in the form of a concave utility-index curve. This type of curve is usually associated with risk-aversion. In other words, anyone whose individual preferences take the form of a concave utility-index curve is a risk avoider; this type of person will usually try to avoid any large gamble in return for a relatively smaller return with certainty. For example, let us assume that an individual is faced with a problem where there is a 50/50 chance of either a $400 profit or a $400 loss. An alternative choice is simply to "do nothing," i.e., expecting no monetary return, either gain or loss. As far as the expected monetary value is concerned, both decisions yield a zero return because the expected value from the first and second courses of action are, respectively,

$$E_1 = \$400 \times 0.5 + (-\$400) \times 0.5$$
$$= 0,$$
$$E_2 = 0.$$

Utility-index curve B, however, indicates that this person prefers to "do nothing," as opposed to taking a risk ($400 loss), since the expected utility index for the first decision is less than the expected utility index for the second decision ("do nothing"), or

$$E(U)_1 = \sum_{i=1}^{2} P(X_i) \cdot U_i$$

$$= 0.5 \times U(+400) + 0.5 \times U(-400)$$
$$= 0.5 \times 0.965 + 0.5 \times 0.380 \quad \text{(points } d \text{ and } e)$$
$$= 0.6725,$$

$$E(U)_2 = \sum_{i=1}^{1} P(X_i) \cdot U_i$$

$$= 1 \times U(0)$$
$$= 1 \times 0.82 \quad \text{(point } f)$$
$$= 0.82.$$

Although the expected monetary return for both decisions is identical ($0), the expected utility index for the first decision is lower than that of the alternative decision ("do nothing"). This implies that the individual desires to avoid any tasks that are accompanied by risk. This type of person is called a "risk avoider."

Convex utility-index curve. When the utility-index curve is increasing at an increasing rate, such as curve C in Figure 9.2, utility is plotted as a convex utility-index curve. When an individual's preferences take the form of a convex utility-index curve, the person is said to be a gambler, or a "risk taker." In other words, the individual prefers a large gain or a large loss, with an equal chance for a small monetary return (gain or loss) with certainty. Let us assume that one has an equal chance of making the choice that results in a $500 profit or $500 loss; the alternative choice again is to "do nothing," with a zero monetary return. The expected monetary return for both decisions should be the same, or zero:

$$E_1 = \$500 \times 0.5 + (-\$500) \times 0.5$$
$$= \$0,$$
$$E_2 = \$0.$$

According to utility-index curve C, however, this type of person definitely prefers the first decision to the alternative choice of "doing nothing," since the expected utility index for the first decision is higher than that for the second, or

$$E(U)_1 = U(+\$500) \times 0.5 + U(-\$500) \times 0.5$$
$$= 0.71 \times 0.5 + 0.02 \times 0.5 \quad \text{(points } g \text{ and } h)$$
$$= 0.355 + 0.010$$
$$= 0.365,$$
$$E(U)_2 = 1.0 \times U(0)$$
$$= 1.0 \times 0.20 \quad \text{(point } i)$$
$$= 0.21.$$

The fact that an individual prefers the first decision, which involves risk, to an alternative decision indicates that he is apparently a risk preferer. In this case, it is clearly seen from curve C in Figure 9.2 that an increase in the utility index is very small around the origin ($-\$600$). The rate of increase in the utility index is, however, more than proportional as the size of the monetary return becomes larger. For example, there is only a 0.02 index increase (from 0.0 to 0.02) in the range $-\$600$ to $-\$500$. On the other hand, the size of the utility index increases by 0.3 (from 0.7 to 1.0) as the monetary return increases from $500 to $600.

Mixed utility-index curve. Not all utility-index curves have typical shapes like A, B, or C. Someone may exhibit a utility-index curve like curve D in

Figure 9.3 / Mixed utility curve.

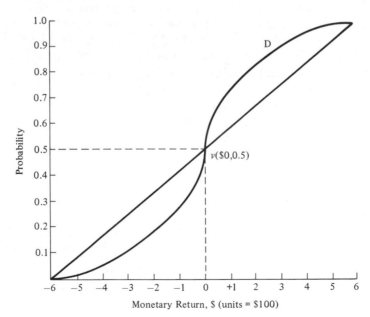

Monetary Return, $ (units = $100)

Figure 9.3. Curve D indicates that part of the utility curve is convex and part is concave. The point of inflection (v) is called an "aspiration level," or a "satisfying point," of monetary return. To the left of point v, a person would like to gamble rather than accept a small return with certainty. To the right of point v, however, the individual rejects a gamble and prefers a small return with certainty.

It is not uncommon to find this type of mixed utility-index curve. For example, a gambler who recently lost $500 may try to recover the lost money by taking a risk or betting a larger sum of money, but just enough to recover his previous loss. After he recovers his loss, however, he may again become conservative toward risk and may prefer a small return with some degree of certainty to a larger gain or loss with an associated high risk.

9.6 / Assessment of Subjective Probabilities

Up to this point we have discussed the computation of objective probabilities based on the long-run frequency. We took it for granted that the decision maker somehow obtained the subjective probabilities; however, we did not discuss how a decision maker could assess the subjective probabilities in the decision-making process.

As mentioned in Chapter 2, the validity of an objective probability is based on two conditions: (1) There is sufficient historical evidence from which a reliable probability distribution concerning the states of nature can be established; and (2) the identical conditions prevail in different time periods. In the absence of these two assumptions, the accuracy of an objective probability becomes unreliable.

Many students of Bayesian decision theory mistakenly assume that subjective probabilities can be assessed on purely a judgmental basis without logic. This is not necessarily the case; the purpose of this section is to discuss the process of assessing logically subjective probabilities in the decision-making process under conditions of uncertainty.

Assessing subjective probabilities—discrete random variable

As previously mentioned, the decision-making process under conditions of uncertainty is analogous to a gambling process; in this analogy the decision maker represents the "gambler." However, unlike card games, lotteries, or dice, in a real-life situation the decision makers often must use subjective rather than objective probabilities.

If we define decision makers under conditions of uncertainty as gamblers, then we may consider a subjective probability as the decision-makers' "betting odd." For example, a bet of $3.00 to $1.00 that the National League will beat the American League in the World Series can be translated into subjective probability terms such as that the odds are 3 to 1, or 0.75, that the National League will win the World Series.

To take another example, suppose that an automobile company is contemplating the introduction of a new compact car into the market. If the sales are successful, the company would make a substantial profit; on the other hand, the failure of the new car model means a further decline of the company's liquidity, which is already at a very low level. In assessing the problem, the decision maker considers the state of the economy for the next five years as being the most critical factor in determining the success or failure of the proposed business venture. Therefore, the decision maker needs certain information concerning the projected state of the economy; e.g., whether the economy will expand for the next five years or remain at essentially the same level. In this situation, it is unrealistic to attempt to compute and use objective probabilities.

First, there are no historical incidences from which probabilities can be objectively computed. Secondly, even if there were historical data, the current conditions or operational environment are noticeably different. The only practical solution in this type of problem is to use subjective probabilities in conjunction with the traditional long-run frequency type that we used previously.

Let us consider a hypothetical lottery where any random value can be drawn from a box containing 1,000 tickets. Some tickets are marked with an "S," indicating success, while others are marked with an "F" for failure. In determining the course of action, the decision maker will ask what mixture (combination) of S and F tickets will make him indifferent concerning the selection of a course of action—in this case, to market or not to market the new car. Suppose that 800 S tickets and 200 F tickets will make him indifferent as to the selection of a course of action. This can be translated as a probability of 0.8 that the economy will expand for the next five years, and 0.2 that the economy will remain static. From these assessments, we can then arrive at an optimum decision under conditions of uncertainty.

Assessing subjective probabilities—continuous random variable

As explained in Chapter 4, in many cases the random variables being considered are in a continuous rather than discrete form. For example, the level of demand for a particular product is normally continuous in nature. In this example, it is very difficult (if not impossible) to assess the probability of demand for a particular amount when it is only one of many possible levels of demand; e.g., 1,000 per month, 2,000 per month, and so on. A logical and relatively simple way of solving this type of problem is through the use of cumulative probabilities of the random variables in which the particular random variable in question falls.

The assessment of cumulative probabilities of random variables is similar to the procedures previously discussed in Section 9.3 to determine the utility index; i.e., a series of subjective questions is asked pertaining to possible demand levels. From these data, a cumulative probability function can then be derived.

EXAMPLE 9.2 / The state government has asked the administrator of a local health-care agency to assess the likelihood of various daily patient numbers in the event that a new hospital in a particular town is built. The assessment and subsequent report concerning expected inpatient use of the new hospital will provide the data required for the state government to determine whether a new hospital should be built at the proposed site or in a different location.

SOLUTION: In an effort to assess the likelihood of various daily patient numbers, the agency statistician assigned to analyze the problem had the following question-and-answer session with the administrator[2]:

Q. In your opinion, what is the possible range of daily patient numbers?
A. Based on my experience, the numbers would range from a low of more than 100 patients to a maximum of 500 daily patients.
Q. What would be the most likely number in these ranges?
A. Approximately 300.
Q. May I assume, then, that the chance is 50/50 that the daily inpatient numbers may be above or below 300?
A. You may.
Q. What is a likely average for the below-300 range?
A. I would say it would be approximately 260 daily patients.
Q. Let's go further down. Can you give me the split numbers (median numbers) below 220?
A. You are now getting into a very tenuous area. If I had to make a guess, I would have to estimate an average 150.
Q. Could you now give me some information about your estimate of daily patient numbers in excess of 300? For example, where would you assess the split of patients number above 300?
A. With little doubt, it would be around 350 per day.
Q. Can you make a similar statement for patients above the 350 level?
A. Yes. I believe that there is a 50/50 chance that the split level would be 385.
Q. One last question. How about between 385 and 500?
A. I would say 440 would be a logical assessment.

From the above conversation between the administrator and the statistician, we can construct a judgmental or subjective cumulative probability; this is shown in Table 9.5.

---

[2] This procedure is identical to the construction of a utility index, discussed in Section 9.3.

Table 9.5 / Cumulative Probabilities for
Daily Patient Numbers

Daily patient numbers, split point	Cumulative probabilities
100	0
150	0.0625
220	0.125
260	0.250
300	0.500
350	0.750
385	0.875
440	0.9375
500	1.0000

For example, according to the questions and answers in the preceding conversation, the median value (split numbers) of daily inpatients below 300 is 260. Therefore, the cumulative probability that, on any given day, the patient numbers will be less than or equal to 260 is $0.5 \times 0.5 = 0.25$. For the split number of 220, the administrator considers a 220 daily-patient number as a median value for daily inpatients of between 100 and 260.

Figure 9.4 / Cumulative probability for daily patient number.

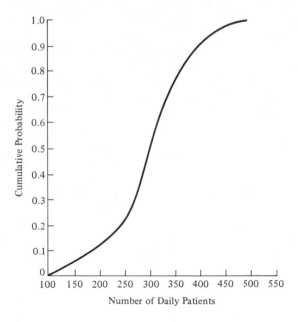

Cumulative Probability

Number of Daily Patients

Therefore, the probability that the daily patient numbers will be less than or equal to 220 is $0.5 \times 0.5 \times 0.5 = 0.125$. The remaining cumulative probabilities were similarily computed.

Using the data in Table 9.5, we can draw a cumulative probability chart depicting the daily patient numbers that are shown in Figure 9.4. The horizontal axis indicates the daily patient numbers and the vertical axis shows the cumulative probabilities of daily patient numbers. The decision maker might wish to establish subjective probabilities for intervals of daily patient numbers ranging from 100 to 500; he would develop these data in increments of, say, 50 patients. Table 9.6 shows the subjective probability associated with each interval.

Column (1) in Table 9.6 lists all possible daily patient numbers, in intervals of 50 patients. Column (2) represents the midpoint of each interval, and column (3) indicates the cumulative probabilities of the upper limit of each interval; these latter data were taken from Figure 9.4. For example: For daily patient numbers of 150 (upper limit of the first interval), locate the daily patient number of 150 on the horizontal axis. Then draw a straight line toward the vertical axis (cumulative probabilities) and find the probability that corresponds to 150 patients. Figure 9.4 indicates that the probability is 0.06 that the daily patient numbers will be at the 150 level in any single day. The remaining values in column (3) were similarly determined.

The values in column (4) show the probability of daily patient numbers for each interval. For example: According to column (3), the cumulative probability is 0.12 that the daily patient number will be less than or equal to 200, but we have just determined that the probability of daily patient

Table 9.6 / Expected Daily Patient Numbers

Daily patient number interval (1)	Midpoint (2)	Cumulative probabilities of patient numbers at or below upper limit (from Figure 9.4) (3)	Probability of patient interval (4)	Expected patient numbers (5) = (2) × (4)
100–150	125	0.06	0.06	7.5
150–200	175	0.12	0.06	10.5
200–250	225	0.24	0.12	27.0
250–300	275	0.50	0.26	71.5
300–350	325	0.75	0.25	81.25
350–400	375	0.90	0.15	56.25
400–450	425	0.96	0.06	25.5
450–500	475	1.00	0.04	19.0
				298.5

numbers, being less than or equal to 150 (upper limit value of the next interval), is 0.06. Therefore, the probability that the daily patient numbers will be between 150 and 200 is $0.12 - 0.06 = 0.06$. The remaining probabilities of daily patient numbers in column (4) were similarly computed. Column (5) shows the expected daily patient numbers. According to column (5), the administrator of the agency should expect about 299 patients per day if the hospital is built on the proposed site.

9.7 / Various Applications of Utility Theory

A few more illustrations may help the student to better understand the substance of this chapter.

EXAMPLE 9.3 / Mr. Johnson, a risk avoider, has the following utility function concerning monetary return (W) between $-\$20$ and $+\$20$,

$$U(W) = W - 0.5W^2,$$

where $-\$20 \leq W \leq +\$20$.

Mr. Johnson is given a chance in a lottery where the monetary prizes are dependent upon the outcomes of the random variable W. This variable has the following probability distribution:

W	$P(W)$
20	0.1
10	0.3
0	0.2
-10	0.3
-20	0.1
	1.0

Assuming that the utility of the lottery is equal to the expected utility of its component prizes, should Mr. Johnson play the lottery?

SOLUTION: Let

$U(L)$ = utility of the lottery

Then

$$U(L) = E[U(W)]$$
$$= E(W - 0.5W^2)$$
$$= E(W) - 0.5E(W^2),$$

where

$$E(W) = \sum P(W_i) \cdot W,$$
$$E(W^2) = \sum P(W_i) \cdot W^2.$$

Table 9.7 / Lottery Utility for Mr. Johnson

W (1)	W^2 (2)	$P(W_i)$ (3)	$E(W) = \sum P(W_i) \cdot W$ (4) = (3) × (1)	$E(W^2) = \sum P(W_i) \cdot W^2$ (5) = (3) × (2)
20	400	0.1	2	40
10	100	0.3	3	30
0	0	0.2	0	0
−10	100	0.3	−3	30
−20	400	0.1	−2	40
		1.0	0	140

From the given data, Table 9.7 can be constructed. Accordingly,

$$E(W) = 0,$$
$$E(W^2) = 140.$$

Therefore,

$$U(L) = E(W) - 0.5E(W^2)$$
$$= 0 - 0.5(140)$$
$$= -70.$$

Since the expected utility from the lottery is negative, Mr. Johnson should not play the game.

EXAMPLE 9.4 / Mr. Jones is an investor who prefers risk with the possibility of large capital returns rather than small capital appreciation with certainty. He has the following utility function for a monetary return between a $100,000 gain and a $200,000 loss:

$$U(W) = 3W + 0.01W^2; \quad -\$200,000 \le W \le +\$100,000.$$

Currently, Mr. Jones is investigating a company that is engaged in oil drilling in Alaska. If oil is discovered, the monetary return will be $100,000; if oil is not found, he faces a potential loss of $200,000. The *Wall Street Wire Service* rates the chance of success for the oil discovery as 0.6. Should Mr. Jones invest in the oil company?

SOLUTION:

$$E[U(W)] = E(3W + 0.01W^2)$$
$$= 3E(W) + 0.01E(W^2).$$

We construct Table 9.8 using the given data. Accordingly,

$$E(W) = -20,$$
$$E(W^2) = 22,000.$$

Table 9.8 / Investment Utility for Mr. Jones (units $= \$1{,}000$)

Events	W_i	W_i^2	$P(W_i)$	$E(W) = \sum P(W) \cdot W$	$E(W^2) = \sum P(W) \cdot W^2$
Success	100	10,000	0.6	60	6,000
Failure	-200	40,000	0.4	-80	16,000
			1.0	-20	22,000

Therefore,

$$\begin{aligned}
E[U(W)] &= 3E(W) + 0.01E(W^2) \\
&= 3(-20) + 0.01(22{,}000) \\
&= -60 + 220 \\
&= 160.
\end{aligned}$$

Since the expected utility index shows a positive value, Mr. Jones should invest in the oil company.[3]

EXAMPLE 9.5 / Mr. Benedict has the following utility function for a specific amount of money, W:

$$U(W) = 2W - 0.03W^2; \qquad -\$5 \le W \le +\$10.$$

Evaluate the following three cases and rank them according to his utility preference.

Case (1). A lottery whose prizes depend on the value of the variable X as follows:

$$W = X - 3,$$

where X is a random variable in a binomial probability distribution with the probability of success (P) being 0.4 in 30 trials.

Case (2). A lottery whose prizes depend on the value of X, defined as follows:

$$W = X - 1,$$

where X is now a random variable in a Poisson probability distribution with a mean value of 5.

Case (3). A tax-free U.S. saving bond of $10.00.

SOLUTION: Case (1) solution:

$$\begin{aligned}
U(W) &= 2W - 0.03W^2, \\
E[U(W)] &= E(2W - 0.03W^2) \\
&= 2E(W) - 0.03(W^2),
\end{aligned}$$

[3] It is interesting to note that, according to EMV theory, the expected return is a $20,000 loss. Since Mr. Jones is a risk preferer, however, he would disregard the results obtained by using EMV theory, since the expected utility index shows a positive result.

where

$$E(W) = E(X - 3)$$
$$= E(X) - E(3)$$
$$= np - 3 \quad \text{(according to Equation 3.13)}$$
$$= 30 \times 0.4 - 3$$
$$= 12 - 3$$
$$= 9,$$

and[4]

$$E(W^2) = \sigma^2(W) + [E(W)]^2.$$

But

$$\sigma^2(W) = \sigma^2(X - 3) = \sigma^2(X) - \sigma^2(3)$$
$$= \sigma^2(X) - 0 \quad \text{(the variance of a constant is zero)}$$
$$- npq \quad \text{(according to Equation 3.14)}$$
$$= 30 \times 0.4 \times 0.6 = 7.2.$$

Therefore,

$$E(W^2) = \sigma^2(W) + [E(W)]^2$$
$$= 7.2 + (9)^2$$
$$= 7.2 + 81$$
$$= 88.2.$$

Accordingly,

$$E[U(W)] \, 2E(W) - 0.03E(W^2)$$
$$= 2(9) - 0.03(88.2)$$
$$= 18 - 2.646$$
$$= 15.354.$$

Case (2) solution:

$$E(W) = E(X - 1)$$
$$= E(X) - E(1)$$
$$= 5 - 1$$
$$= 4,$$

and

$$E(W^2) = \sigma^2(W) + [E(W)]^2$$

---

[4] From Equation 3.8, the variance of a binomial distribution (σ^2) is $\sigma^2(X) = \sum(X - E)^2 \cdot P(X) = E(X^2) - [E(X)]^2$. Therefore, $\sigma^2(W) = E(W^2) - [E(W)]^2$ and $E(W^2) = \sigma^2(W) + [E(W)]^2$.

where

$$\sigma^2(W) = \sigma^2(X - 1)$$
$$= \sigma^2(X) - 0$$
$$= E(X) \quad \text{(the variance of a Poisson distribution is the}$$
$$\text{same as its mean; see Equations 3.18 and 3.19),}$$
$$= 5.0$$

Therefore,

$$E(W^2) = \sigma^2(W) + [E(W)]^2$$
$$= 5 + (4)^2$$
$$= 5 + 16$$
$$= 21.$$

Accordingly,

$$E[U(W)] = 2E(W) - 0.03E(W^2)$$
$$= 2 \times 4 - 0.03(+21)$$
$$= 8 - 0.63$$
$$= 7.37.$$

Case (3) solution:

$$E[U(W)] = E(2W - 0.03W^2)$$
$$= E[2 \times 10 - 0.03(10)^2]$$
$$= E(2 \times 10 - 0.03 \times 100)$$
$$= E(20 - 3)$$
$$= E(17)$$
$$= 17.$$

Therefore, the ranking of the utility choices in this problem would be (3), (1), and (2), in that order, since the expected utility for problems (1), (2), and (3) are, respectively, 15.354, 7.37, and 17.

9.8 / Summary

The use of EMV theory as a decision-making tool under conditions of uncertainty contains two basic assumptions that may restrict its applicability to the decision-making process. These two restrictive assumptions are (1) a linear utility of monetary reward, and (2) no external constraints imposed on the decision maker.

Under conditions where EMV theory is not an appropriate decision-making criterion, a utility scale can be used. A utility index is a measure of a condition that involves a risk.

The following summary should help students to understand the contents of this chapter:

(1) The optimum decision using utility-index theory is the one that yields the highest expected utility index among various alternative decisions.

(2) Inherent in utility-index theory are the following assumptions: subjectivity, preference, transitivity, critical probability, indifference and preference of higher probability.

(3) There are mainly three different scales that are used to measure a utility index: (a) nominal, (b) ordinal, and (c) cardinal.

(4) When a person's individual preferences exhibit a linear utility function, a change in the amount of the utility index is proportional to the change in the amount of expected monetary return. Accordingly, under these conditions, maximization of both the expected monetary value and the utility index produces identical results.

(5) When the utility curve increases at a decreasing rate as the monetary reward increases, a concave utility-index curve results. If a person's preferences take the form of a concave utility curve, the individual is said to be a "risk avoider"; i.e., the person prefers a small monetary return with certainty to a gamble.

(6) When the utility curve increases at an increasing rate as the monetary reward increases, a convex utility curve results. When an individual's preferences take the form of a convex utility curve, he is said to be a "risk preferer"; in other words, the person prefers the possibility of a large return with some degree of risk rather than a small return with certainty.

(7) It is not uncommon to see part of a utility curve as convex and the other part of the curve as concave. This usually indicates that the value of the monetary return changes as the monetary situation varies; i.e., one changes from a risk preferer at one extreme to a risk avoider at the other extreme. The point of inflection is called the "aspiration level" or "satisfying point."

EXERCISES

1. Discuss carefully the differences between EMV theory and utility theory as each relates to the decision-making process under conditions of uncertainty.

2. Explain the use of the St. Petersburg paradox in the decision-making process under conditions of uncertainty.

3. "When an individual has a linear utility-index curve, it does not matter whether he uses utility theory or EMV theory in determining the optimum decision under conditions of uncertainty." Explain.

4. Discuss the meaning of the inflection point as it relates to the decision-making process under conditions of uncertainty.

5. George Minter, who has not been involved in an automobile accident for the past ten years, computed the total amount of the automobile insurance premium he expects to pay next year. He also calculated the theoretical amount of compensation

that would have been paid to him by the insurance company if he had incurred accidents in this same period of time. The resulting net monetary return from the two events turned out to be negative. Accordingly, Mr. Minter concluded that he does not need automobile insurance. Do you agree with Mr. Minter's logic?

6. Explain why many companies pay large insurance premiums for protection from fire or theft even though the probability of occurrence of these events is very low. Explain your answer in terms of monetary value theory and utility theory.

7. The Rutanda Electronics Company has one plant which is currently worth about $10,000,000. The fire department estimates that there is a 2% chance that a company like Rutanda will be completely destroyed by fire. The company currently has a $5,000,000 reserve fund set aside for an emergency situation. The insurance company estimates the fire insurance premium for minimum fire coverage of the plant as $300,000 per year. Should the company purchase the insurance policy? Explain your answer using EMV and utility theory.

8. Chris Rozychi, a junior at George Washington University, prepared a list of possible ways to spend his summer vacation and assigned a corresponding utility value to each choice, as shown below:

Options	Utility index
A. Go to summer school	45
B. Go to Europe	100
C. Part-time work	80
D. Full-time work	50
E. Do nothing	30

Do you think Chris prefers option A for certainty over
(a) a 10% and 90% chance of getting options B and E, respectively?
(b) an even chance of getting options C and D, respectively?
(c) an 80% and 20% chance of getting options E and C, respectively?

9. Mr. Sanson and Mr. Pinton exhibit in the accompanying table the utility indexes for a certain amount of monetary return (Y).

Monetary return, $	Utility index	
(Y)	Sanson	Pinton
+50,000	70	70
+40,000	70	41
+30,000	68	24
+20,000	64	16
+10,000	60	10
0	52	7
-10,000	44	5
-20,000	34	2
-30,000	20	1
-40,000	0	0

(a) Draw individual utility curves for Mr. Sanson and Mr. Pinton.
(b) What kind of utility curve do these data represent?
(c) Is either man a risk preferer, risk avoider, or a risk neutral investor, if any?
(d) Explain the logic of having different utility indexes for the same amount of monetary return.

10. Suppose that you are given two choices: (a) placing $10,000 in a bank with a projected yield of $550 per year and a corresponding utility index of 10, or (b) investing the same amount of money with an even chance of either making a $1,000 gain or a $100 loss. The corresponding utility indexes for gain and loss in this situation are 50 and − 10, respectively. What is the optimum decision? Solve this problem, using both EMV and utility theory.

11. Mr. Bill Menner is considering the purchase of a medium-sized grocery store. He estimates that the chances for making an annual profit of + $20,000, + $40,000, and − $30,000 are 0.3, 0.5, and 0.2, respectively. The corresponding utility indexes for these three events are 20, 28, and − 100.
(a) Using EMV theory, compute the optimum decision.
(b) Compute the optimum decision, using utility theory.
(c) Is Mr. Menner a risk preferer, risk avoider, or a risk neutral investor?

12. Mr. Jackson is considering the purchase of ten shares of Consolidated Grocery Company stock. In terms of personal utility, Mr. Jackson has assigned a utility index of − 70 for a $100 gain and an index of − 200 for a zero gain. Furthermore, he is indifferent between a gain of $150 for certain and the following chances: 0.8 chance of a $100 gain and a 0.2 chance of no gain at all. What is his utility index for a $150 gain?

13. An individual has a utility index of 10 for a profit of $1,000 and an index of 0 for a loss of $500. Suppose further that he is indifferent between a gain of $100 for certain and the following chances: 0.5 chance of a $1,000 profit and a 0.5 chance of a $500 loss. What is the utility index for a $100 gain?

14. Mr. Englewood assigned a personal utility index of 300 for a $10,000 profit and 10 for a $100 profit on a given business venture. He is indifferent between the possibility of a $100 profit with certainty and a 0.5 chance of a $10,000 profit plus a 0.5 chance of $20,000 loss. What is his utility index for the potential $20,000 loss?

15. Consider yourself as a risk avoider with the following utility function for concerning a possible monetary return (Z):

$$U(Z) = Z - 0.001Z^2,$$

where − $20 \le Z \le$ $50. You are offered the opportunity to participate in a contest with the following expected return (Z):

Return (Z), $	Probability $P(Z_i)$
50	0.1
30	0.1
10	0.2
0	0.2
− 10	0.2
− 20	0.1
− 50	0.1

Assuming that the utility of the contest is equal to the expected utility of the prizes, should you play the game? Is your answer based on utility theory different from a decision based on EMV theory?

16. A major airline company negotiated a contract with the Protective Metal Research Company of Los Angeles to develop a special metal detective device within a year for use in detecting potential hijackers. If the device passes the initial research and development (R & D) tests and is subsequently mass produced, the company would make a profit of $500,000. If, on the other hand, the prototype item fails the initial R & D test, the company would not be awarded a follow-up contract to mass-produce the detection device, and would therefore suffer a $1,000,000 loss for development costs, including opportunity loss. The management assigned a 70% chance of success and a 30% chance of failure for the project. After an intensive analysis of several pilot tests, the management of Protective determined that the chance of success within a year is only 40%. (Hint: Assume 40% success as a conditional probability.)

    (a) Assuming that the company has a neutral utility curve, what is the prior and posterior optimum decision concerning success or failure of the project?

    (b) Would the prior and posterior optimum decisions change if the company's previous neutral position switched to the following utility function (unit: $ million)?

    $$U(W) = W - 0.05W^3,$$

    where W is a monetary return.

17. For a given amount of money (N), Mr. Shaffley has the following utility function:

    $$U(N) = 0.5N + 0.1N^2$$

    where $-\$40 \le N \le \$20$. Evaluate the following three cases and rank them in order of preference according to the utility function given above:

    (a) A contest where the prizes depend upon the following value of x:

    $$N = x - 2,$$

    where x represents a random variable in a binomial probability distribution with the probability of success (P) being 0.5 in 20 trials.

    (b) A lottery in which the potential monetary reward depends on a following value of m:

    $$N = m - 5,$$

    where m is a random variable in a Poisson probability distribution with a mean of 10.

    (c) A treasury bond that yields a tax-free $5.00.

SUGGESTED READING

Bierman, Harold, Jr.; Bonini, Charles P.; and Hauserman, Warren H. *Quantitative Analysis for Business Decisions* (3d ed.). Homewood, Ill.: Richard D. Irwin, 1973, chapter 17.

Dyckman, T. R.; Smidt, S.; and McAdams, A. K. *Management Decision Making under Uncertainty*. New York: Macmillan, 1969, chapter 11.

Friedman, M., and Savage, L. J. "The Utility Analysis of Choices Involving Risk," *Journal of Political Economy* (August, 1948), pp. 279–304.

Friedman, M., and Savage, L.J. "The Expected Utility Hypothesis and the Measure-ability of Utility," *Journal of Political Economy*, Vol. 60 (December 1952), pp. 463–474.

Halter, Albert N., and Dean, Gerald W. *Decisions under Uncertainty with Research Application.* Cincinnati: South-Western Publishing, 1971, chapters 3 and 5.

Hays, William L., and Winkler, Robert L. *Statistics: Probability, Inference, and Decision.* New York: Holt, Rinehart and Winston, 1971, chapter 9, sections 7 through 9.

Morgan, Bruce W. *An Introduction to Bayesian Statistical Decision Process.* Englewood Cliffs, N.J.: Prentice-Hall, 1968, chapter 6.

Raiffa, Howard. *Decision Analysis: Introductory Lectures on Choices under Uncertainty.* Reading, Mass.: Addison-Wesley, 1968, chapter 4.

Schlaifer, Robert. *Probability and Statistics for Business Decisions.* New York: McGraw-Hill, 1959, chapter 2, sections 1, 2, 3, and 6.

Thompson, Gerald E. *Statistics for Decisions—An Elementary Introduction.* Boston: Little, Brown, 1972, chapter 9.

Von Neumann, J., and Morgenstern, O. *Theory of Games and Economic Behavior.* Princeton, N.J.: Princeton University Press, 1944, chapter 6.

10 / Classical Decision Theory

10.1 / Purpose

The main purpose of this book is to discuss modern decision theory concepts and procedures. However, no book on decision theory is complete without at least some explanation of, and comparison with, classical decision theory. In fact, this chapter shows that there are significant differences in the procedures used to analyze problems under conditions of uncertainty, using modern versus classical decision theory approaches. In addition to a discussion of classical decision theory, we briefly examine both the similarities and differences between these two decision theory concepts, and outline some of the advantages of using modern decision theory compared to the use of classical decision theory.

Our discussion of classical decision theory is limited to two subjects: estimation of population parameters and the testing of hypotheses. It is through these two subjects that a direct comparison of classical and modern decision theory is normally made.

10.2 / Estimate of the Population Parameter

Solving problems associated with estimating population values is crucial in many aspects of business and economics. Decision makers can derive very few optimum decisions without having at least some knowledge of the population value. However, it is almost impossible (from a practical standpoint) in most cases to examine the entire population in order to obtain the appropriate statistical parameters (mean, standard deviation, etc.) on population; this is mainly due to the time and cost of obtaining such data. Therefore, most decision makers use samples to estimate the population parameters.

The use of samples, however, imposes a major analytical problem that must be addressed; namely, determining the size of the error associated with a particular sample, simply referred to as the "sampling error." Accordingly, statistical methods have been developed to control the sampling error when estimating population parameters. If the decision maker is able to control or at least identify the size of the sampling error, he has some knowledge of the error margin in the estimation process. Indeed, one of the primary objectives of this section is to explain some of the basic statistical procedures used to estimate population parameters when samples are used.

Case with a large sample

Let us reexamine the case of the Goodstone Tire Manufacturing Company (Example 4.8).

EXAMPLE 10.1 / Suppose that a local cab company purchased 100 tires on a trial basis from the Goodstone Tire Company for its new fleet of cabs. After six months of usage, the cab company records showed that the mean $(\bar{X})$ usage for the sample tires was approximately 38,000 miles, with a standard deviation (S) of 10,000 miles. From the given data, estimate the population mean based on a 95% confidence level. Is the company's experience of 38,000 average miles significantly different from the Goodstone's claim of 40,000 miles?

SOLUTION: The basic question to be resolved in this problem is whether the difference in mileage between 40,000 (advertised or claimed) and 38,000 (actual) is due to pure chance or a significantly different population mean. In order to answer this question, it is necessary to determine or to estimate the degree to which the ranges within the population mean (μ) can vary due to pure chance or sampling error, as opposed to an actual change in the population mean. If the value of μ falls outside these acceptable error ranges, we can say that the company's claim of a 40,000-mile tire is not justified.

From the given information, $\bar{X} = 38,000$ miles, $S = 10,000$ miles, and $n = 100$ tires, the standard error of sample mean is

$$S_{\bar{x}} = \frac{S}{\sqrt{n}} = \frac{10,000}{\sqrt{100}} = 1,000 \text{ miles.}$$

If we let $E =$ error range from the mean, which is equal to $(S_{\bar{x}})(Z_{\alpha})$, then the sample mean $(\bar{X})$ could vary from the population mean (μ) by

$$\bar{X} - E \leq \mu \leq \bar{X} + E,$$

or

$$\bar{X} - (S_{\bar{x}})(Z_{\alpha}) \leq \mu \leq \bar{X} + (S_{\bar{x}})(Z_{\alpha}), \tag{10.1}$$

where $Z =$ standard normal deviate at the significance level of α. It follows, therefore, that from the given data of $\bar{X} = 38,000$, $S_{\bar{x}} = 1,000$, and $Z_{\alpha = 0.05} = 1.96$, and by Equation 10.1, the population mean varies by

$$38,000 - (1,000)(1.96) \leq \mu \leq 38,000 + (1,000)(1.96),$$
$$36,040 \leq \mu \leq 39,960.$$

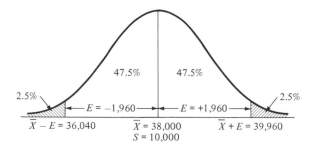

Figure 10.1 / Goodstone Tire
Manufacturing Company.

In essence, we are saying that, according to the error ranges (E) given above, we are 95% confident that the population mean of the Goodstone tires ranges from 36,040 miles to 39,960 miles. Since there is only a 2.5% chance that the population mean exceeds 39,960 miles (there is also a 2.5% chance that the population mean is less than 36,040 miles), we must conclude that the Goodstone's claim of 40,000 miles per average tire cannot be justified. The results of this discussion are shown in Figure 10.1.

Case with a small sample

If the sample size in a problem is relatively smaller (e.g., less than 30), the size of the sampling error tends to increase, since $S_{\bar{x}} = S/\sqrt{n}$. Under these circumstances, the sampling distribution usually does not exhibit exactly a normal distribution; therefore, we cannot obtain the value for Z for our desired level of confidence from the Appendix Table G.

In those cases where the sample size is small, a new type of probability distribution must be introduced. This new probability distribution is called "Student's t distribution," and is derived as follows:

$$t = \frac{(\bar{X} - \mu)\sqrt{n}}{S} \tag{10.2}$$

The t distribution has a relative frequency curve that is both bell-shaped and symmetrical; this curve is illustrated in Figure 10.2.

The only parameter that determines the shape of t distribution curve is the degree of freedom (df) assigned to the problem; this is given by $n - 1$. As in the case of a normal probability distribution, the t values corresponding to the degree of freedom and level of significance are given in Appendix Table J.

EXAMPLE 10.2 / Let us reexamine the Goodstone Tire Company case again. Suppose that the local cab company purchased 25 tires on a trial

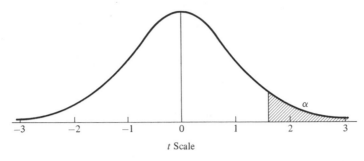

Figure 10.2 / Student's t distribution.

basis from the Goodstone Tire Company. After six months of usage, the sample mean ($\bar{X}$) was recorded as 38,000 miles, with an associated standard deviation of 10,000 miles. Estimate the population mean at a 95% confidence level.

SOLUTION: From the given information, where $\bar{X}$ = 38,000 miles, S = 10,000 miles, n = 25 tires,

$$S_{\bar{x}} = \frac{S}{\sqrt{n}} = \frac{10,000}{\sqrt{25}} = 2,000 \text{ miles,}$$

From Appendix Table J (df = 25 − 1 = 24), $t_{\alpha=0.05}$ = 2.064. According to Equation 10.1,

$$\bar{X} - (S_{\bar{x}})(t_{\alpha=0.05}) \le \mu \le \bar{X} + (S_{\bar{x}})(t_{\alpha=0.05}),$$

$$38,000 - \left(\frac{10,000}{\sqrt{25}}\right)(2.064) \le \mu \le 38,000 + \left(\frac{10,000}{\sqrt{25}}\right)(2.064),$$

$$33,872 \le \mu \le 42,128.$$

As shown above, when the sample size is small, the error range tends to increase. The significance of this difference will be explained in a subsequent section.

Necessary sample size

From the previous two sections, we can see that the sampling error can be effectively controlled by the sample size. Since the sample size and the level of significance (α) are subject to the control of the decision maker, it is also possible that the error range (E) can be controlled in the process of population estimation.

From Equation 10.1, it was shown that

$$E = \pm(S_{\bar{x}})(Z_{\alpha}),$$

or

$$E = \pm(S/\sqrt{n})(Z_\alpha).$$

Multiplying both sides of the equation by $\sqrt{n}$ and then dividing by E, we obtain

$$\sqrt{n} = \frac{(S)(Z_\alpha)}{E}.$$

By squaring both sides, the equation becomes

$$n = \left[\frac{(S)(Z_\alpha)}{E}\right]^2 \tag{10.3}$$

In Equation 10.3, n represents the sample size required to change (either increase or decrease) the predetermined error range (E).

EXAMPLE 10.3 / Suppose that the cab company in Example 10.1 wants to reduce the error range (E) from the current 1,960 miles to 1,000 miles. How many samples should the company take to accomplish this purpose? (Use a 95% confidence level.)

SOLUTION: In this case, where $S = 10,000$ miles, $Z = 1.96$, and $E = 1,000$ miles, the optimum sample size required to reduce the range from 1,960 to 1,000 miles is computed to be

$$n = \left[\frac{(10,000)(1.96)}{1,000}\right]^2$$

$$= (19.6)^2 = 384.16 \equiv 384 \text{ tires.}$$

The cab company therefore needs approximately 384 sample tires to reduce the error range from the current 1,960 to 1,000 miles. As the reader may also have noticed, the sample size had to be increased by almost four times (i.e., from 100 to 384 tires) to reduce the range (E) by only approximately 50%, or from 1,960 to 1,000 miles.

10.3 / Testing Hypotheses

In Section 10.2, we saw how samples could be used to estimate the population parameters. In this section, we discuss another method of using samples to make statistical inferences concerning population parameters; i.e., we use samples to decide between two mutually exclusive courses of action.

Examples of mutually exclusive actions include such everyday decisions as determining whether a certain process produces more or less than 10% defective parts, or determining whether your car will get more or less than

20 miles per gallon of gasoline, and many other practical situations involving judgment.

Decision makers usually arrive at a final decision after analyzing the test results of two mutually exclusive courses of action. For example, the decision maker may choose to ship his products as they are currently being produced, if the test results indicate that the shipment contains less than or as many as 10% defective parts. On the other hand, the shipment may have to be delayed or totally reprocessed if the sample indicates a defective rate greater than 10%. In another example, you may determine to buy a particular compact car if test results confirm the manufacturer's claim that the car will actually get 20 miles to the gallon.

When the decision maker uses hypotheses testing as the method for determining the statistical inference of the population value, his first step is to determine whether or not a population parameter is in fact equal to some predetermined or prescribed value.

Let us again use the Goodstone Tire Manufacturing Company case to illustrate this concept. Suppose that the management of Goodstone considers its manufacturing process to be under control if the mean mileage of tires ranges from 38,000 to 42,000 miles; on the other hand, the company considers the process to be out of control if the mean is greater than 42,000 miles or less than 38,000 miles. The management of the company wishes to test the hypothesis that the true mean (μt) in this problem is 40,000 miles as compared to another hypothesis that the mean is not 40,000 miles. The former hypothesis is known as a "null" hypothesis and is generally denoted by the symbol H_0; the latter is referred to as the alternative hypothesis and is denoted by H_1.

In the Goodstone case, the company will accept the null hypothesis (H_0) and reject the alternative hypothesis (H_1) if $38,000 \leq \mu \leq 42,000$. Conversely, the company should accept H_1 and consequently reject H_0 if $\mu > 42,000$ or $\mu < 38,000$ miles.

Errors associated with hypothesis testing

There are two major types of errors that may be committed in the process of making statistical inference based on hypothesis testing. First, a decision maker may consider the process to be out of control—i.e., the true mean is outside the acceptable range boundaries (which will subsequently lead to the rejection of the null hypothesis)—when, in fact, the situation is really within acceptable control limits. This type of situation can occur, for example, in the Goodstone case if the true mean is greater than 42,000 or less than 38,000 miles, whereas the sample mean is between 38,000 and 42,000 miles.

Secondly, the opposite situation may occur: the decision maker may conclude that the process is within the control boundaries when, in fact, the situation is really out of control. This condition could happen in the Goodstone case if, for example, the true mean is not 40,000 miles, but the sample mean ranges from 38,000 to 42,000 miles.

Table 10.1 / Description of Type I and
Type II Errors for the
Goodstone Company

	H_0 is *true* ($\mu = 40,000$)	H_0 is *false* ($\mu \neq 40,000$)
Accept H_0	Correct decision	Type II error
Reject H_0	Type I error	Correct decision

The first kind of statistical error described above is called a "Type I" error, and the second kind is referred to as a "Type II" error. Table 10.1 summarizes the salient features of both Type I and Type II errors.

In general, a Type I error is committed if H_0 is true but is subsequently rejected. The probability of committing a Type I error is denoted by the Greek letter α (alpha). Conversely, a Type II error is committed if the hypothesis is false but is accepted as the true state of nature; the probability of committing a Type II error is denoted by the Greek letter β (beta).

As the reader has probably surmised, there is often a definite adverse effect or penalty associated with a "bad" decision, which results in either a Type I or Type II error. For example, the economic penalty of a Type I error may result in a decision by the Goodstone Company to shut down production in order to search for a nonexistent mean. On the other hand, the economic penalty of a Type II error could be that undesirable products are continuously allowed to be produced. In either case, the potential economic losses are severe. Let us examine this situation in further detail.

Two-tailed test versus one-tailed test

Depending upon the nature of the problem being analyzed, the decision maker is interested in evaluating either both ends of the distribution (two-tailed test) or only one end of the distribution (one-tailed test).

Two-tailed test. In a two-tailed test, the decision maker is concerned with both ends of distribution. For example, the management of the Goodstone Tire Company wants to ensure that it produces quality tires in terms of mileage so that its customers will not eventually reject its products; therefore, the company is interested in testing only the lower end of distribution (see Figure 10.1). On the other hand, the company wants to control the upper end of the distribution because it does not want to sell superior tires at too low a price. Therefore, in order to develop a successful product, the company intends to control both ends of the distribution.

EXAMPLE 10.4 / Suppose that the Goodstone Company examined a sample of 100 tires. The results of the sample showed a 38,000 average mileage life per tire, with a corresponding standard deviation of 10,000 miles. Assume that the Goodstone Company wanted to test the null hypothesis of $\mu = 40,000$ miles against an alternative hypothesis that $\mu \neq 40,000$ miles. Assume further that a critical value of 0.05 (i.e., $\alpha = 0.05$) was used in this test. Should the company accept the null hypothesis or the alternative hypothesis?

SOLUTION: The procedures to be followed in a two-tail test are as follows:

(1) Determine the hypotheses under consideration.

$H_0: \mu = 40,000$ miles,

$H_1: \mu \neq 40,000$ miles,

$\alpha = 0.05$ (Type I error).

(2) Establish the decision criteria.

Accept H_0 if $-1.96 \leq Z \leq 1.96$

Reject H_0 if $Z > 1.96$ or $Z < -1.96$.

(3) Compute the actual value of Z. Given that $\bar{X} = 38,000$ miles, $S = 10,000$ miles, $n = 100$, and $S_x = S/\sqrt{n} = 10,000/\sqrt{100} = 1,000$ miles, then

$$Z = \frac{\bar{X} - \mu}{S_{\bar{x}}} = \frac{38,000 - 40,000}{1,000} = -2.$$

(4) Analyze the test results. Since $-2 < -1.96$, the company should reject the null hypothesis that the true mean of the tire is 40,000 miles. In other words, the difference of 2,000 miles between the sample mean (38,000 miles) and the hypothetical or claimed mean (40,000 miles) is too large to attribute to pure chance.

The preceding solution is shown graphically in Figure 10.3, where the shaded areas represent the regions of rejection applicable in this problem.

One-tail test. In many instances, due to government regulations, financial or production constraints, or a host of other (normally externally induced) reasons, only one side or tail of a normal distribution needs to be evaluated in order to select an appropriate decision. Examples of cases that are amenable to a one-tail test include the following: the manufacture of medical items where minimum health standards must be met; the meeting of minimum net-weight requirements in a canned food item; or the maximum weight that may be carried in a small boat to meet safety standards. In each of these examples, owing to a predetermined restriction on one end of the normal distribution, an analysis needs to be performed on only one side of the distribution.

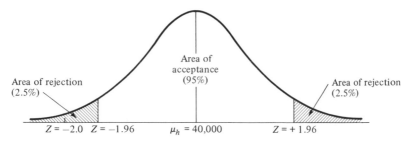

Figure 10.3 / Goodstone Tire Company: two-tail test.

EXAMPLE 10.5 / Suppose, in the Goodstone Company case, consumers will probably not be too concerned if a tire lasts more than 40,000 miles; a customer would probably become very concerned, however, if the tire wore out in less than 40,000 miles. Accordingly, from a practical viewpoint, the appropriate method to be used in this situation is a single-tail test. In this case, we are concerned with only the left tail of the curve; i.e., $\mu < 40,000$ miles.

SOLUTION:

(1) Determine the hypotheses under consideration.

$H_0: \mu = 40,000$ miles,

$H_1: \mu < 40,000$ miles,

$\alpha = 0.05$ (Type I error).

(2) Establish the decision criteria.

Accept H_0 if $Z \geq -1.64$ (from Appendix Table G, the Z value of a one-tail test is ± 1.64).

Reject H_0 if $Z < -1.64$.

(3) Compute the actual value of Z. Given that $\bar{X} = 38,000$ miles, $S = 10,000$ miles, $n = 100$, and $S_{\bar{x}} = 1,000$ miles, then

$$Z = \frac{38,000 - 40,000}{1,000} = -2.0.$$

(4) Analyze the test results. Since $-2 < -1.64$, the company should reject the null hypothesis that the true mean concerning tire life is 40,000 miles. The preceding solution is shown graphically in Figure 10.4.

It is interesting to note that in either the two-tail or one-tail test (and contrary to the results obtained above for $\alpha = 0.05$), the null hypothesis

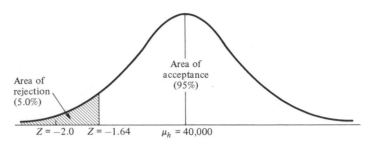

Figure 10.4 / Goodstone Tire Company: one-tail test.

would be accepted if the critical level was 0.01 ($\alpha = 0.01$); this is true because the Z values at $\alpha = 0.01$ are 2.58 in a two-tail test and 3.09 in a one-tail test. Both values are, of course, greater than the actual value of -2.

Determination of α and β

As the reader may have already noticed, the probability that a Type I error will occur is determined by the decision maker's choice concerning the amount of risk that one is willing to accept; e.g., a 5% risk or a 1% risk. If the individual elects to accept a 5% chance, or risk, that the resulting statistical inference will be wrong; accordingly, the size of the corresponding Type I error is 5%. In other words, the size of the Type I error (or alpha error) is always equal to the value of α established by an individual in the decision-making process. The size of the potential Type II error (beta error) varies, however, depending upon the values of the true and hypothetical mean.

EXAMPLE 10.6 / Suppose the Goodstone Tire Company claims that, as a result of a sample test of 100 tires, its radial tires have a mean (μ_h) of 40,000 miles and a corresponding standard deviation of 10,000 miles. Using a two-tail test, where $\alpha = 0.05$, we would accept the null hypothesis of $\mu = 40,000$ miles, as opposed to the selection of the alternative hypothesis of $\mu \neq 40,000$ miles, if $-1.96 \leq Z \leq 1.96$; this is illustrated in Figure 10.5, where the value of E was estimated by

$$E = (S_{\bar{x}})(Z_\alpha) = \frac{S}{\sqrt{n}}(Z_\alpha)$$

$$= \left(\frac{10,000}{\sqrt{100}}\right)(1.96)$$

$$= 1,960 \text{ miles.}$$

Suppose, however, that the true mean (μ_t) actually turns out to be 41,000 miles. In this case, the company obviously committed a Type II error; this

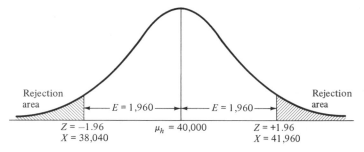

Figure 10.5 / Goodstone Tire Company. The value of E was estimated by $E = (S\bar{x})(Z_\alpha) = (S/\sqrt{n})(Z_\alpha)$.

is concluded because the company accepted the false hypothesis that $\mu = 40{,}000$ miles, while in reality the true mean (μ_t) was 41,000 miles. Based upon these data, what is the probability of the Type II error?

SOLUTION: Figure 10.6 illustrates the area of a Type II error when the true mean is 41,000 miles. Based upon the null hypothesis, the results from the sample would show a mean ranging from 38,040 miles to 41,960 miles in order for the original null hypothesis of $\mu_h = 40{,}000$ miles to be accepted as an approximation of the true mean. This is shown in curve A of Figure 10.6 (curve A was reproduced from Figure 10.3). If the true mean is 41,000 miles, however, the entire area within the boundary of 38,040 to 41,960 miles (except for a thin line of the 41,000th mile itself) in curve A is false. Therefore, if the null hypothesis is accepted, a Type II error would be committed.

Figure 10.6 / Type II error when $\mu_t = 41{,}000$ miles, $\mu_h = 40{,}000$ miles.

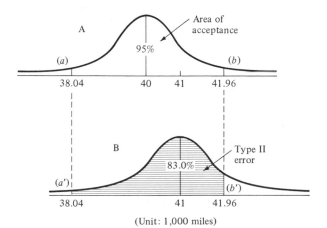

275

To estimate the size of the Type II error, we draw a new normal curve, B, based upon the true mean of $\mu_t = 41{,}000$ miles and $S = 10{,}000$ miles. This new curve is drawn immediately below the original normal curve in Figure 10.6. Straight lines are then drawn from the original boundary, points a and b in curve A to the corresponding points a' and b' in curve B. The area applicable to a Type II error is indicated by the shaded area in curve B.

The computation of the probability of the shaded area (or a Type II error) located in curve B of Figure 10.6 is rather straightforward. From Equation 4.12, the probability of the shaded area (β) is determined by

$$\beta = P_{a'}\left(Z = \frac{X - \mu_t}{S_{\bar{x}}}\right) + P_{b'}\left(Z = \frac{X - \mu_t}{S_{\bar{x}}}\right)$$

$$= P_{a'}\left(Z = \frac{41{,}960 - 41{,}000}{1{,}000}\right) + P_{b'}\left(Z = \frac{38{,}040 - 41{,}000}{1{,}000}\right)$$

$$= P(Z = 0.96) + P(Z = -2.96)$$

$$= 0.3315 + 0.4985 = 0.8300.$$

The probability of having a Type II error occur is therefore 83% when the true mean is 41,000 miles and the hypothetical mean is estimated at 40,000 miles.

On the other hand, suppose that the true mean is actually 39,500 miles rather than the hypothetical mean of 40,000 miles. The probability of having a Type II error occur in this case is similarly computed, as shown in Figure 10.7. Accordingly,

$$\beta = P_{a'}\left(Z = \frac{38{,}040 - 39{,}500}{1{,}000}\right)$$

$$= + P_{b'}\left(Z = \frac{41{,}960 - 39{,}500}{1{,}000}\right)$$

$$= P(Z = -1.46) + P(Z = 2.46)$$

$$= 0.4279 + 0.4931 = 0.9210.$$

In other words, the probability of a Type II error is 92.1% when the true mean is 39,500 miles and the hypothetical mean is estimated to be 40,000 miles.

Table 10.2 illustrates several examples of different sizes of Type II errors when the true mean varies from a hypothetical mean of 40,000 miles. The third column $(1 - \beta)$ indicates the probabilities of making a correct decision when the true mean is different from the hypothetical mean. If we plot the probabilities contained in the second column (error β) against the corresponding changes in the population parameter (μ_t), we obtain what is called an operating characteristic curve (OC curve), shown in Figure 10.8.

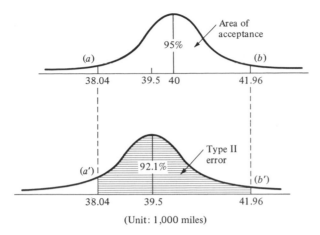

(Unit: 1,000 miles)

Figure 10.7 / Type II error when $\mu_h = 39,500$ miles, and $\mu_h = 40,000$ miles.

By using the OC curve, we are able to estimate (through an interpolation process) the probability of committing a Type II error for any given true mean (μ_t) value other than those already given in Table 10.2. For example, the probability of committing a Type II error for a true mean of 38,800 miles is estimated to be 0.76 from Figure 10.8.

From the data contained in the Table 10.2 and Figure 10.8, it can be seen that the probability of committing a Type II error is at its highest point when the value of the true mean changes only slightly from the hypothetical mean of 40,000 miles; and, conversely, the probabilities of committing a Type II error gradually decrease as the differences between the

Table 10.2 / Type II Errors for Different
Values of True Mean and
$\mu_h = 40,000$ Miles

True mean (μ_t)	Type II error (β)	Correct decision $1 - \beta$
41,500	0.6769	0.3231
41,000	0.8300	0.1700
40,500	0.9210	0.0790
40,000	0.0000	0.0500*
39,500	0.9210	0.0790
39,000	0.8300	0.1700
38,500	0.6769	0.3231

* Type I error.

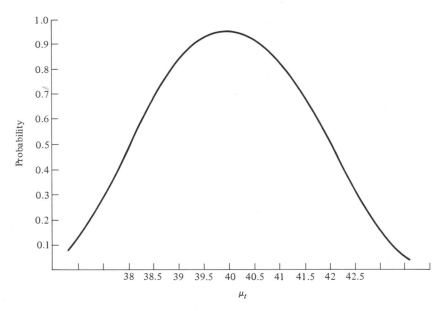

Figure 10.8 / OC curve for the Goodstone Tire Company when $\mu_h = 40{,}000$ miles.

true mean and the hypothetical mean increase. In other words, when the true mean is significantly different from the hypothetical mean, there is less chance for the decision maker to accept a false hypothesis.

Hypothesis testing is often used as an important decision criterion in the decision-making process. In this respect, the probability of committing a Type I error is compared against the likelihood of committing a Type II error. Of course several other factors do affect the ultimate decision that is reached. For example, the degree of optimism and pessimism on the part of the decision maker will affect the ultimate decision that is reached.

10.4 / Relationship between Modern Decision Theory and Classical Decision Theory

As shown in the preceding section it is evident that classical decision theory, unlike modern decision theory, does not explicitly consider the monetary or other utility consequences inherent in the decision model. In classical decision theory, the seriousness of committing errors is measured in terms of probability values and not in terms of monetary or utility loss, as in modern decision theory. It may also be apparent in comparing these two decision theories that the modern approach seems to be concerned with making "decisions" based on an exhaustive course of actions, whereas classical decision theory seems to be oriented more toward just drawing "conclusions" or "inferences" based on two mutually exclusive courses of action. Since the decision maker who is using a modern decision-theory approach is forced to consider all possible alternatives in order to arrive at an optimum decision under conditions of uncertainty, this type of analysis is normally considered a more effective method of problem solving.

The basic approach used in classical decision theory is to differentiate between Type I and Type II errors, and then to measure the risk of committing each of these two types of errors in terms of the probability of occurrence. However, a close analysis of the decision-making process reveals a similarity underlying the principles on which these two approaches are based. We can use a Type I and Type II error chart to explain these points. Table 10.3 was reproduced with a few modifications from Table 10.1.

In Table 10.3, let a_1 = course of action that accepts the H_0 hypothesis; a_2 represents the alternative that accepts the H_1 hypothesis. It follows, therefore, that the probability of committing a Type I error is $\alpha = P(a_2|H_0)$, and the probability of committing a Type II error is $\beta = P(a_1|H_1)$.

According to classical decision theory, the optimum decision is to select a course of action that will minimize the probability of committing a Type II error, given the probability that a Type I error occurs. In other words, according to classical decision theory, the best decision rule is derived from the function that will minimize $P(a_1|H_1)$, subject to the restriction $P(a_2|H_0) = \alpha$.

Error chart and opportunity loss. Let us now discuss Table 10.3 from a modern decision-theory viewpoint. According to modern decision theory, the optimum decision is the course of action that yields either the maximum expected monetary value (utility) or the least opportunity loss. In this respect, an error chart such as Table 10.3 can be rearranged into an opportunity-loss form from which an optimum decision can be derived.

EXAMPLE 10.7 / The A & S Supermarket in Toledo, Ohio, is considering buying 1,000 cases of fresh produce. Based on past experience, the manager of the supermarket expects either that 30% or 60% of the produce received will be in an unsalable condition. The cost of the produce is $5.00 per case, with a retail selling price of $10.00 per case; i.e., a $5.00 profit per case (or 100% markup) is expected. Furthermore, unsold produce cannot be

Table 10.3 / Modified Error Chart

	Accept H_0 (accept action a_1)	Accept H_1 (accept action a_2)
$H_0: \mu = \mu_0$ is true	$1 - \alpha$	$P(a_2|H_0) = \alpha$
$H_1: \mu \neq \mu_0$ is true	$P(a_1|H_1) = \beta$	$1 - \beta$

Table 10.4 / Profit/Loss for
A & S Supermarket

	To buy (a_1)	Not to buy (a_2)
$H_0: P = 0.3$ is true	2,000	0
$H_1: P = 0.6$ is true	$-1,000$	0

returned to the wholesaler for credit. Therefore, if 30% of the case contains spoiled products, the profit would be

$$1,000 \times 0.7 \times \$5.00 - 1,000 \times 0.3 \times \$5.00 = \$2,000.$$

On the other hand, if the case contains 60% spoiled products, the profit would be

$$1,000 \times 0.4 \times \$5.00 - 1,000 \times 0.6 \times \$5.00 = -\$1,000.$$

SOLUTION: Let a_1 = decision to buy the produce, a_2 = decision not to buy the produce, and P = proportion of spoiled produce per case. Then the preceding profit and loss data can be rearranged into Table 10.4.

Suppose that Table 10.4 is a profit matrix. It is not too difficult to generate an opportunity-loss table. This is illustrated in Table 10.5.

If the case contains only 0.3 (30%) spoiled produce, the optimum decision is to select course of action a_1, with its corresponding $2,000 profit and no opportunity loss. On the other hand, if the decision maker decides not to buy (a_2), although $P = 0.3$, then the opportunity loss of this decision is $2,000.

Profit functions. The profit function for the A & S problem is as follows:

$$a_1 = 1,000(1 - P) \times \$5.00 - 1,000P \times \$5.00$$
$$= 5,000 - 10,000P,$$
$$a_2 = 0.$$

Table 10.5 / Opportunity-Loss Table for
A & S Supermarket

	Proportion of spoiled produce (P)	Course of actions	
		To buy (a_1)	Not to buy (a_2)
State of nature	0.3	0	2,000
State of nature	0.6	1,000	0

Break-even value. According to Equation 7.7, the break-even value of P (proportion of spoiled produce) is

$$P_k = \frac{0 - 5{,}000}{-10{,}000 - 0} = 0.5.$$

In other words, if the produce is unsalable 50% of the time, the manager is indifferent as to which course of action (a_1 or a_2) is selected because the expected value at this rate would be identical, or

Profit for $a_1 = 5{,}000 - 10{,}000(0.5) = 0$

Profit for $a_2 = 0$.

Opportunity loss functions. It is now possible to compute the opportunity loss functions for the A & S Supermarket. Let $L = $ opportunity-loss function; it then follows that for action a_1:

$$L_1 = 0 \qquad \text{(if } P < 0.5\text{)},$$
$$L_1 = E(a_2) - E(a_1)$$
$$= -5{,}000 + 10{,}000E(P) \qquad \text{(if } P > 0.5\text{)}.$$

and for action a_2:

$$L_2 = 0 \qquad \text{(if } P > 0.5\text{)},$$
$$L_2 = E(a_1) - E(a_2)$$
$$= 5{,}000 - 10{,}000E(P) \qquad \text{(if } P < 0.5\text{)}.$$

10.5 / Typical Application of Classical Theory

Let us once again use the Goodstone Tire Company example to illustrate the classical decision concept.

EXAMPLE 10.8 / Suppose that a local cab company purchased 100 tires on a trial basis from both the Goodstone Tire Company and the United Tire Company. The cab company plans to sign a long-term contract for tires with the company whose tires yield the greatest average mileage life. Upon completion of an extended road test, the results obtained are tabulated as in Table 10.6. What is the optimum decision?

SOLUTION: Let us solve this problem by following the decision-making steps outlined in Chapter 1.

(1) Objective: To reduce operating costs by signing a long-term contract with the company whose tires yield the greatest average mileage life.

(2) Alternative courses of action: To sign the contract with either the Goodstone Company or United Tire Company.

Table 10.6 / Sampling Results for Tire Life

Goodstone Company			United Company		
Mileage	Frequency	Probability	Mileage	Frequency	Probability
38,000	20	0.20	37,000	5	0.05
39,000	40	0.40	38,000	15	0.15
40,000	30	0.30	39,000	30	0.30
41,000	5	0.05	40,000	40	0.40
42,000	5	0.05	41,000	10	0.10
	100	1.00		100	1.00

(3) States of nature: The range of variation of tires for the two companies. In our current example (from Table 10.6), this variation ranges from 38,000 miles to 42,000 miles for the Goodstone Company, and from 37,000 miles to 41,000 miles for the United Company.

(4) Probabilities: The likelihoods concerning the states of nature were obtained from the given sample frequencies shown in Table 10.6.

(5) Payoffs: Since there are no cost or revenue data provided, this problem does not require the listing of payoffs.

(6) Expected values: The expected value concerning the average tire life for each company is computed from Equation 3.3:

$$E = \sum P(X_i) \cdot X_i.$$

Accordingly,

$$E(\text{Goodstone}) = 38{,}000 \times 0.20 + 39{,}000 \times 0.40 + 40{,}000$$
$$\times 0.30 + 41{,}000 \times 0.05 + 42{,}000 \times 0.05$$
$$= 39{,}350 \text{ miles},$$
$$E(\text{United}) = 37{,}000 \times 0.05 + 38{,}000 \times 0.15 + 39{,}000$$
$$\times 0.30 + 40{,}000 \times 0.40 + 41{,}000 \times 0.10$$
$$= 39{,}350 \text{ miles}.$$

As the data indicate, the expected values of the tire life for both companies are identical, i.e., 39,350 miles.

(7) Optimum decision: If we assume that the prices of tires are the same for both companies, and if the decision maker elects to use the expected monetary value as the sole decision criterion, the optimum decision in this problem is to purchase the tires from *either* the Goodstone Company or the United Company. The decision maker could, of course, expand the decision criteria to suit any specific conditions that may be desired. For example, if he has established a minimum tolerance level of 38,000 miles as a decision criterion, the optimum decision in this case will

be to buy the tires from the Goodstone Company, since the least mileage experienced from a Goodstone tire is 38,000 miles, while the minimum mileage obtained from a United tire is 37,000 miles.

10.6 / Advantages of Modern Decision Theory

In general, the use of the modern decision theory approach to solve problems has several advantages over the use of classical decision theory procedures. Some of these advantages include the following: providing for several alternatives; making uncertainty explicit in the form and subsequent use of a subjective probability distribution; and including provisions for the use of both prior knowledge and current sampling information in the solution of problems. The use of the expected monetary value concept along with utility theory provides the decision maker a clear choice under conditions of uncertainty concerning the selection of an alternative or course of action which best fits the conditions and objectives of the situation being analyzed. Since the selection of an optimum decision will vary, depending upon the environment and circumstances surrounding the decision maker, it is apparent that modern decision theory, which utilizes a subjective probability concept, provides a much more flexible decision tool than that of the classical approach.

The concepts and procedures encompassed by modern decision theory are relatively new and are therefore currently limited to a rather small number of practical applications. Nevertheless, there is clear evidence that a growing number of theoreticians as well as practitioners are convinced that the use of modern decision models is greatly increasing the ability of the decision maker to deal with and perhaps solve the complex problems that face us in the dynamic world in which we live.[1]

10.7 / Summary

This chapter has been concerned with a discussion of classical decision theory, and the following items represent its highlights.

(1) Classical decision theory usually employs two statistical procedures: (a) estimation of population parameters, and (b) testing of hypotheses.

(2) The determination of a population parameter is a statistical procedure that estimates the population value of both the mean and standard deviation through the use of samples and a predetermined level of confidence.

(3) Hypotheses testing is a statistical procedure to confirm or disprove the predetermined population parameters.

---

[1] For a detailed, practical application of modern decision theory, see Joseph W. Newmann, *Management Application of Decision Theory.* (New York: Harper & Row, 1973).

(4) A Type I error is committed when one rejects a hypothesis that is in fact true. The economic consequences of a Type I error may be an unnecessary closedown of a production process in an effort to search for a nonexistent population parameter.

(5) A Type II error occurs when the decision maker accepts a hypothesis that is in fact false. The economic consequence of a Type II error is normally the continued production of inferior products.

(6) The decision maker may control the error range (E) by varying the sample size (n):

$$n = \left[\frac{(S)(Z_\alpha)}{E} \right]^2$$

(7) In classical decision theory, the optimum decision rule is to minimize the probability of a Type II error (β), provided a given level of Type I error (α).

(8) Although there is some relationship between classical decision theory and modern decision theory, modern decision theory is a more explicit decision-making process, especially in the areas of uncertainties, monetary profit or loss, or utility and all alternative courses of action.

EXERCISES

1. Discuss the major advantages that modern decision theory has over classical decision theory.

2. What are the potential economic penalties of Type I and Type II errors?

3. Explain how the concepts of Type I and Type II errors are used as decision criteria.

4. "As the likelihood of a Type I error increases the probability that a Type II error will occur decreases." Do you agree? Why or why not?

5. The probability of committing a Type II error decreases as the difference between the true mean and the hypothetical mean increases, while the value of the EVPI decreases as the difference between the break-even value (μ_k) and the mean (μ) increases. Explain the significance of these two statements.

6. When may the tolerance level be used as a decision criterion?

7. The Manchester Chemical Company of St. Louis is currently using 150 gallons per day of a chemical called Chloline 100, with an associated standard deviation of 20 gallons per day. Since the supply of Chloline 100 is limited and the only supplier of this chemical is located in South America, the inventory control of Chloline 100 is very important. In view of this fact, the usage of Chloline 100 has been carefully monitored for the past 25 working days; during this sample time period, the mean usage of Chloline 100 was determined to be 160 gallons per day.
   (a) Estimate the population mean of Chloline 100, using a 95% confidence level.
   (b) Using the answer derived in (a), do you believe that the mean usage of Chloline 100 is still 150 gallons per day?

8. By determining the average number of miles that his car travels per gallon of gas, Mr. Turner claims that he can tell whether his car needs to be tuned. If his car is properly tuned, he will normally get 15 miles per gallon, with a standard deviation of 2 miles. After 16 fill-ups last month, he found that his actual gas consumption rate was 14 miles per gallon.
   (a) With a 95% confidence level, test the hypothesis that the average miles per gallon is still 15 miles.
   (b) What are the economic penalties for Type I and Type II errors in this case?
   (c) How many samples are required to reduce the range cited in (a) by 50%?
   (d) Suppose that the true mean is 13 miles per gallon. Using a 95% confidence level, what is the probability of committing a Type II error?

SUGGESTED READING

Costis, Harry G. *Statistics for Business.* Columbus, Ohio: Charles E. Merrill, 1972, chapters 14 through 17.

Fellner, William. *Probability and Profit.* Homewood, Ill.: Richard D. Irwin, 1965, chapter 3.

Freund, John E., and Williams, Frank J. *Elementary Business Statistics* (2d ed.). Englewood Cliffs, N. J.: Prentice-Hall, 1972, chapters 9 and 10.

Halter, Albert N., and Dean, Gerald W. *Decisions under Uncertainty with Research Applications.* Cincinnati, Ohio: South-Western Publishing, 1971, chapters 12 and 14.

Hamburg, Morris. *Statistical Analysis for Decision Making.* New York: Harcourt, Brace & World, 1970, chapter 17.

Newman, Joseph W. *Management Applications of Decision Theory.* New York: Harper & Row, 1973, chapter 11.

Raiffa, Howard. *Decision Analysis: Introductory Lectures on Choices under Uncertainty.* Reading, Mass.: Addison-Wesley, 1968, chapter 10.

Schlaifer, Robert. *Analysis of Decisions under Uncertainty.* New York: McGraw-Hill, 1969, chapters 1 and 2.

Specer, Milton H. *Managerial Economics.* Homewood, Ill.: Richard D. Irwin, 1968, chapter 1.

Thompson, Gerald E. *Statistics for Decisions.* Boston: Little, Brown, 1972, chapter 2.

APPENDIX

Tables

A / Six-Place Logarithms of Base 10

N.	0	1	2	3	4	5	6	7	8	9	D.
100	000000	000434	000868	001301	001734	002166	002598	003029	003461	003891	432
1	4321	4751	5181	5609	6038	6466	6894	7321	7748	8174	428
2	8600	9026	9451	9876	010300	010724	011147	011570	011993	012415	424
3	012837	013259	013680	014100	4521	4940	5360	5779	6197	6616	420
4	7033	7451	7868	8284	8700	9116	9532	9947	020361	020775	416
105	021189	021603	022016	022428	022841	023252	023664	024075	4486	4896	412
6	5306	5715	6125	6533	6942	7350	7757	8164	8571	8978	408
7	9384	9789	030195	030600	031004	031408	031812	032216	032619	033021	404
8	033424	033826	4227	4628	5029	5430	5830	6230	6629	7028	400
9	7426	7825	8223	8620	9017	9414	9811	040207	040602	040998	397
110	041393	041787	042182	042576	042969	043362	043755	044148	044540	044932	393
1	5323	5714	6105	6495	6885	7275	7664	8053	8442	8830	390
2	9218	9606	9993	050380	050766	051153	051538	051924	052309	052694	386
3	053078	053463	053846	4230	4613	4996	5378	5760	6142	6524	383
4	6905	7286	7666	8046	8426	8805	9185	9563	9942	060320	379
115	060698	061075	061452	061829	062206	062582	062958	063333	063709	4083	376
6	4458	4832	5206	5580	5953	6326	6699	7071	7443	7815	373
7	8186	8557	8928	9298	9668	070038	070407	070776	071145	071514	370
8	071882	072250	072617	072985	073352	3718	4085	4451	4816	5182	366
9	5547	5912	6276	6640	7004	7368	7731	8094	8457	8819	363
120	079181	079543	079904	080266	080626	080987	081347	081707	082067	082426	360
1	082785	083144	083503	3861	4219	4576	4934	5291	5647	6004	357
2	6360	6716	7071	7426	7781	8136	8490	8845	9198	9552	355
3	9905	090258	090611	090963	091315	091667	092018	092370	092721	093071	352
4	093422	3772	4122	4471	4820	5169	5518	5866	6215	6562	349
125	6910	7257	7604	7951	8298	8644	8990	9335	9681	100026	346
6	100371	100715	101059	101403	101747	102091	102434	102777	103119	3462	343
7	3804	4146	4487	4828	5169	5510	5851	6191	6531	6871	341
8	7210	7549	7888	8227	8565	8903	9241	9579	9916	110253	338
9	110590	110926	111263	111599	111934	112270	112605	112940	113275	3609	335
130	113943	114277	114611	114944	115278	115611	115943	116276	116608	116940	333
1	7271	7603	7934	8265	8595	8926	9256	9536	9915	120245	330
2	120574	120903	121231	121560	121888	122216	122544	122871	123198	3525	328
3	3852	4178	4504	4830	5156	5481	5806	6131	6456	6781	325
4	7105	7429	7753	8076	8399	8722	9045	9368	9690	130012	323
135	130334	130655	130977	131298	131619	131939	132260	132580	132900	3219	321
6	3539	3858	4177	4496	4814	5133	5451	5769	6086	6403	318
7	6721	7037	7354	7671	7987	8303	8618	8934	9249	9564	316
8	9879	140194	140508	140822	141136	141450	141763	142076	142389	142702	314
9	143015	3327	3639	3951	4263	4574	4885	5196	5507	5818	311
140	146128	146438	146748	147058	147367	147676	147985	148294	148603	148911	309
1	9219	9527	9835	150142	150449	150756	151063	151370	151676	151982	307
2	152288	152594	152900	3205	3510	3815	4120	4424	4728	5032	305
3	5336	5640	5943	6246	6549	6852	7154	7457	7759	8061	303
4	8362	8664	8965	9266	9567	9868	160168	160469	160769	161068	301
145	161368	161667	161967	162266	162564	162863	3161	3460	3758	4055	299
6	4353	4650	4947	5244	5541	5838	6134	6430	6726	7022	297
7	7317	7613	7908	8203	8497	8792	9086	9380	9674	9968	295
8	170555	170848	171141	171434	171726	172019	172311	172603	172895		293
9	3186	3478	3769	4060	4351	4641	4932	5222	5512	5802	291
150	176091	176381	176670	176959	177248	177536	177825	178113	178401	178689	289
1	8977	9264	9552	9839	180126	180413	180699	180986	181272	181558	287
2	181844	182129	182415	182700	2985	3270	3555	3839	4123	4407	285
3	4691	4975	5259	5542	5825	6108	6391	6674	6956	7239	283
4	7521	7803	8084	8366	8647	8928	9209	9490	9771	190051	281
155	190332	190612	190892	191171	191451	191730	192010	192289	192567	2846	279
6	3125	3403	3681	3959	4237	4514	4792	5069	5346	5623	278
7	5900	6176	6453	6729	7005	7281	7556	7832	8107	8382	276
8	8657	8932	9206	9481	9755	200029	200303	200577	200850	201124	274
9	201397	201670	201943	202216	202488	2761	3033	3305	3577	3848	272
N.	0	1	2	3	4	5	6	7	8	9	D.

Table A (Continued)

N.	0	1	2	3	4	5	6	7	8	9	D.
160	204120	204391	204663	204934	205204	205475	205746	206016	206286	206556	271
1	6826	7096	7365	7634	7904	8173	8441	8710	8979	9247	269
2	9515	9783	210051	210319	210586	210853	211121	211388	211654	211921	267
3	212188	212454	2720	2986	3252	3518	3783	4049	4314	4579	266
4	4844	5109	5373	5638	5902	6166	6430	6694	6957	7221	264
165	7484	7747	8010	8273	8536	8798	9060	9323	9585	9846	262
6	220108	220370	220631	220892	221153	221414	221675	221936	222196	222456	261
7	2716	2976	3236	3496	3755	4015	4274	4533	4792	5051	259
8	5309	5568	5826	6084	6342	6600	6858	7115	7372	7630	258
9	7887	8144	8400	8657	8913	9170	9426	9682	9938	230193	256
170	230449	230704	230960	231215	231470	231724	231979	232234	232488	232742	255
1	2996	3250	3504	3757	4011	4264	4517	4770	5023	5276	253
2	5528	5781	6033	6285	6537	6789	7041	7292	7544	7795	252
3	8046	8297	8548	8799	9049	9299	9550	9800	240050	240300	250
4	240549	240799	241048	241297	241546	241795	242044	242293	2541	2790	249
175	3038	3286	3534	3782	4030	4277	4525	4772	5019	5266	248
6	5513	5759	6006	6252	6499	6745	6991	7237	7482	7728	246
7	7973	8219	8464	8709	8954	9198	9443	9687	9932	250176	245
8	250420	250664	250908	251151	251395	251638	251081	252125	252358	2610	243
9	2853	3096	3338	3580	3822	4064	4306	4548	4790	5031	242
180	255273	255514	255755	255996	256237	256477	256718	256958	257198	257439	241
1	7679	7918	8158	8398	8637	8377	9116	9355	9594	9833	239
2	260071	260310	260548	260787	261025	261263	261501	261739	261976	262214	238
3	2451	2688	2925	3162	3399	3636	3373	4109	4346	4582	237
4	4818	5054	5290	5525	5761	5996	6232	6467	6702	6937	235
185	7172	7406	7641	7875	8110	8344	8578	8812	9046	9279	234
6	9513	9746	9930	270213	270446	270679	270912	271144	271377	271609	233
7	271842	272074	272306	2538	2770	3001	3233	3454	3696	3927	232
8	4158	4389	4620	4850	5031	5311	5542	5772	6002	6232	230
9	6462	6692	6921	7151	7380	7609	7838	8067	8296	8525	229
190	278754	278982	279211	279439	279667	279895	280123	280351	280578	280806	228
1	281033	281251	281488	281715	281942	282169	2396	2622	2849	3075	227
2	3301	3527	3753	3979	4205	4431	4556	4882	5107	5332	226
3	5557	5782	6007	6232	6456	6631	6905	7130	7354	7578	225
4	7802	8026	8249	8473	8696	8920	9143	9366	9589	9812	223
195	290035	290257	290480	290702	290925	291147	291369	291591	291813	292034	222
6	2256	2478	2699	2920	3141	3333	3384	3804	4025	4246	221
7	4466	4537	4907	5127	5347	5567	5787	6007	6226	6446	220
8	6665	6834	7104	7323	7542	7761	7979	8198	8416	8635	219
9	8853	9071	8289	9507	9725	9943	300161	300378	300595	300813	218
200	301030	301247	301464	301631	301898	302114	302331	302547	302764	302980	217
1	3196	3412	3628	3844	4059	4275	4491	4706	4921	5136	216
2	5351	5566	5781	5996	6211	6425	6639	6854	7068	7282	215
3	7496	7710	7924	8137	8351	8564	8778	8991	9204	9417	213
4	9630	9843	310056	310268	310481	310693	310906	311118	311330	311542	212
205	311754	311966	2177	2389	2600	2812	3023	3234	3445	3656	211
6	3867	4078	4289	4499	4710	4920	5130	5340	5551	5760	210
7	5970	6180	6390	6599	6809	7018	7227	7436	7646	7854	209
8	8063	8272	8481	8689	8898	9106	9314	9522	9730	9938	208
9	320146	320354	320562	320769	320977	321184	321391	321598	321805	322012	207
210	322219	322426	322633	322839	323046	323252	323458	323665	323871	324077	206
1	4282	4488	4694	4899	5105	5310	5516	5721	5926	6131	205
2	6336	6541	6745	6950	7155	7359	7563	7767	7972	8176	204
3	8380	8583	8787	8991	9194	9398	9601	9805	330008	330211	203
4	330414	330617	330819	331022	331225	331427	331630	331832	2034	2236	202
215	2438	2640	2842	3044	3246	3447	3649	3850	4051	4253	202
6	4454	4655	4856	5057	5257	5458	5658	5859	6059	6260	201
7	6460	6660	6860	7060	7260	7459	7659	7858	8058	8257	200
8	8456	8656	8855	9054	9253	9451	9650	9849	340047	340246	199
9	340444	340642	340841	341039	341237	341435	341632	341830	2028	2225	198
N.	0	1	2	3	4	5	6	7	8	9	D.

Table A (Continued)

N.	0	1	2	3	4	5	6	7	8	9	D.
220	342423	342620	342817	343014	343212	343409	343606	343802	343999	344196	197
1	4392	4589	4785	4981	5178	5374	5570	5766	5962	6157	196
2	6353	6549	6744	6939	7135	7330	7525	7720	7915	8110	195
3	8305	8500	8694	8889	9083	9278	9472	9666	9860	350054	194
4	350248	350442	350636	350829	351023	351216	351410	351603	351796	1989	193
225	2183	2375	2568	2761	2954	3147	3339	3532	3724	3916	193
6	4108	4301	4493	4685	4876	5068	5260	5452	5643	5834	192
7	6026	6217	6408	6599	6790	6981	7172	7363	7554	7744	191
8	7935	8125	8316	8506	8696	8886	9076	9266	9456	9646	190
9	9835	360025	360215	360404	360593	360783	360972	361161	361350	361539	189
230	361728	361917	362105	362294	362482	362671	362859	363048	363236	363424	188
1	3612	3800	3988	4176	4363	4551	4739	4926	5113	5301	188
2	5488	5675	5862	6049	6236	6423	6610	6796	6983	7169	187
3	7356	7542	7729	7915	8101	8287	8473	8659	8845	9030	186
4	9216	9401	9587	9772	9958	370143	370328	370513	370698	370883	185
235	371068	371253	371437	371622	371806	1991	2175	2360	2544	2728	184
6	2912	3096	3280	3464	3647	3831	4015	4198	4382	4565	184
7	4748	4932	5115	5298	5481	5664	5846	6029	6212	6394	183
8	6577	6759	6942	7124	7306	7488	7670	7852	8034	8216	182
9	8398	8580	8761	8943	9124	9306	9487	9668	9849	380030	181
240	380211	380392	380573	380754	380934	381115	381296	381476	381656	381837	181
1	2017	2197	2377	2557	2737	2917	3097	3277	3456	3636	180
2	3815	3995	4174	4353	4533	4712	4891	5070	5249	5428	179
3	5606	5785	5964	6142	6321	6499	6677	6856	7034	7212	178
4	7390	7568	7746	7923	8101	8279	8456	8634	8811	8989	178
245	9166	9343	9520	9698	9875	390051	390228	390405	390582	390759	177
6	390935	391112	391288	391464	391641	1817	1993	2169	2345	2521	176
7	2697	2873	3048	3224	3400	3575	3751	3926	4101	4277	176
8	4452	4627	4802	4977	5152	5326	5501	5676	5850	6025	175
9	6199	6374	6548	6722	6896	7071	7245	7419	7592	7766	174
250	397940	398114	398287	398461	398634	398808	398981	399154	399328	399501	173
1	9674	9847	400020	400192	400365	400538	400711	400883	401056	401228	173
2	401401	401573	1745	1917	2089	2261	2433	2605	2777	2949	172
3	3121	3292	3464	3635	3807	3978	4149	4320	4492	4663	171
4	4834	5005	5176	5346	5517	5688	5858	6029	6199	6370	171
255	6540	6710	6881	7051	7221	7391	7561	7731	7901	8070	170
6	8240	8410	8579	8749	8918	9087	9257	9426	9595	9764	169
7	9933	410102	410271	410440	410609	410777	410964	411114	411283	411451	169
8	411620	1788	1956	2124	2293	2461	2629	2796	2964	3132	168
9	3300	3467	3635	3803	3970	4137	4305	4472	4639	4806	167
260	414973	415140	415307	415474	415641	415808	415974	416141	416308	416474	167
1	6641	6807	6973	7139	7306	7472	7638	7804	7970	8135	166
2	8301	8467	8633	8798	8964	9129	9295	9460	9625	9791	165
3	9956	420121	420286	420451	420616	420781	420945	421110	421275	421439	165
4	421604	1768	1933	2097	2261	2426	2590	2754	2918	3082	164
265	3246	3410	3574	3737	3901	4065	4228	4392	4555	4718	164
6	4882	5045	5208	5371	5534	5697	5860	6023	6186	6349	163
7	6511	6674	6836	6999	7161	7324	7486	7648	7811	7973	162
8	8135	8297	8459	8621	8783	8944	9106	9268	9429	9591	162
9	9752	9914	430075	430236	430398	430559	430720	430881	431042	431203	161
270	431364	431525	431685	431846	432007	432167	432328	432488	432649	432809	161
1	2959	3130	3290	3450	3610	3770	3930	4090	4249	4409	160
2	4569	4729	4888	5048	5207	5367	5526	5685	5844	6004	159
3	6163	6322	6481	6640	6799	6957	7116	7275	7433	7592	159
4	7751	7909	8067	8226	8384	8542	8701	8859	9017	9175	158
275	9333	9491	9648	9806	9964	440122	440279	440437	440594	440752	158
6	440909	441066	441224	441381	441538	1695	1852	2009	2166	2323	157
7	2480	2637	2793	2950	3106	3263	3419	3576	3732	3889	157
8	4045	4201	4357	4513	4669	4825	4981	5137	5293	5449	156
9	5604	5760	5915	6071	6226	6382	6537	6692	6848	7003	155
N	0	1	2	3	4	5	6	7	8	9	D.

Table A (Continued)

N.	0	1	2	3	4	5	6	7	8	9	D.
280	447158	447313	447468	447623	447778	447933	448088	448242	448397	448552	155
1	8706	8861	9015	9170	9324	9478	9633	9787	9941	450095	154
2	450249	450403	450557	450711	450865	451018	451172	451326	451479	1633	154
3	1786	1940	2093	2247	2400	2553	2706	2859	3012	3165	153
4	3318	3471	3624	3777	3930	4082	4235	4387	4540	4692	153
285	4845	4997	5150	5302	5454	5606	5758	5910	6062	6214	152
6	6366	6518	6670	6821	6973	7125	7276	7428	7579	7731	152
7	7882	8033	8184	8336	8487	8638	8789	8940	9091	9242	151
8	9392	9543	9694	9345	9995	460146	460296	460447	460597	460748	151
9	460898	461048	461198	461348	461499	1649	1799	1948	2098	2248	150
290	462398	462548	462697	462847	462997	463146	463296	463445	463594	463744	150
1	3893	4042	4191	4340	4490	4639	4788	4936	5085	5234	149
2	5383	5532	5680	5329	5977	6126	6274	6423	6571	6719	149
3	6868	7016	7164	7312	7460	7608	7756	7904	8052	8200	148
4	8347	8495	8643	8790	8938	9085	9233	9380	9527	9675	148
295	9822	9969	470116	470263	470410	470557	470704	470851	470998	471145	147
6	471292	471438	1585	1732	1878	2025	2171	2318	2464	2610	146
7	2756	2903	3049	3195	3341	3487	3633	3779	3925	4071	146
8	4216	4362	4508	4653	4799	4944	5090	5235	5381	5526	146
9	5671	5816	5962	6107	6252	6397	6542	6687	6832	6976	145
300	477121	477266	477411	477555	477700	477844	477989	478133	478278	478422	145
1	8566	8711	8855	8999	9143	9287	9431	9575	9719	9863	144
2	480007	480151	480294	480438	480582	480725	480869	481012	481156	481299	144
3	1443	1586	1729	1872	2016	2159	2302	2445	2588	2731	143
4	2874	3016	3159	3302	3445	3587	3730	3872	4015	4157	143
305	4300	4442	4585	4727	4869	5011	5153	5295	5437	5579	142
6	5721	5863	6005	6147	6289	6430	6572	6714	6855	6997	142
7	7138	7280	7421	7563	7704	7845	7986	8127	8269	8410	141
8	8551	8692	8833	8974	9114	9255	9396	9537	9677	9818	141
9	9958	490099	490239	490380	490520	490661	490801	490941	491081	491222	140
310	491362	491502	491642	491782	491922	492062	492201	492341	492481	492621	140
1	2760	2900	3040	3179	3319	3458	3597	3737	3876	4015	139
2	4155	4294	4433	4572	4711	4850	4939	5128	5267	5406	139
3	5544	5683	5822	5960	6099	6238	6376	6515	6653	6791	139
4	6930	7068	7206	7344	7483	7621	7759	7897	8035	8173	138
315	8311	8448	8586	8724	8862	8999	9137	9275	9412	9550	138
6	9687	9824	9962	500099	500236	500374	500511	500648	500785	500922	137
7	501059	501196	501333	1470	1607	1744	1880	2017	2154	2291	137
8	2427	2564	2700	2837	2973	3109	3246	3382	3518	3655	136
9	3791	3927	4063	4199	4335	4471	4607	4743	4878	5014	136
320	505150	505286	505421	505557	505693	505828	505964	506099	506234	506370	136
1	6505	6640	6776	6911	7046	7181	7316	7451	7586	7721	135
2	7856	7991	8126	8260	8395	8530	8664	8799	8934	9068	135
3	9203	9337	9471	9606	9740	9874	510009	510143	510277	510411	134
4	510545	510679	510813	510947	511081	511215	1349	1482	1616	1750	134
325	1883	2017	2151	2284	2418	2551	2684	2818	2951	3084	133
6	3218	3351	3484	3617	3750	3883	4016	4149	4282	4415	133
7	4548	4681	4813	4946	5079	5211	5344	5476	5609	5741	133
8	5874	6006	6139	6271	6403	6535	6668	6800	6932	7064	132
9	7196	7328	7460	7592	7724	7855	7987	8119	8251	8382	132
330	518514	518646	518777	518909	519040	519171	519303	519434	519566	519697	131
1	9828	9959	520090	520221	520353	520484	520615	520745	520876	521007	131
2	521138	521269	1400	1530	1661	1792	1922	2053	2183	2314	131
3	2444	2575	2705	2835	2966	3096	3226	3356	3486	3616	130
4	3746	3876	4006	4136	4266	4396	4526	4656	4785	4915	130
335	5045	5174	5304	5434	5563	5693	5822	5951	6081	6210	129
6	6339	6469	6598	6727	6856	6985	7114	7243	7372	7501	129
7	7630	7759	7888	8016	8145	8274	8402	8531	8660	8788	129
8	8917	9045	9174	9302	9430	9559	9687	9815	9943	530072	128
9	530200	530328	530456	530584	530712	530840	530968	531096	531223	1351	128
N.	0	1	2	3	4	5	6	7	8	9	D.

Table A (Continued)

N.	0	1	2	3	4	5	6	7	8	9	D.
340	531479	531607	531734	531862	531990	532117	532245	532372	532500	532627	128
1	2754	2882	3009	3136	3264	3391	3518	3645	3772	3899	127
2	4026	4153	4280	4407	4534	4661	4787	4914	5041	5167	127
3	5294	5421	5547	5674	5800	5927	6053	6180	6306	6432	126
4	6558	6685	6811	6937	7063	7189	7315	7441	7567	7693	126
345	7819	7945	8071	8197	8322	8448	8574	8699	8825	8951	126
6	9076	9202	9327	9452	9578	9703	9829	9954	540079	540204	125
7	540329	540455	540580	540705	540830	540955	541080	541205	1330	1454	125
8	1579	1704	1829	1953	2078	2203	2327	2452	2576	2701	125
9	2825	2950	3074	3199	3323	3447	3571	3696	3820	3944	124
350	544068	544192	544316	544440	544564	544688	544812	544936	545060	545183	124
1	5307	5431	5555	5678	5802	5925	6049	6172	6296	6419	124
2	6543	6666	6789	6913	7036	7159	7282	7405	7529	7652	123
3	7775	7898	8021	8144	8267	8389	8512	8635	8758	8881	123
4	9003	9126	9249	9371	9494	9616	9739	9861	9984	550106	123
355	550228	550351	550473	550595	550717	550840	550962	551084	551206	1328	122
6	1450	1572	1694	1816	1938	2060	2181	2303	2425	2547	122
7	2668	2790	2911	3033	3155	3276	3398	3519	3640	3762	12'
8	3883	4004	4126	4247	4368	4489	4610	4731	4852	4973	121
9	5094	5215	5336	5457	5578	5699	5820	5940	6061	6182	121
360	556303	556423	556544	556664	556785	556905	557026	557146	557267	557387	120
1	7507	7627	7748	7868	7988	8108	8228	8349	8469	8589	120
2	8709	8829	8948	9068	9188	9308	9428	9548	9667	9787	120
3	9907	560026	560146	560265	560385	560504	560624	560743	560863	560982	119
4	561101	1221	1340	1459	1578	1698	1817	1936	2055	2174	119
365	2293	2412	2531	2650	2769	2887	3006	3125	3244	3362	119
6	3481	3600	3718	3837	3955	4074	4192	4311	4429	4548	119
7	4666	4784	4903	5021	5139	5257	5376	5494	5612	5730	118
8	5848	5966	6084	6202	6320	6437	6555	6673	6791	6909	118
9	7026	7144	7262	7379	7497	7614	7732	7849	7967	8084	118
370	568202	568319	568436	568554	568671	568788	568905	569023	569140	569257	117
1	9374	9491	9608	9725	9842	9959	570076	570193	570309	570426	117
2	570543	570660	570776	570893	571010	571126	1243	1359	1476	1592	117
3	1709	1825	1942	2058	2174	2291	2407	2523	2639	2755	116
4	2872	2988	3104	3220	3336	3452	3568	3684	3800	3915	116
375	4031	4147	4263	4379	4494	4610	4726	4841	4957	5072	116
6	5188	5303	5419	5534	5650	5765	5880	5996	6111	6226	115
7	6341	6457	6572	6687	6802	6917	7032	7147	7252	7377	115
8	7492	7607	7722	7836	7951	8066	8181	8295	8410	8525	115
9	8639	8754	8868	8983	9097	9212	9326	9441	9555	9669	114
380	579784	579898	580012	580126	580241	580355	580469	580583	580697	580811	114
1	580925	581039	1153	1267	1381	1495	1608	1722	1836	1950	114
2	2063	2177	2291	2404	2518	2631	2745	2858	2972	3085	114
3	3199	3312	3426	3539	3652	3765	3879	3992	4105	4218	113
4	4331	4444	4557	4670	4783	4896	5009	5122	5235	5348	113
385	5461	5574	5686	5799	5912	6024	6137	6250	6362	6475	113
6	6587	6700	6812	6925	7037	7149	7262	7374	7486	7599	112
7	7711	7823	7935	8047	8160	8272	8384	8496	8608	8720	112
8	8832	8944	9056	9167	9279	9391	9503	9615	9726	9838	112
9	9950	590061	590173	590284	590396	590507	590619	590730	590842	590953	112
390	591065	591176	591287	591399	591510	591621	591732	591843	591955	592066	111
1	2177	2288	2399	2510	2621	2732	2843	2954	3064	3175	111
2	3286	3397	3508	3618	3729	3840	3950	4061	4171	4282	111
3	4393	4503	4614	4724	4834	4945	5055	5165	5276	5380	110
4	5496	5606	5717	5827	5937	6047	6157	6267	6377	6487	110
395	6597	6707	6817	6927	7037	7146	7256	7366	7476	7586	110
6	7695	7805	7914	8024	8134	8243	8353	8462	8572	8681	110
7	8791	8900	9009	9119	9228	9337	9446	9556	9665	9774	109
8	9883	9992	600101	600210	600319	600428	600537	600646	600755	600864	109
9	600973	601082	1191	1299	1408	1517	1625	1734	1843	1951	109
N.	0	1	2	3	4	5	6	7	8	9	D.

Table A (Continued)

N.	0	1	2	3	4	5	6	7	8	9	D.
400	602060	602169	602277	602386	602494	602603	602711	602819	602928	603036	108
1	3144	3253	3361	3469	3577	3686	3794	3902	4010	4118	108
2	4226	4334	4442	4550	4658	4766	4874	4982	5089	5197	108
3	5305	5413	5521	5628	5736	5844	5951	6059	6166	6274	108
4	6381	6489	6596	6704	6811	6919	7026	7133	7241	7348	107
405	7455	7562	7669	7777	7884	7991	8098	8205	8312	8419	107
6	8526	8633	8740	8847	8954	9061	9167	9274	9381	9488	107
7	9594	9701	9808	9914	610021	610128	610234	610341	610447	610554	107
8	610660	610767	610873	610979	1086	1192	1298	1405	1511	1617	106
9	1723	1829	1936	2042	2148	2254	2360	2466	2572	2678	106
410	612784	612890	612996	613102	613207	613313	613419	613525	613630	613736	106
1	3842	3947	4053	4159	4264	4370	4475	4581	4686	4792	106
2	4897	5003	5108	5213	5319	5424	5529	5634	5740	5845	105
3	5950	6055	6160	6265	6370	6476	6581	6686	6790	6895	105
4	7000	7105	7210	7315	7420	7525	7629	7734	7839	7943	105
415	8048	8153	8257	8362	8466	8571	8676	8780	8884	8989	105
6	9093	9198	9302	9406	9511	9615	9719	9824	9928	620032	104
7	620136	620240	620344	620448	620552	620656	620760	620864	620968	1072	104
8	1176	1280	1384	1488	1592	1695	1799	1903	2007	2110	104
9	2214	2318	2421	2525	2628	2732	2835	2939	3042	3146	104
420	623249	623353	623456	623559	623663	623766	623869	623973	624076	624179	103
1	4282	4385	4488	4591	4695	4798	4901	5004	5107	5210	103
2	5312	5415	5518	5621	5724	5827	5929	6032	6135	6238	103
3	6340	6443	6546	6648	6751	6853	6956	7058	7161	7263	103
4	7366	7468	7571	7673	7775	7878	7980	8082	8185	8287	102
425	8389	8491	8593	8695	8797	8900	9002	9104	9206	9308	102
6	9410	9512	9613	9715	9817	9919	630021	630123	630224	630326	102
7	630428	630530	630631	630733	630835	630936	1038	1139	1241	1342	102
8	1444	1545	1647	1748	1849	1951	2052	2153	2255	2356	101
9	2457	2559	2660	2761	2862	2963	3064	3165	3266	3367	101
430	633468	633569	633670	633771	633872	633973	634074	634175	634276	634376	101
1	4477	4578	4679	4779	4880	4981	5081	5182	5283	5383	101
2	5484	5584	5685	5785	5886	5986	6087	6187	6287	6388	100
3	6488	6588	6688	6789	6889	6989	7089	7189	7290	7390	100
4	7490	7590	7690	7790	7890	7990	8090	8190	8290	8389	100
435	8489	8589	8689	8789	8888	8988	9088	9188	9287	9387	100
6	9486	9586	9686	9785	9889	9984	640084	640183	640283	640382	99
7	640481	640581	640680	640779	640879	640978	1077	1177	1276	1375	99
8	1474	1573	1672	1771	1871	1970	2069	2168	2267	2366	99
9	2465	2563	2662	2761	2860	2959	3058	3156	3255	3354	99
440	643453	643551	643650	643749	643847	643946	644044	644143	644242	644340	98
1	4439	4537	4636	4734	4832	4931	5029	5127	5226	5324	98
2	5422	5521	5619	5717	5815	5913	6011	6110	6208	6306	98
3	6404	6502	6600	6698	6796	6894	6992	7089	7197	7285	98
4	7383	7481	7579	7676	7774	7872	7969	8067	8165	8262	98
445	8360	8458	8555	8653	8750	8848	8945	9043	9140	9237	97
6	9335	9432	9530	9627	9724	9821	9919	650016	650113	650210	97
7	650308	650405	650502	650599	650696	650793	650890	0987	1084	1181	97
8	1278	1375	1472	1569	1666	1762	1859	1956	2053	2150	97
9	2246	2343	2440	2536	2633	2730	2826	2923	3019	3116	97
450	653213	653309	653405	653502	653598	653695	653791	653888	653984	654080	96
1	4177	4273	4369	4465	4562	4658	4754	4850	4946	5042	96
2	5138	5235	5331	5427	5523	5619	5715	5810	5906	6002	96
3	6098	6194	6290	6386	6482	6577	6673	6769	6864	6960	96
4	7056	7152	7247	7343	7438	7534	7629	7725	7820	7916	96
455	8011	8107	8202	8298	8393	8488	8584	8679	8774	8870	95
6	8965	9060	9155	9250	9346	9441	9536	9631	9726	9821	95
7	9916	660011	660106	660201	660296	660391	660486	660581	660676	660771	95
8	660865	0960	1055	1150	1245	1339	1434	1529	1623	1718	95
9	1813	1907	2002	2096	2191	2286	2380	2475	2569	2663	95
N.	0	1	2	3	4	5	6	7	8	9	D.

Table A (Continued)

N.	0	1	2	3	4	5	6	7	8	9	D.
460	662758	662852	662947	663041	663135	663230	663324	663418	663512	663607	94
1	3701	3795	3889	3983	4078	4172	4266	4360	4454	4548	94
2	4642	4736	4830	4924	5018	5112	5206	5299	5393	5487	94
3	5581	5675	5769	5862	5956	6050	6143	6237	6331	6424	94
4	6518	6612	6705	6799	6892	6986	7079	7173	7266	7360	94
465	7453	7546	7640	7733	7826	7920	8013	8106	8199	8293	93
6	8386	8479	8572	8665	8759	8852	8945	9038	9131	9224	93
7	9317	9410	9503	9596	9689	9782	9875	9967	670060	670153	93
8	670246	670339	670431	670524	670617	670710	670802	670895	0988	1080	93
9	1173	1265	1358	1451	1543	1636	1728	1821	1913	2005	93
470	672098	672190	672283	672375	672467	672560	672652	672744	672836	672929	92
1	3021	3113	3205	3297	3390	3482	3574	3666	3758	3850	92
2	3942	4034	4126	4218	4310	4402	4494	4586	4677	4769	92
3	4861	4953	5045	5137	5228	5320	5412	5503	5595	5687	92
4	5778	5870	5962	6053	6145	6236	6328	6419	6511	6602	92
475	6694	6785	6876	6968	7059	7151	7242	7333	7424	7516	91
6	7607	7698	7789	7881	7972	8063	8154	8245	8336	8427	91
7	8518	8609	8700	8791	8882	8973	9064	9155	9246	9337	91
8	9428	9519	9610	9700	9791	9882	9973	680063	680154	680245	91
9	680336	680426	680517	680607	680698	680789	680879	0970	1060	1151	91
480	681241	681332	681422	681513	681603	681693	681784	681874	681964	682055	90
1	2145	2235	2326	2416	2506	2596	2686	2777	2867	2957	90
2	3047	3137	3227	3317	3407	3497	3587	3677	3767	3857	90
3	3947	4037	4127	4217	4307	4396	4486	4576	4666	4756	90
4	4845	4935	5025	5114	5204	5294	5383	5473	5563	5652	90
485	5742	5831	5921	6010	6100	6189	6279	6368	6458	6547	89
6	6636	6726	6815	6904	6994	7033	7172	7261	7351	7440	89
7	7529	7618	7707	7796	7886	7975	8064	8153	8242	8331	89
8	8420	8509	8598	8687	8776	8865	8953	9042	9131	9220	89
9	9309	9398	9486	9575	9664	9753	9841	9930	690019	690107	89
490	690196	690285	690373	690462	690550	690639	690728	690816	690905	690993	89
1	1081	1170	1258	1347	1435	1524	1612	1700	1789	1877	88
2	1965	2053	2142	2230	2318	2406	2494	2583	2671	2759	88
3	2847	2935	3023	3111	3199	3287	3375	3463	3551	3639	88
4	3727	3815	3903	3991	4078	4166	4254	4342	4430	4517	88
495	4605	4693	4781	4868	4956	5044	5131	5219	5307	5394	88
6	5482	5569	5657	5744	5832	5919	6007	6094	6182	6269	87
7	6356	6444	6531	6618	6706	6793	6880	6968	7055	7142	87
8	7229	7317	7404	7491	7578	7665	7752	7839	7926	8014	87
9	8101	8188	8275	8362	8449	8535	8622	8709	8796	8883	87
500	698970	699057	699144	699231	699317	699404	699491	699578	699664	699751	87
1	9838	9924	700011	700098	700184	700271	700358	700444	700531	700617	87
2	700704	700790	0877	0963	1050	1136	1222	1309	1395	1482	86
3	1568	1654	1741	1827	1913	1999	2085	2172	2258	2344	86
4	2431	2517	2603	2689	2775	2861	2947	3033	3119	3205	86
505	3291	3377	3463	3549	3635	3721	3807	3893	3979	4065	86
6	4151	4236	4322	4408	4494	4579	4665	4751	4837	4922	86
7	5008	5094	5179	5265	5350	5436	5522	5607	5693	5778	86
8	5864	5949	6035	6120	6206	6291	6376	6462	6547	6632	85
9	6718	6803	6888	6974	7059	7144	7229	7315	7400	7485	85
510	707570	707655	707740	707826	707911	707996	708081	708166	708251	708336	85
1	8421	8506	8591	8676	8761	8846	8931	9015	9100	9185	85
2	9270	9355	9440	9524	9609	9694	9779	9863	9948	710033	85
3	710117	710202	710287	710371	710456	710540	710625	710710	710794	0879	85
4	0963	1048	1132	1217	1301	1385	1470	1554	1639	1723	84
515	1807	1892	1976	2060	2144	2229	2313	2397	2481	2566	84
6	2650	2734	2818	2902	2986	3070	3154	3238	3323	3407	84
7	3491	3575	3659	3742	3826	3910	3994	4078	4162	4246	84
8	4330	4414	4497	4581	4665	4749	4833	4916	5000	5084	84
9	5167	5251	5335	5418	5502	5586	5669	5753	5836	5920	84
N.	0	1	2	3	4	5	6	7	8	9	D.

Table A (Continued)

N.	0	1	2	3	4	5	6	7	8	9	D.
520	716003	716087	716170	716254	716337	716421	716504	716588	716671	716754	83
1	6838	6921	7004	7088	7171	7254	7338	7421	7504	7587	83
2	7671	7754	7837	7920	8003	8086	8169	8253	8336	8419	83
3	8502	8585	8668	8751	8834	8917	9000	9083	9165	9248	83
4	9331	9414	9497	9580	9663	9745	9828	9911	9994	720077	83
525	720159	720242	720325	720407	720490	720573	720655	720738	720821	0903	83
6	0986	1068	1151	1233	1316	1398	1481	1563	1646	1728	82
7	1811	1893	1975	2058	2140	2222	2305	2387	2469	2552	82
8	2634	2716	2798	2881	2963	3045	3127	3209	3291	3374	82
9	3456	3538	3620	3702	3784	3866	3948	4030	4112	4194	82
530	724276	724358	724440	724522	724604	724685	724767	724849	724931	725013	82
1	5095	5176	5258	5340	5422	5503	5585	5667	5748	5830	82
2	5912	5993	6075	6156	6238	6320	6401	6483	6564	6646	82
3	6727	6809	6890	6972	7053	7134	7216	7297	7379	7460	81
4	7541	7623	7704	7785	7866	7948	8029	8110	8191	8273	81
535	8354	8435	8516	8597	8678	8759	8841	8922	9003	9084	81
6	9165	9246	9327	9408	9439	9570	9651	9732	9813	9893	81
7	9974	730055	730136	730217	730298	730378	730459	730540	730621	730702	81
8	730782	0863	0944	1024	1105	1186	1266	1347	1428	1508	81
9	1589	1669	1750	1830	1911	1991	2072	2152	2233	2313	81
540	732394	732474	732555	732635	732715	732796	732876	732956	733037	733117	80
1	3197	3278	3358	3438	3518	3598	3679	3759	3839	3919	80
2	3999	4079	4160	4240	4320	4400	4480	4560	4640	4720	80
3	4800	4880	4960	5040	5120	5200	5279	5359	5439	5519	80
4	5599	5679	5759	5838	5918	5998	6078	6157	6237	6317	80
545	6397	6476	6556	6635	6715	6795	6874	6954	7034	7113	80
6	7193	7272	7352	7431	7511	7590	7670	7749	7829	7908	79
7	7987	8067	8146	8225	8305	8384	8463	8543	8622	8701	79
8	8781	8860	8939	9018	9097	9177	9256	9335	9414	9493	79
9	9572	9651	9731	9810	9889	9968	740047	740126	740205	740284	79
550	740363	740442	740521	740600	740678	740757	740836	740915	740994	741073	79
1	1152	1230	1309	1388	1467	1546	1624	1703	1782	1860	79
2	1939	2018	2096	2175	2254	2332	2411	2499	2568	2647	79
3	2725	2804	2882	2961	3039	3118	3196	3275	3353	3431	78
4	3510	3588	3667	3745	3823	3902	3980	4058	4136	4215	78
555	4293	4371	4449	4528	4606	4684	4762	4840	4919	4997	78
6	5075	5153	5231	5309	5387	5465	5543	5621	5699	5777	78
7	5855	5933	6011	6089	6167	6245	6323	6401	6479	6556	78
8	6634	6712	6790	6868	6945	7023	7101	7179	7256	7334	78
9	7412	7489	7567	7645	7722	7800	7878	7955	8033	8110	78
560	748188	748266	748343	748421	748498	748576	748653	748731	748808	748885	77
1	8963	9040	9118	9195	9272	9350	9427	9504	9582	9659	77
2	9736	9814	9891	9968	750045	750123	750200	750277	750354	750431	77
3	750508	750586	750663	750740	0817	0894	0971	1048	1125	1202	77
4	1279	1356	1433	1510	1587	1664	1741	1818	1895	1972	77
565	2048	2125	2202	2279	2356	2433	2509	2586	2663	2740	77
6	2816	2893	2970	3047	3123	3200	3277	3353	3430	3506	77
7	3583	3660	3736	3813	3889	3966	4042	4119	4195	4272	77
8	4348	4425	4501	4578	4654	4730	4807	4883	4960	5036	76
9	5112	5189	5265	5341	5417	5494	5570	5646	5722	5799	76
570	755875	755951	756027	756103	756180	756256	756332	756408	756484	756560	76
1	6636	6712	6788	6864	6940	7016	7092	7168	7244	7320	76
2	7396	7472	7548	7624	7700	7775	7851	7927	8003	8079	76
3	8155	8230	8306	8382	8458	8533	8609	8685	8761	8836	76
4	8912	8988	9063	9139	9214	9290	9366	9441	9517	9592	76
575	9668	9743	9819	9894	9970	760045	760121	760196	760272	760347	75
6	760422	760498	760573	760649	760724	0799	0875	0950	1025	1101	75
7	1176	1251	1326	1402	1477	1552	1627	1702	1778	1853	75
8	1928	2003	2078	2153	2228	2303	2378	2453	2529	2604	75
9	2679	2754	2829	2904	2978	3053	3128	3203	3278	3353	75
N.	0	1	2	3	4	5	6	7	8	9	D.

Table A (Continued)

N.	0	1	2	3	4	5	6	7	8	9	D.
580	763428	763503	763578	763653	763727	763802	763877	763952	764027	764101	75
1	4176	4251	4326	4400	4475	4550	4624	4699	4774	4848	75
2	4923	4998	5072	5147	5221	5296	5370	5445	5520	5594	75
3	5669	5743	5818	5892	5966	6041	6115	6190	6264	6338	74
4	6413	6487	6562	6636	6710	6785	6859	6933	7007	7082	74
585	7156	7230	7304	7379	7453	7527	7601	7675	7749	7823	74
6	7898	7972	8046	8120	8194	8268	8342	8416	8490	8564	74
7	8638	8712	8786	8860	8934	9008	9082	9156	9230	9303	74
8	9377	9451	9525	9599	9673	9746	9820	9894	9968	770042	74
9	770115	770189	770263	770336	770410	770484	770557	770631	770705	0778	74
590	770852	770926	770999	771073	771146	771220	771293	771367	771440	771514	74
1	1587	1661	1734	1808	1881	1955	2028	2102	2175	2248	73
2	2322	2395	2468	2542	2615	2688	2762	2835	2908	2981	73
3	3055	3128	3201	3274	3348	3421	3494	3567	3640	3713	73
4	3786	3860	3933	4006	4079	4152	4225	4298	4371	4444	73
595	4517	4590	4663	4736	4809	4882	4955	5028	5100	5173	73
6	5246	5319	5392	5465	5538	5610	5683	5756	5829	5902	73
7	5974	6047	6120	6193	6265	6338	6411	6483	6556	6629	73
8	6701	6774	6846	6919	6992	7064	7137	7209	7282	7354	73
9	7427	7499	7572	7644	7717	7789	7862	7934	8006	8079	72
600	778151	778224	778296	778368	778441	778513	778585	778658	778730	778802	72
1	8874	8947	9019	9091	9163	9236	9308	9380	9452	9524	72
2	9596	9669	9741	9813	9885	9957	780029	780101	780173	780245	72
3	780317	780389	780461	780533	780605	780677	0749	0821	0893	0965	72
4	1037	1109	1181	1253	1324	1396	1468	1540	1612	1684	72
605	1755	1827	1899	1971	2042	2114	2186	2258	2329	2401	72
6	2473	2544	2616	2688	2759	2831	2902	2974	3046	3117	72
7	3189	3260	3332	3403	3475	3546	3618	3689	3761	3832	71
8	3904	3975	4046	4118	4189	4261	4332	4403	4475	4546	71
9	4617	4689	4760	4831	4902	4974	5045	5116	5187	5259	71
610	785330	785401	785472	785543	785615	785686	785757	785828	785899	785970	7:
1	6041	6112	6183	6254	6325	6396	6467	6538	6609	6680	71
2	6751	6822	6893	6964	7035	7106	7177	7248	7319	7390	71
3	7460	7531	7602	7673	7744	7815	7885	7956	8027	8098	71
4	8168	8239	8310	8381	8451	8522	8593	8663	8734	8804	71
615	8875	8946	9016	9087	9157	9228	9299	9369	9440	9510	71
6	9581	9651	9722	9792	9863	9933	790004	790074	790144	790215	70
7	790285	790356	790426	790496	790567	790637	0707	0778	0848	0918	70
8	0988	1059	1129	1199	1269	1340	1410	1480	1550	1620	70
9	1691	1761	1831	1901	1971	2041	2111	2181	2252	2322	70
620	792392	792462	792532	792602	792672	792742	792812	792882	792952	793022	70
1	3092	3162	3231	3301	3371	3441	3511	3581	3651	3721	70
2	3790	3860	3930	4000	4070	4139	4209	4279	4349	4418	70
3	4488	4558	4627	4697	4767	4836	4906	4976	5045	5115	70
4	5185	5254	5324	5393	5463	5532	5602	5672	5741	5811	70
625	5880	5949	6019	6088	6158	6227	6297	6366	6436	6505	69
6	6574	6644	6713	6782	6852	6921	6990	7060	7129	7198	69
7	7268	7337	7406	7475	7545	7614	7683	7752	7821	7890	69
8	7960	8029	8098	8167	8236	8305	8374	8443	8513	8582	69
9	8651	8720	8789	8858	8927	8996	9065	9134	9203	9272	69
630	799341	799409	799478	799547	799616	799685	799754	799823	799892	799961	69
1	800029	800098	800167	800236	800305	800373	800442	800511	800580	800648	69
2	0717	0786	0854	0923	0992	1061	1129	1198	1266	1335	69
3	1404	1472	1541	1609	1678	1747	1815	1884	1952	2021	69
4	2089	2158	2226	2295	2363	2432	2500	2568	2637	2705	68
635	2774	2842	2910	2979	3047	3116	3184	3252	3321	3389	68
6	3457	3525	3594	3662	3730	3798	3867	3935	4003	4071	68
7	4139	4208	4276	4344	4412	4480	4548	4616	4685	4753	68
8	4821	4889	4957	5025	5093	5161	5229	5297	5365	5433	68
9	5501	5569	5637	5705	5773	5841	5908	5976	6044	6112	68
N.	0	1	2	3	4	5	6	7	8	9	D.

Table A (Continued)

N.	C	1	2	3	4	5	6	7	8	9	D.
640	806180	806248	806316	806384	806451	806519	806587	806655	806723	806790	68
1	6858	6926	6994	7061	7129	7197	7264	7332	7400	7467	68
2	7535	7603	7670	7738	7806	7873	7941	8008	8076	8143	68
3	8211	8279	8346	8414	8481	8549	8616	8684	8751	8818	67
4	8886	8953	9021	9088	9156	9223	9290	9358	9425	9492	67
645	9560	9627	9694	9762	9829	9896	9964	810031	810098	810165	67
6	810233	810300	810367	810434	810501	810569	810636	0703	0770	0837	67
7	0904	0971	1039	1106	1173	1240	1307	1374	1441	1508	67
8	1575	1642	1709	1776	1843	1910	1977	2044	2111	2178	67
9	2245	2312	2379	2445	2512	2579	2646	2713	2780	2847	67
650	812913	812980	813047	813114	813181	813247	813314	813381	813448	813514	67
1	3581	3648	3714	3781	3848	3914	3981	4048	4114	4181	67
2	4248	4314	4381	4447	4514	4581	4647	4714	4780	4847	67
3	4913	4980	5046	5113	5179	5246	5312	5378	5445	5511	66
4	5578	5644	5711	5777	5843	5910	5976	6042	6109	6175	66
655	6241	6308	6374	6440	6506	6573	6639	6705	6771	6838	66
6	6904	6970	7036	7102	7169	7235	7301	7367	7433	7499	66
7	7565	7631	7698	7764	7830	7896	7962	8028	8094	8160	66
8	8226	8292	8358	8424	8490	8556	8622	8688	8754	8820	66
9	8885	8951	9017	9083	9149	9215	9281	9346	9412	9478	66
660	819544	819610	819676	819741	819807	819873	819939	820004	820070	820136	66
1	820201	820267	820333	820399	820464	820530	820596	0661	0727	0792	66
2	0858	0924	0989	1055	1120	1186	1251	1317	1382	1448	66
3	1514	1579	1645	1710	1775	1841	1905	1972	2037	2103	65
4	2168	2233	2299	2364	2430	2495	2560	2626	2691	2756	65
665	2822	2887	2952	3018	3083	3148	3213	3279	3344	3409	65
6	3474	3539	3605	3670	3735	3800	3865	3930	3996	4061	65
7	4126	4191	4256	4321	4386	4451	4516	4581	4646	4711	65
8	4776	4841	4906	4971	5036	5101	5166	5231	5296	5361	65
9	5426	5491	5556	5621	5686	5751	5815	5880	5945	6010	65
670	826075	826140	826204	826269	826334	826399	826464	826528	826593	826658	65
1	6723	6787	6852	6917	6981	7046	7111	7175	7240	7305	65
2	7369	7434	7499	7563	7628	7692	7757	7821	7886	7951	65
3	8015	8080	8144	8209	8273	8338	8402	8467	8531	8595	64
4	8660	8724	8789	8853	8918	8982	9046	9111	9175	9239	64
675	9304	9368	9432	9497	9561	9625	9690	9754	9818	9882	64
6	9947	830011	830075	830139	830204	830268	830332	830396	830460	830525	64
7	830589	0653	0717	0781	0845	0909	0973	1037	1102	1166	64
8	1230	1294	1358	1422	1486	1550	1614	1678	1742	1806	64
9	1870	1934	1998	2062	2126	2189	2253	2317	2381	2445	64
680	832509	832573	832637	832700	832764	832828	832892	832956	833020	833083	64
1	3147	3211	3275	3338	3402	3466	3530	3593	3657	3721	64
2	3784	3848	3912	3975	4039	4103	4166	4230	4294	4357	64
3	4421	4484	4548	4611	4675	4739	4802	4866	4929	4993	64
4	5056	5120	5183	5247	5310	5373	5437	5500	5564	5627	63
685	5691	5754	5817	5881	5944	6007	6071	6134	6197	6261	63
6	6324	6387	6451	6514	6577	6641	6704	6767	6830	6894	63
7	6957	7020	7083	7146	7210	7273	7336	7399	7462	7525	63
8	7588	7652	7715	7778	7841	7904	7967	8030	8093	8156	63
9	8219	8282	8345	8408	8471	8534	8597	8660	8723	8786	63
690	838849	838912	838975	839038	839101	839164	839227	839289	839352	839415	63
1	9478	9541	9604	9667	9729	9792	9855	9918	9981	840043	63
2	840106	840169	840232	840294	840357	840420	840482	840545	840608	0671	63
3	0733	0796	0859	0921	0984	1046	1109	1172	1234	1297	63
4	1359	1422	1485	1547	1610	1672	1735	1797	1860	1922	63
695	1985	2047	2110	2172	2235	2297	2360	2422	2484	2547	62
6	2609	2672	2734	2796	2859	2921	2983	3046	3108	3170	62
7	3233	3295	3357	3420	3482	3544	3606	3669	3731	3793	62
8	3855	3918	3930	4042	4104	4166	4229	4291	4353	4415	62
9	4477	4539	4601	4664	4726	4788	4850	4912	4974	5036	62
N.	0	1	2	3	4	5	6	7	8	9	D.

Table A (Continued)

N.	0	1	2	3	4	5	6	7	8	9	D.
700	845098	845160	845222	845284	845346	845408	845470	845532	845594	845656	62
1	5718	5780	5842	5904	5966	6028	6090	6151	6213	6275	62
2	6337	6399	6461	6523	6585	6646	6708	6770	6832	6894	62
3	6955	7017	7079	7141	7202	7264	7326	7388	7449	7511	62
4	7573	7634	7696	7758	7819	7881	7943	8004	8066	8128	62
705	8189	8251	8312	8374	8435	8497	8559	8620	8682	8743	62
6	8805	8866	8928	8989	9051	9112	9174	9235	9297	9358	61
7	9419	9481	9542	9604	9665	9726	9788	9849	9911	9972	61
8	850033	850095	850156	850217	850279	850340	850401	850462	850524	850585	61
9	0646	0707	0769	0830	0891	0952	1014	1075	1136	1197	61
710	851258	851320	851381	851442	851503	851564	851625	851686	851747	851809	61
1	1870	1931	1992	2053	2114	2175	2236	2297	2358	2419	61
2	2480	2541	2602	2663	2724	2785	2846	2907	2968	3029	61
3	3090	3150	3211	3272	3333	3394	3455	3516	3577	3637	61
4	3698	3759	3820	3881	3941	4002	4063	4124	4185	4245	61
715	4306	4367	4428	4488	4549	4610	4670	4731	4792	4852	61
6	4913	4974	5034	5095	5156	5216	5277	5337	5398	5459	61
7	5519	5580	5640	5701	5761	5822	5882	5943	6003	6064	61
8	6124	6185	6245	6306	6356	6427	6487	6548	6608	6668	60
9	6729	6789	6850	6910	6970	7031	7091	7152	7212	7272	60
720	857332	857393	857453	857513	857574	857634	857694	857755	857815	857875	60
1	7935	7995	8056	8116	8176	8236	8297	8357	8417	8477	60
2	8537	8597	8657	8718	8778	8838	8898	8958	9018	9078	60
3	9138	9198	9258	9318	9379	9439	9499	9559	9619	9679	60
4	9739	9799	9859	9918	9978	860038	860098	860158	860218	860278	60
725	860338	860398	860458	860518	860578	0637	0697	0757	0817	0877	60
6	0937	0996	1056	1116	1176	1236	1295	1355	1415	1475	60
7	1534	1594	1654	1714	1773	1833	1893	1952	2012	2072	60
8	2131	2191	2251	2310	2370	2430	2489	2549	2608	2668	60
9	2728	2787	2847	2906	2966	3025	3085	3144	3204	3263	60
730	863323	863382	863442	863501	863561	863620	863680	863739	863799	863858	59
1	3917	3977	4036	4096	4155	4214	4274	4333	4392	4452	59
2	4511	4570	4630	4689	4748	4808	4867	4926	4985	5045	59
3	5104	5163	5222	5282	5341	5400	5459	5519	5578	5637	59
4	5696	5755	5814	5874	5933	5992	6051	6110	6169	6228	59
735	6287	6346	6405	6465	6524	6583	6642	6701	6760	6819	59
6	6878	6937	6996	7055	7114	7173	7232	7291	7350	7409	59
7	7467	7526	7585	7644	7703	7762	7821	7880	7939	7998	59
8	8056	8115	8174	8233	8292	8350	8409	8468	8527	8586	59
9	8644	8703	8762	8821	8879	8938	8997	9056	9114	9173	59
740	869232	869290	869349	869408	869466	869525	869584	869642	869701	869760	59
1	9818	9877	9935	9994	870053	870111	870170	870228	870287	870345	59
2	870404	870462	870521	870579	0638	0696	0755	0813	0872	0930	58
3	0989	1047	1106	1164	1223	1281	1339	1398	1456	1515	58
4	1573	1631	1690	1748	1806	1865	1923	1981	2040	2098	58
745	2156	2215	2273	2331	2389	2448	2506	2564	2622	2681	58
6	2739	2797	2855	2913	2972	3030	3088	3146	3204	3262	58
7	3321	3379	3437	3495	3553	3611	3669	3727	3785	3844	58
8	3902	3960	4018	4076	4134	4192	4250	4308	4366	4424	58
9	4482	4540	4598	4656	4714	4772	4830	4888	4945	5003	58
750	875061	875119	875177	875235	875293	875351	875409	875466	875524	875582	58
1	5640	5698	5756	5813	5871	5929	5987	6045	6102	6160	58
2	6218	6276	6333	6391	6449	6507	6564	6622	6680	6737	58
3	6795	6853	6910	6968	7026	7083	7141	7199	7256	7314	58
4	7371	7429	7487	7544	7602	7659	7717	7774	7832	7889	58
755	7947	8004	8062	8119	8177	8234	8292	8349	8407	8464	57
6	8522	8579	8637	8694	8752	8809	8866	8924	8981	9039	57
7	9096	9153	9211	9268	9325	9383	9440	9497	9555	9612	57
8	9669	9726	9784	9841	9898	9956	880013	880070	880127	880185	57
9	880242	880299	880356	880413	880471	880528	0585	0642	0699	0756	57
N.	0	1	2	3	4	5	6	7	8	9	D.

Table A (Continued)

N.	0	1	2	3	4	5	6	7	8	9	D.
760	880814	880871	880928	880985	881042	881099	881156	881213	881271	881328	57
1	1385	1442	1499	1556	1613	1670	1727	1784	1841	1898	57
2	1955	2012	2069	2126	2183	2240	2297	2354	2411	2468	57
3	2525	2581	2638	2695	2752	2809	2866	2923	2980	3037	57
4	3093	3150	3207	3264	3321	3377	3434	3491	3548	3605	57
765	3661	3718	3775	3832	3888	3945	4002	4059	4115	4172	57
6	4229	4285	4342	4399	4455	4512	4569	4625	4682	4739	57
7	4795	4852	4909	4965	5022	5078	5135	5192	5248	5305	57
8	5361	5418	5474	5531	5587	5644	5700	5757	5813	5870	57
9	5926	5983	6039	6096	6152	6209	6265	6321	6378	6434	56
770	886491	886547	886604	886660	886716	886773	886829	886C85	886942	886998	56
1	7054	7111	7167	7223	7280	7336	7392	7449	7505	7561	56
2	7617	7674	7730	7786	7842	7898	7955	8011	8067	8123	56
3	8179	8236	8292	8348	8404	8460	8516	8573	8629	8685	56
4	8741	8797	8853	8909	8965	9021	9077	9134	9190	9246	56
775	9302	9358	9414	9470	9526	9582	9638	9694	9750	9806	56
6	9862	9918	9974	890030	890086	890141	890197	890253	890309	890365	56
7	890421	890477	890533	0589	0645	0700	0756	0812	0868	0924	56
8	0980	1035	1091	1147	1203	1259	1314	1370	1426	1482	56
9	1537	1593	1649	1705	1760	1816	1872	1928	1983	2039	56
780	892095	892150	892206	892262	892317	892373	892429	892484	892540	892595	56
1	2651	2707	2762	2818	2873	2929	2985	3040	3096	3151	56
2	3207	3262	3318	3373	3429	3484	3540	3595	3651	3706	56
3	3762	3817	3873	3928	3984	4039	4094	4150	4205	4261	55
4	4316	4371	4427	4482	4538	4593	4648	4704	4759	4814	55
785	4870	4925	4980	5036	5091	5146	5201	5257	5312	5367	55
6	5423	5478	5533	5588	5644	5699	5754	5809	5864	5920	55
7	5975	6030	6085	6140	6195	6251	6306	6361	6416	6471	55
8	6526	6581	6636	6692	6747	6802	6857	6912	6967	7022	55
9	7077	7132	7187	7242	7297	7352	7407	7462	7517	7572	55
790	897627	897682	897737	897792	897847	897902	897957	898012	898067	898122	55
1	8176	8231	8286	8341	8396	8451	8506	8561	8615	8670	55
2	8725	8780	8835	8890	8944	8999	9054	9109	9164	9218	55
3	9273	9328	9383	9437	9492	9547	9602	9656	9711	9766	55
4	9821	9875	9930	9985	900039	900094	900149	900203	900258	900312	55
795	900367	900422	900476	900531	0586	0640	0695	0749	0804	0859	55
6	0913	0968	1022	1077	1131	1186	1240	1295	1349	1404	55
7	1458	1513	1567	1622	1676	1731	1785	1840	1894	1948	54
8	2003	2057	2112	2166	2221	2275	2329	2384	2438	2492	54
9	2547	2601	2655	2710	2764	2818	2873	2927	2981	3036	54
800	903090	903144	903199	903253	903307	903361	903416	903470	903524	903578	54
1	3633	3687	3741	3795	3849	3904	3958	4012	4066	4120	54
2	4174	4229	4283	4337	4391	4445	4499	4553	4607	46C1	54
3	4716	4770	4824	4878	4932	4986	5040	5094	5148	5202	54
4	5256	5310	5364	5418	5472	5526	5580	5634	5688	5742	54
805	5796	5850	5904	5958	6012	6066	6119	6173	6227	6281	54
6	6335	6389	6443	6497	6551	6604	6658	6712	6766	6820	54
7	6874	6927	6981	7035	7089	7143	7196	7250	7304	7358	54
8	7411	7465	7519	7573	7626	7680	7734	7787	7841	7895	54
9	7949	8002	8056	8110	8163	8217	8270	8324	8378	8431	54
810	908485	908539	908592	908646	908699	908753	908807	908860	908914	908967	54
1	9021	9074	9128	9181	9235	9289	9342	9396	9449	9503	54
2	9556	9610	9663	9716	9770	9823	9877	9930	9984	910037	53
3	910091	910144	910197	910251	910304	910358	910411	910464	910518	0571	53
4	0624	0678	0731	0784	0838	0891	0944	0998	1051	1104	53
815	1158	1211	1264	1317	1371	1424	1477	1530	1584	1637	53
6	1690	1743	1797	1850	1903	1956	2009	2063	2116	2169	53
7	2222	2275	2328	2381	2435	2488	2541	2594	2647	2700	53
8	2753	2806	2859	2913	2966	3019	3072	3125	3178	3231	53
9	3284	3337	3390	3443	3496	3549	3602	3655	3708	3761	53
N.	0	1	2	3	4	5	6	7	8	9	D.

Table A (Continued)

N.	0	1	2	3	4	5	6	7	8	9	D.
820	913814	913867	913920	913973	914026	914079	914132	914184	914237	914290	53
1	4343	4396	4449	4502	4555	4608	4660	4713	4766	4819	53
2	4872	4925	4977	5030	5083	5136	5189	5241	5294	5347	53
3	5400	5453	5505	5558	5611	5664	5716	5769	5822	5875	53
4	5927	5980	6033	6085	6138	6191	6243	6296	6349	6401	53
825	6454	6507	6559	6612	6664	6717	6770	6822	6875	6927	53
6	6980	7033	7085	7138	7190	7243	7295	7348	7400	7453	53
7	7506	7558	7611	7663	7716	7768	7820	7873	7925	7978	52
8	8030	8083	8135	8188	8240	8293	8345	8397	8450	8502	52
9	8555	8607	8659	8712	8764	8816	8869	8921	8973	9026	52
830	919078	919130	919183	919235	919287	919340	919392	919444	919496	919549	52
1	9601	9653	9706	9758	9810	9862	9914	9967	920019	920071	52
2	920123	920176	920228	920280	920332	920384	920436	920489	0541	0593	52
3	0645	0697	0749	0801	0853	0906	0958	1010	1062	1114	52
4	1166	1218	1270	1322	1374	1426	1478	1530	1582	1634	52
835	1686	1738	1790	1842	1894	1946	1998	2050	2102	2154	52
6	2206	2258	2310	2362	2414	2466	2518	2570	2622	2674	52
7	2725	2777	2829	2881	2933	2985	3037	3089	3140	3192	52
8	3244	3296	3348	3399	3451	3503	3555	3607	3658	3710	52
9	3762	3814	3865	3917	3969	4021	4072	4124	4176	4228	52
840	924279	924331	924383	924434	924486	924538	924589	924641	924693	924744	52
1	4796	4848	4899	4951	5003	5054	5106	5157	5209	5261	52
2	5312	5364	5415	5467	5518	5570	5621	5673	5725	5776	52
3	5828	5879	5931	5982	6034	6085	6137	6188	6240	6291	51
4	6342	6394	6445	6497	6548	6600	6651	6702	6754	6805	51
845	6857	6908	6959	7011	7062	7114	7165	7216	7268	7319	51
6	7370	7422	7473	7524	7576	7627	7678	7730	7781	7832	51
7	7883	7935	7986	8037	8088	8140	8191	8242	8293	8345	51
8	8396	8447	8498	8549	8601	8652	8703	8754	8805	8857	51
9	8908	8959	9010	9061	9112	9163	9215	9266	9317	9368	51
850	929419	929470	929521	929572	929623	929674	929725	929776	929827	929879	51
1	9930	9981	930032	930083	930134	930185	930236	930287	930338	930389	51
2	930440	930491	0542	0592	0643	0694	0745	0796	0847	0898	51
3	0949	1000	1051	1102	1153	1204	1254	1305	1356	1407	51
4	1458	1509	1560	1610	1661	1712	1763	1814	1865	1915	51
855	1966	2017	2068	2118	2169	2220	2271	2322	2372	2423	51
6	2474	2524	2575	2626	2677	2727	2778	2829	2879	2930	51
7	2981	3031	3082	3133	3183	3234	3285	3335	3386	3437	51
8	3487	3538	3589	3639	3690	3740	3791	3841	3892	3943	51
9	3993	4044	4094	4145	4195	4246	4296	4347	4397	4448	51
860	934498	934549	934599	934650	934700	934751	934801	934852	934902	934953	50
1	5003	5054	5104	5154	5205	5255	5306	5356	5406	5457	50
2	5507	5558	5608	5658	5709	5759	5809	5860	5910	5960	50
3	6011	6061	6111	6162	6212	6262	6313	6363	6413	6463	50
4	6514	6564	6614	6665	6715	6765	6815	6865	6916	6966	50
865	7016	7066	7117	7167	7217	7267	7317	7367	7418	7468	50
6	7518	7568	7618	7668	7718	7769	7819	7869	7919	7969	50
7	8019	8069	8119	8169	8219	8269	8320	8370	8420	8470	50
8	8520	8570	8620	8670	8720	8770	8820	8870	8920	8970	50
9	9020	9070	9120	9170	9220	9270	9320	9369	9419	9469	50
870	939519	939569	939619	939669	939719	939769	939819	939869	939918	939968	50
1	940018	940068	940118	940168	940218	940267	940317	940367	940417	940467	50
2	0516	0566	0616	0666	0716	0765	0815	0865	0915	0964	50
3	1014	1064	1114	1163	1213	1263	1313	1362	1412	1462	50
4	1511	1561	1611	1660	1710	1760	1809	1859	1909	1958	50
875	2008	2058	2107	2157	2207	2256	2306	2355	2405	2455	50
6	2504	2554	2603	2653	2702	2752	2801	2851	2901	2950	50
7	3000	3049	3099	3148	3198	3247	3297	3346	3396	3445	49
8	3495	3544	3593	3643	3692	3742	3791	3841	3890	3939	49
9	3989	4038	4088	4137	4186	4236	4285	4335	4384	4433	49
N.	0	1	2	3	4	5	6	7	8	9	D.

Table A (Continued)

N.	0	1	2	3	4	5	6	7	8	9	D.
880	944483	944532	944581	944631	944680	944729	944779	944828	944877	944927	49
1	4976	5025	5074	5124	5173	5222	5272	5321	5370	5419	49
2	5469	5518	5567	5616	5665	5715	5764	5813	5862	5912	49
3	5961	6010	6059	6108	6157	6207	6256	6305	6354	6403	49
4	6452	6501	6551	6600	6649	6698	6747	6796	6845	6894	49
885	6943	6992	7041	7090	7140	7189	7238	7287	7336	7385	49
6	7434	7483	7532	7581	7630	7679	7728	7777	7826	7875	49
7	7924	7973	8022	8070	8119	8168	8217	8266	8315	8364	49
8	8413	8462	8511	8560	8609	8657	8706	8755	8804	8853	49
9	8902	8951	8999	9048	9097	9146	9195	9244	9292	9341	49
890	949390	949439	949488	949536	949585	949634	949683	949731	949780	949829	49
1	9878	9926	9975	950024	950073	950121	950170	950219	950267	950316	49
2	950365	950414	950462	0511	0560	0608	0657	0706	0754	0803	49
3	0851	0900	0949	0997	1046	1095	1143	1192	1240	1289	49
4	1338	1386	1435	1483	1532	1580	1629	1677	1726	1775	49
895	1823	1872	1920	1969	2017	2066	2114	2163	2211	2260	48
6	2308	2356	2405	2453	2502	2550	2599	2647	2696	2744	48
7	2792	2841	2889	2938	2986	3034	3083	3131	3180	3228	48
8	3276	3325	3373	3421	3470	3518	3566	3615	3663	3711	48
9	3760	3808	3856	3905	3953	4001	4049	4098	4146	4194	48
900	954243	954291	954339	954387	954435	954484	954532	954580	954628	954677	48
1	4725	4773	4821	4869	4918	4966	5014	5062	5110	5158	48
2	5207	5255	5303	5351	5399	5447	5495	5543	5592	5640	48
3	5688	5736	5784	5832	5880	5928	5976	6024	6072	6120	48
4	6168	6216	6265	6313	6361	6409	6457	6505	6553	6601	48
905	6649	6697	6745	6793	6840	6888	6936	6984	7032	7080	48
6	7128	7176	7224	7272	7320	7368	7416	7464	7512	7559	48
7	7607	7655	7703	7751	7799	7847	7894	7942	7990	8038	48
8	8086	8134	8181	8229	8277	8325	8373	8421	8468	8516	48
9	8564	8612	8659	8707	8755	8803	8850	8898	8946	8994	48
910	959041	959089	959137	959185	959232	959280	959328	959375	959423	959471	48
1	9518	9566	9614	9661	9709	9757	9804	9852	9900	9947	48
2	9995	960042	960090	960138	960185	960233	960280	960328	960376	960423	48
3	960471	0518	0566	0613	0661	0709	0756	0804	0851	0899	48
4	0946	0994	1041	1089	1136	1184	1231	1279	1326	1374	48
915	1421	1469	1516	1563	1611	1658	1706	1753	1801	1848	47
6	1895	1943	1990	2038	2085	2132	2180	2227	2275	2322	47
7	2369	2417	2464	2511	2559	2606	2653	2701	2748	2795	47
8	2843	2890	2937	2985	3032	3079	3126	3174	3221	3268	47
9	3316	3363	3410	3457	3504	3552	3599	3646	3693	3741	47
920	963788	963835	963882	963929	963977	964024	964071	964118	964165	964212	47
1	4260	4307	4354	4401	4448	4495	4542	4590	4637	4684	47
2	4731	4778	4825	4872	4919	4966	5013	5061	5108	5155	47
3	5202	5249	5296	5343	5390	5437	5484	5531	5578	5625	47
4	5672	5719	5766	5813	5860	5907	5954	6001	6048	6095	47
925	6142	6189	6236	6283	6329	6376	6423	6470	6517	6564	47
6	6611	6658	6705	6752	6799	6845	6892	6939	6986	7033	47
7	7080	7127	7173	7220	7267	7314	7361	7408	7454	7501	47
8	7548	7595	7642	7688	7735	7782	7829	7875	7922	7969	47
9	8016	8062	8109	8156	8203	8249	8296	8343	8390	8436	47
930	968483	968530	968576	968623	968670	968716	968763	968810	968856	968903	47
1	8950	8996	9043	9090	9136	9183	9229	9276	9323	9369	47
2	9416	9463	9509	9556	9602	9649	9695	9742	9789	9835	47
3	9882	9928	9975	970021	970068	970114	970161	970207	970254	970300	47
4	970347	970393	970440	0486	0533	0579	0626	0672	0719	0765	46
935	0812	0858	0904	0951	0997	1044	1090	1137	1183	1229	46
6	1276	1322	1369	1415	1461	1508	1554	1601	1647	1693	46
7	1740	1786	1832	1879	1925	1971	2018	2064	2110	2157	46
8	2203	2249	2295	2342	2388	2434	2481	2527	2573	2619	46
9	2666	2712	2758	2804	2851	2897	2943	2989	3035	3082	46
N.	0	1	2	3	4	5	6	7	8	9	D.

Table A (Continued)

N.	0	1	2	3	4	5	6	7	8	9	D.
940	973128	973174	973220	973266	973313	973359	973405	973451	973497	973543	46
1	3590	3636	3682	3728	3774	3820	3866	3913	3959	4005	46
2	4051	4097	4143	4189	4235	4281	4327	4374	4420	4466	46
3	4512	4558	4604	4650	4696	4742	4788	4834	4880	4926	46
4	4972	5018	5064	5110	5156	5202	5248	5294	5340	5386	46
945	5432	5478	5524	5570	5616	5662	5707	5753	5799	5845	46
6	5891	5937	5983	6029	6075	6121	6167	6212	6258	6304	46
7	6350	6396	6442	6488	6533	6579	6625	6671	6717	6763	46
8	6808	6854	6900	6946	6992	7037	7083	7129	7175	7220	46
9	7266	7312	7358	7403	7449	7495	7541	7586	7632	7678	46
950	977724	977769	977815	977861	977906	977952	977998	978043	978089	978135	46
1	8181	8226	8272	8317	8363	8409	8454	8500	8546	8591	46
2	8637	8683	8728	8774	8819	8865	8911	8956	9002	9047	46
3	9093	9138	9184	9230	9275	9321	9366	9412	9457	9503	46
4	9548	9594	9639	9685	9730	9776	9821	9867	9912	9958	46
955	980003	980049	980094	980140	980185	980231	980276	980322	980367	980412	45
6	0458	0503	0549	0594	0640	0685	0730	0776	0821	0867	45
7	0912	0957	1003	1048	1093	1139	1184	1229	1275	1320	45
8	1366	1411	1456	1501	1547	1592	1637	1683	1728	1773	45
9	1819	1864	1909	1954	2000	2045	2090	2135	2181	2226	45
960	982271	982316	982362	982407	982452	982497	982543	982588	982633	982678	45
1	2723	2769	2814	2859	2904	2949	2994	3040	3085	3130	45
2	3175	3220	3265	3310	3356	3401	3446	3491	3536	3581	45
3	3626	3671	3716	3762	3807	3852	3897	3942	3987	4032	45
4	4077	4122	4167	4212	4257	4302	4347	4392	4437	4482	45
965	4527	4572	4617	4662	4707	4752	4797	4842	4887	4932	45
6	4977	5022	5067	5112	5157	5202	5247	5292	5337	5382	45
7	5426	5471	5516	5561	5606	5651	5696	5741	5786	5830	45
8	5875	5920	5965	6010	6055	6100	6144	6189	6234	6279	45
9	6324	6369	6413	6458	6503	6548	6593	6637	6682	6727	45
970	986772	986817	986861	986906	986951	986996	987040	987085	987130	987175	45
1	7219	7264	7309	7353	7398	7443	7488	7532	7577	7622	45
2	7666	7711	7756	7800	7845	7890	7934	7979	8024	8068	45
3	8113	8157	8202	8247	8291	8336	8381	8425	8470	8514	45
4	8559	8604	8648	8693	8737	8782	8826	8871	8916	8960	45
975	9005	9049	9094	9138	9183	9227	9272	9316	9361	9405	45
6	9450	9494	9539	9583	9628	9672	9717	9761	9806	9850	44
7	9995	9939	9983	990028	990072	990117	990161	990206	990250	990294	44
8	990339	990383	990428	0472	0516	0561	0605	0650	0694	0738	44
9	0783	0827	0871	0916	0960	1004	1049	1093	1137	1192	44
980	991226	991270	991315	991359	991403	991448	991492	991536	991580	991625	44
1	1669	1713	1758	1802	1846	1890	1935	1979	2023	2067	44
2	2111	2156	2200	2244	2288	2333	2377	2421	2465	2509	44
3	2554	2598	2642	2686	2730	2774	2819	2863	2907	2951	44
4	2995	3039	3083	3127	3172	3216	3260	3304	3348	3392	44
985	3436	3480	3524	3568	3613	3657	3701	3745	3789	3833	44
6	3877	3921	3965	4009	4053	4097	4141	4185	4229	4273	44
7	4317	4361	4405	4449	4493	4537	4581	4625	4669	4713	44
8	4757	4801	4845	4889	4933	4977	5021	5065	5108	5152	44
9	5196	5240	5284	5328	5372	5416	5460	5504	5547	5591	44
990	995635	995679	995723	995767	995811	995854	995898	995942	995986	996030	44
1	6074	6117	6161	6205	6249	6293	6337	6380	6424	6468	44
2	6512	6555	6599	6643	6687	6731	6774	6818	6862	6906	44
3	6949	6993	7037	7080	7124	7168	7212	7255	7299	7343	44
4	7386	7430	7474	7517	7561	7605	7648	7692	7736	7779	44
995	7823	7867	7910	7954	7998	8041	8085	8129	8172	8216	44
6	8259	8303	8347	8390	8434	8477	8521	8564	8608	8652	44
7	8695	8739	8782	8826	8869	8913	8956	9000	9043	9087	44
8	9131	9174	9218	9261	9305	9348	9392	9435	9479	9522	44
9	9565	9609	9652	9696	9739	9783	9826	9870	9913	9957	43
N.	0	1	2	3	4	5	6	7	8	9	D.

B / Squares, Square Roots, and Reciprocal 1–1.000

N	N^2	$\sqrt{N}$	$\sqrt{10N}$	$1/N$	N	N^2	$\sqrt{N}$	$\sqrt{10N}$	$1/N$ .0
					50	2 500	7.071 068	22.36068	2000000
1	1	1.000 000	3.162 278	1.0000000	51	2 601	7.141 428	22.58318	1960784
2	4	1.414 214	4.472 136	.5000000	52	2 704	7.211 103	22.80351	1923077
3	9	1.732 051	5.477 226	.3333333	53	2 809	7.280 110	23.02173	1886792
4	16	2.000 000	6.324 555	.2500000	54	2 916	7.348 469	23.23790	1851852
5	25	2.236 068	7.071 068	.2000000	55	3 025	7.416 198	23.45208	1818182
6	36	2.449 490	7.745 967	.1666667	56	3 136	7.483 315	23.66432	1785714
7	49	2.645 751	8.366 600	.1428571	57	3 249	7.549 834	23.87467	1754386
8	64	2.828 427	8.944 272	.1250000	58	3 364	7.615 773	24.08319	1724138
9	81	3.000 000	9.486 833	.1111111	59	3 481	7.681 146	24.28992	1694915
10	100	3.162 278	10.00000	.1000000	60	3 600	7.745 967	24.49490	1666667
11	121	3.316 625	10.48809	.09090909	61	3 721	7.810 250	24.69818	1639344
12	144	3.464 102	10.95445	.08333333	62	3 844	7.874 008	24.89980	1612903
13	169	3.605 551	11.40175	.07692308	63	3 969	7.937 254	25.09980	1587302
14	196	3.741 657	11.83216	.07142857	64	4 096	8.000 000	25.29822	1562500
15	225	3.872 983	12.24745	.06666667	65	4 225	8.062 258	25.49510	1538462
16	256	4.000 000	12.64911	.06250000	66	4 356	8.124 038	25.69047	1515152
17	289	4.123 106	13.03840	.05882353	67	4 489	8.185 353	25.88436	1492537
18	324	4.242 641	13.41641	.05555556	68	4 624	8.246 211	26.07681	1470588
19	361	4.358 899	13.78405	.05263158	69	4 761	8.306 624	26.26785	1449275
20	400	4.472 136	14.14214	.05000000	70	4 900	8.366 600	26.45751	1428571
21	441	4.582 576	14.49138	.04761905	71	5 041	8.426 150	26.64583	1408451
22	484	4.690 416	14.83240	.04545455	72	5 184	8.485 281	26.83282	1388889
23	529	4.795 832	15.16575	.04347826	73	5 329	8.544 004	27.01851	1369863
24	576	4.898 979	15.49193	.04166667	74	5 476	8.602 325	27.20294	1351351
25	625	5.000 000	15.81139	.04000000	75	5 625	8.660 254	27.38613	1333333
26	676	5.099 020	16.12452	.03846154	76	5 776	8.717 798	27.56810	1315789
27	729	5.196 152	16.43168	.03703704	77	5 929	8.774 964	27.74887	1298701
28	784	5.291 503	16.73320	.03571429	78	6 084	8.831 761	27.92848	1282051
29	841	5.385 165	17.02939	.03448276	79	6 241	8.888 194	28.10694	1265823
30	900	5.477 226	17.32051	.03333333	80	6 400	8.944 272	28.28427	1250000
31	961	5.567 764	17.60682	.03225806	81	6 561	9.000 000	28.46050	1234568
32	1 024	5.656 854	17.88854	.03125000	82	6 724	9.055 385	28.63564	1219512
33	1 089	5.744 563	18.16590	.03030303	83	6 889	9.110 434	28.80972	1204819
34	1 156	5.830 952	18.43909	.02941176	84	7 056	9.165 151	28.98275	1190476
35	1 225	5.916 080	18.70829	.02857143	85	7 225	9.219 544	29.15476	1176471
36	1 296	6.000 000	18.97367	.02777778	86	7 396	9.273 618	29.32576	1162791
37	1 369	6.082 763	19.23538	.02702703	87	7 569	9.327 379	29.49576	1149425
38	1 444	6.164 414	19.49359	.02631579	88	7 744	9.380 832	29.66479	1136364
39	1 521	6.244 998	19.74842	.02564103	89	7 921	9.433 981	29.83287	1123596
40	1 600	6.324 555	20.00000	.02500000	90	8 100	9.486 833	30.00000	1111111
41	1 681	6.403 124	20.24846	.02439024	91	8 281	9.539 392	30.16621	1098901
42	1 764	6.480 741	20.49390	.02380952	92	8 464	9.591 663	30.33150	1086957
43	1 849	6.557 439	20.73644	.02325581	93	8 649	9.643 651	30.49590	1075269
44	1 936	6.633 250	20.97618	.02272727	94	8 836	9.695 360	30.65942	1063830
45	2 025	6.708 204	21.21320	.02222222	95	9 025	9.746 794	30.82207	1052632
46	2 116	6.782 330	21.44761	.02173913	96	9 216	9.797 959	30.98387	1041667
47	2 209	6.855 655	21.67948	.02127660	97	9 409	9.848 858	31.14482	1030928
48	2 304	6.928 203	21.90890	.02083333	98	9 604	9.899 495	31.30495	1020408
49	2 401	7.000 000	22.13594	.02040816	99	9 801	9.949 874	31.46427	1010101
50	2 500	7.071 068	22.36068	.02000000	100	10 000	10.00000	31.62278	1000000

Table B (Continued)

N	N²	√N	√10N	1/N .0
100	10 000	10.00000	31.62278	10000000
101	10 201	10.04988	31.78050	09900990
102	10 404	10.09950	31.93744	09803922
103	10 609	10.14889	32.09361	09708738
104	10 816	10.19804	32.24903	09615385
105	11 025	10.24695	32.40370	09523810
106	11 236	10.29563	32.55764	09433962
107	11 449	10.34408	32.71085	09345794
108	11 664	10.39230	32.86335	09259259
109	11 881	10.44031	33.01515	09174312
110	12 100	10.48809	33.16625	09090909
111	12 321	10.53565	33.31666	09009009
112	12 544	10.58301	33.46640	08928571
113	12 769	10.63015	33.61547	08849558
114	12 996	10.67708	33.76389	08771930
115	13 225	10.72381	33.91165	08695652
116	13 456	10.77033	34.05877	08620690
117	13 689	10.81665	34.20526	08547009
118	13 924	10.86278	34.35113	08474576
119	14 161	10.90871	34.49638	08403361
120	14 400	10.95445	34.64102	08333333
121	14 641	11.00000	34.78505	08264463
122	14 884	11.04536	34.92850	08196721
123	15 129	11.09054	35.07136	08130081
124	15 376	11.13553	35.21363	08064516
125	15 625	11.18034	35.35534	08000000
126	15 876	11.22497	35.49648	07936508
127	16 129	11.26943	35.63706	07874016
128	16 384	11.31371	35.77709	07812500
129	16 641	11.35782	35.91657	07751938
130	16 900	11.40175	36.05551	07692308
131	17 161	11.44552	36.19392	07633588
132	17 424	11.48913	36.33180	07575758
133	17 689	11.53256	36.46917	07518797
134	17 956	11.57584	36.60601	07462687
135	18 225	11.61895	36.74235	07407407
136	18 496	11.66190	36.87818	07352941
137	18 769	11.70470	37.01351	07299270
138	19 044	11.74734	37.14835	07246377
139	19 321	11.78983	37.28270	07194245
140	19 600	11.83216	37.41657	07142857
141	19 881	11.87434	37.54997	07092199
142	20 164	11.91638	37.68289	07042254
143	20 449	11.95826	37.81534	06993007
144	20 736	12.00000	37.94733	06944444
145	21 025	12.04159	38.07887	06896552
146	21 316	12.08305	38.20995	06849315
147	21 609	12.12436	38.34058	06802721
148	21 904	12.16553	38.47077	06756757
149	22 201	12.20656	38.60052	06711409
150	22 500	12.24745	38.72983	06666667

N	N²	√N	√10N	1/N .00
150	22 500	12.24745	38.72983	6666667
151	22 801	12.28821	38.85872	6622517
152	23 104	12.32883	38.98718	6578947
153	23 409	12.36932	39.11521	6535948
154	23 716	12.40967	39.24283	6493506
155	24 025	12.44990	39.37004	6451613
156	24 336	12.49000	39.49684	6410256
157	24 649	12.52996	39.62323	6369427
158	24 964	12.56981	39.74921	6329114
159	25 281	12.60952	39.87480	6289308
160	25 600	12.64911	40.00000	6250000
161	25 921	12.68858	40.12481	6211180
162	26 244	12.72792	40.24922	6172840
163	26 569	12.76715	40.37326	6134969
164	26 896	12.80625	40.49691	6097561
165	27 225	12.84523	40.62019	6060606
166	27 556	12.88410	40.74310	6024096
167	27 889	12.92285	40.86563	5988024
168	28 224	12.96148	40.98780	5952381
169	28 561	13.00000	41.10961	5917160
170	28 900	13.03840	41.23106	5882353
171	29 241	13.07670	41.35215	5847953
172	29 584	13.11488	41.47288	5813953
173	29 929	13.15295	41.59327	5780347
174	30 276	13.19091	41.71331	5747126
175	30 625	13.22876	41.83300	5714286
176	30 976	13.26650	41.95235	5681818
177	31 329	13.30413	42.07137	5649718
178	31 684	13.34166	42.19005	5617978
179	32 041	13.37909	42.30839	5586592
180	32 400	13.41641	42.42641	5555556
181	32 761	13.45362	42.54409	5524862
182	33 124	13.49074	42.66146	5494505
183	33 489	13.52775	42.77850	5464481
184	33 856	13.56466	42.89522	5434783
185	34 225	13.60147	43.01163	5405405
186	34 596	13.63818	43.12772	5376344
187	34 969	13.67479	43.24350	5347594
188	35 344	13.71131	43.35897	5319149
189	35 721	13.74773	43.47413	5291005
190	36 100	13.78405	43.58899	5263158
191	36 481	13.82027	43.70355	5235602
192	36 864	13.85641	43.81780	5208333
193	37 249	13.89244	43.93177	5181347
194	37 636	13.92839	44.04543	5154639
195	38 025	13.96424	44.15880	5128205
196	38 416	14.00000	44.27189	5102041
197	38 809	14.03567	44.38468	5076142
198	39 204	14.07125	44.49719	5050505
199	39 601	14.10674	44.60942	5025126
200	40 000	14.14214	44.72136	5000000

Table B (Continued)

N	N²	√N	√10N	1/N .00	N	N²	√N	√10N	1/N .00
200	40 000	14.14214	44.72136	5000000	250	62 500	15.81139	50.00000	4000000
201	40 401	14.17745	44.83302	4975124	251	63 001	15.84298	50.09990	3984064
202	40 804	14.21267	44.94441	4950495	252	63 504	15.87451	50.19960	3968254
203	41 209	14.24781	45.05552	4926108	253	64 009	15.90597	50.29911	3952569
204	41 616	14.28286	45.16636	4901961	254	64 516	15.93738	50.39841	3937008
205	42 025	14.31782	45.27693	4878049	255	65 025	15.96872	50.49752	3921569
206	42 436	14.35270	45.38722	4854369	256	65 536	16.00000	50.59644	3906250
207	42 849	14.38749	45.49725	4830918	257	66 049	16.03122	50.69517	3891051
208	43 264	14.42221	45.60702	4807692	258	66 564	16.06238	50.79370	3875969
209	43 681	14.45683	45.71652	4784689	259	67 081	16.09348	50.89204	3861004
210	44 100	14.49138	45.82576	4761905	260	67 600	16.12452	50.99020	3846154
211	44 521	14.52584	45.93474	4739336	261	68 121	16.15549	51.08816	3831418
212	44 944	14.56022	46.04346	4716981	262	68 644	16.18641	51.18594	3816794
213	45 369	14.59452	46.15192	4694836	263	69 169	16.21727	51.28353	3802281
214	45 796	14.62874	46.26013	4672897	264	69 696	16.24808	51.38093	3787879
215	46 225	14.66288	46.36809	4651163	265	70 225	16.27882	51.47815	3773585
216	46 656	14.69694	46.47580	4629630	266	70 756	16.30951	51.57519	3759398
217	47 089	14.73092	46.58326	4608295	267	71 289	16.34013	51.67204	3745318
218	47 524	14.76482	46.69047	4587156	268	71 824	16.37071	51.76872	3731343
219	47 961	14.79865	46.79744	4566210	269	72 361	16.40122	51.86521	3717472
220	48 400	14.83240	46.90416	4545455	270	72 900	16.43168	51.96152	3703704
221	48 841	14.86607	47.01064	4524887	271	73 441	16.46208	52.05766	3690037
222	49 284	14.89966	47.11688	4504505	272	73 984	16.49242	52.15362	3676471
223	49 729	14.93318	47.22288	4484305	273	74 529	16.52271	52.24940	3663004
224	50 176	14.96663	47.32864	4464286	274	75 076	16.55295	52.34501	3649635
225	50 625	15.00000	47.43416	4444444	275	75 625	16.58312	52.44044	3636364
226	51 076	15.03330	47.53946	4424779	276	76 176	16.61325	52.53570	3623188
227	51 529	15.06652	47.64452	4405286	277	76 729	16.64332	52.63079	3610108
228	51 984	15.09967	47.74935	4385965	278	77 284	16.67333	52.72571	3597122
229	52 441	15.13275	47.85394	4366812	279	77 841	16.70329	52.82045	3584229
230	52 900	15.16575	47.95832	4347826	280	78 400	16.73320	52.91503	3571429
231	53 361	15.19868	48.06246	4329004	281	78 961	16.76305	53.00943	3558719
232	53 824	15.23155	48.16638	4310345	282	79 524	16.79286	53.10367	3546099
233	54 289	15.26434	48.27007	4291845	283	80 089	16.82260	53.19774	3533569
234	54 756	15.29706	48.37355	4273504	284	80 656	16.85230	53.29165	3521127
235	55 225	15.32971	48.47680	4255319	285	81 225	16.88194	53.38539	3508772
236	55 696	15.36229	48.57983	4237288	286	81 796	16.91153	53.47897	3496503
237	56 169	15.39480	48.68265	4219409	287	82 369	16.94107	53.57238	3484321
238	56 644	15.42725	48.78524	4201681	288	82 944	16.97056	53.66563	3472222
239	57 121	15.45962	48.88763	4184100	289	83 521	17.00000	53.75872	3460208
240	57 600	15.49193	48.98979	4166667	290	84 100	17.02939	53.85165	3448276
241	58 081	15.52417	49.09175	4149378	291	84 681	17.05872	53.94442	3436426
242	58 564	15.55635	49.19350	4132231	292	85 264	17.08801	54.03702	3424658
243	59 049	15.58846	49.29503	4115226	293	85 849	17.11724	54.12947	3412969
244	59 536	15.62050	49.39636	4098361	294	86 436	17.14643	54.22177	3401361
245	60 025	15.65248	49.49747	4081633	295	87 025	17.17556	54.31390	3389831
246	60 516	15.68439	49.59839	4065041	296	87 616	17.20465	54.40588	3378378
247	61 009	15.71623	49.69909	4048583	297	88 209	17.23369	54.49771	3367003
248	61 504	15.74802	49.79960	4032258	298	88 804	17.26268	54.58938	3355705
249	62 001	15.77973	49.89990	4016064	299	89 401	17.29162	54.68089	3344482
250	62 500	15.81139	50.00000	4000000	300	90 000	17.32051	54.77226	3333333

Table B (Continued)

N	N²	√N	√10N	1/N .00	N	N²	√N	√10N	1/N .00
300	90 000	17.32051	54.77226	3333333	350	122 500	18.70829	59.16080	2857143
301	90 601	17.34935	54.86347	3322259	351	123 201	18.73499	59.24525	2849003
302	91 204	17.37815	54.95453	3311258	352	123 904	18.76166	59.32959	2840909
303	91 809	17.40690	55.04544	3300330	353	124 609	18.78829	59.41380	2832861
304	92 416	17.43560	55.13620	3289474	354	125 316	18.81489	59.49790	2824859
305	93 025	17.46425	55.22681	3278689	355	126 025	18.84144	59.58188	2816901
306	93 636	17.49286	55.31727	3267974	356	126 736	18.86796	59.66574	2808989
307	94 249	17.52142	55.40758	3257329	357	127 449	18.89444	59.74948	2801120
308	94 864	17.54993	55.49775	3246753	358	128 164	18.92089	59.83310	2793296
309	95 481	17.57840	55.58777	3236246	359	128 881	18.94730	59.91661	2785515
310	96 100	17.60682	55.67764	3225806	360	129 600	18.97367	60.00000	2777778
311	96 721	17.63519	55.76737	3215434	361	130 321	19.00000	60.08328	2770083
312	97 344	17.66352	55.85696	3205128	362	131 044	19.02630	60.16644	2762431
313	97 969	17.69181	55.94640	3194888	363	131 769	19.05256	60.24948	2754821
314	98 596	17.72005	56.03570	3184713	364	132 496	19.07878	60.33241	2747253
315	99 225	17.74824	56.12486	3174603	365	133 225	19.10497	60.41523	2739726
316	99 856	17.77639	56.21388	3164557	366	133 956	19.13113	60.49793	2732240
317	100 489	17.80449	56.30275	3154574	367	134 689	19.15724	60.58052	2724796
318	101 124	17.83255	56.39149	3144654	368	135 424	19.18333	60.66300	2717391
319	101 761	17.86057	56.48008	3134796	369	136 161	19.20937	60.74537	2710027
320	102 400	17.88854	56.56854	3125000	370	136 900	19.23538	60.82763	2702703
321	103 041	17.91647	56.65686	3115265	371	137 641	19.26136	60.90977	2695418
322	103 684	17.94436	56.74504	3105590	372	138 384	19.28730	60.99180	2688172
323	104 329	17.97220	56.83309	3095975	373	139 129	19.31321	61.07373	2680965
324	104 976	18.00000	56.92100	3086420	374	139 876	19.33908	61.15554	2673797
325	105 625	18.02776	57.00877	3076923	375	140 625	19.36492	61.23724	2666667
326	106 276	18.05547	57.09641	3067485	376	141 376	19.39072	61.31884	2659574
327	106 929	18.08314	57.18391	3058104	377	142 129	19.41649	61.40033	2652520
328	107 584	18.11077	57.27128	3048780	378	142 884	19.44222	61.48170	2645503
329	108 241	18.13836	57.35852	3039514	379	143 641	19.46792	61.56298	2638522
330	108 900	18.16590	57.44563	3030303	380	144 400	19.49359	61.64414	2631579
331	109 561	18.19341	57.53260	3021148	381	145 161	19.51922	61.72520	2624672
332	110 224	18.22087	57.61944	3012048	382	145 924	19.54483	61.80615	2617801
333	110 889	18.24829	57.70615	3003003	383	146 689	19.57039	61.88699	2610966
334	111 556	18.27567	57.79273	2994012	384	147 456	19.59592	61.96773	2604167
335	112 225	18.30301	57.87918	2985075	385	148 225	19.62142	62.04837	2597403
336	112 896	18.33030	57.96551	2976190	386	148 996	19.64688	62.12890	2590674
337	113 569	18.35756	58.05170	2967359	387	149 769	19.67232	62.20932	2583979
338	114 244	18.38478	58.13777	2958580	388	150 544	19.69772	62.28965	2577320
339	114 921	18.41195	58.22371	2949853	389	151 321	19.72308	62.36986	2570694
340	115 600	18.43909	58.30952	2941176	390	152 100	19.74842	62.44998	2564103
341	116 281	18.46619	58.39521	2932551	391	152 881	19.77372	62.52999	2557545
342	116 964	18.49324	58.48077	2923977	392	153 664	19.79899	62.60990	2551020
343	117 649	18.52026	58.56620	2915452	393	154 449	19.82423	62.68971	2544529
344	118 336	18.54724	58.65151	2906977	394	155 236	19.84943	62.76942	2538071
345	119 025	18.57418	58.73670	2898551	395	156 025	19.87461	62.84903	2531646
346	119 716	18.60108	58.82176	2890173	396	156 816	19.89975	62.92853	2525253
347	120 409	18.62794	58.90671	2881844	397	157 609	19.92486	63.00794	2518892
348	121 104	18.65476	58.99152	2873563	398	158 404	19.94994	63.08724	2512563
349	121 801	18.68154	59.07622	2865330	399	159 201	19.97498	63.16645	2506266
350	122 500	18.70829	59.16080	2857143	400	160 000	20.00000	63.24555	2500000

Table B (Continued)

N	N²	√N	√10N	1/N .00	N	N²	√N	√10N	1/N .00
400	160 000	20.00000	63.24555	2500000	450	202 500	21.21320	67.08204	2222222
401	160 801	20.02498	63.32456	2493766	451	203 401	21.23676	67.15653	2217295
402	161 604	20.04994	63.40347	2487562	452	204 304	21.26029	67.23095	2212389
403	162 409	20.07486	63.48228	2481390	453	205 209	21.28380	67.30527	2207506
404	163 216	20.09975	63.56099	2475248	454	206 116	21.30728	67.37952	2202643
405	164 025	20.12461	63.63961	2469136	455	207 025	21.33073	67.45369	2197802
406	164 836	20.14944	63.71813	2463054	456	207 936	21.35416	67.52777	2192982
407	165 649	20.17424	63.79655	2457002	457	208 849	21.37756	67.60178	2188184
408	166 464	20.19901	63.87488	2450980	458	209 764	21.40093	67.67570	2183406
409	167 281	20.22375	63.95311	2444988	459	210 681	21.42429	67.74954	2178649
410	168 100	20.24846	64.03124	2439024	460	211 600	21.44761	67.82330	2173913
411	168 921	20.27313	64.10928	2433090	461	212 521	21.47091	67.89698	2169197
412	169 744	20.29778	64.18723	2427184	462	213 444	21.49419	67.97058	2164502
413	170 569	20.32240	64.26508	2421308	463	214 369	21.51743	68.04410	2159827
414	171 396	20.34699	64.34283	2415459	464	215 296	21.54066	68.11755	2155172
415	172 225	20.37155	64.42049	2409639	465	216 225	21.56386	68.19091	2150538
416	173 056	20.39608	64.49806	2403846	466	217 156	21.58703	68.26419	2145923
417	173 889	20.42058	64.57554	2398082	467	218 089	21.61018	68.33740	2141328
418	174 724	20.44505	64.65292	2392344	468	219 024	21.63331	68.41053	2136752
419	175 561	20.46949	64.73021	2386635	469	219 961	21.65641	68.48357	2132196
420	176 400	20.49390	64.80741	2380952	470	220 900	21.67948	68.55655	2127660
421	177 241	20.51828	64.88451	2375297	471	221 841	21.70253	68.62944	2123142
422	178 084	20.54264	64.96153	2369668	472	222 784	21.72556	68.70226	2118644
423	178 929	20.56696	65.03845	2364066	473	223 729	21.74856	68.77500	2114165
424	179 776	20.59126	65.11528	2358491	474	224 676	21.77154	68.84766	2109705
425	180 625	20.61553	65.19202	2352941	475	225 625	21.79449	68.92024	2105263
426	181 476	20.63977	65.26868	2347418	476	226 576	21.81742	68.99275	2100840
427	182 329	20.66398	65.34524	2341920	477	227 529	21.84033	69.06519	2096436
428	183 184	20.68816	65.42171	2336449	478	228 484	21.86321	69.13754	2092050
429	184 041	20.71232	65.49809	2331002	479	229 441	21.88607	69.20983	2087683
430	184 900	20.73644	65.57439	2325581	480	230 400	21.90890	69.28203	2083333
431	185 761	20.76054	65.65059	2320186	481	231 361	21.93171	69.35416	2079002
432	186 624	20.78461	65.72671	2314815	482	232 324	21.95450	69.42622	2074689
433	187 489	20.80865	65.80274	2309469	483	233 289	21.97726	69.49820	2070393
434	188 356	20.83267	65.87868	2304147	484	234 256	22.00000	69.57011	2066116
435	189 225	20.85665	65.95453	2298851	485	235 225	22.02272	69.64194	2061856
436	190 096	20.88061	66.03030	2293578	486	236 196	22.04541	69.71370	2057613
437	190 969	20.90454	66.10598	2288330	487	237 169	22.06808	69.78539	2053388
438	191 844	20.92845	66.18157	2283105	488	238 144	22.09072	69.85700	2049180
439	192 721	20.95233	66.25708	2277904	489	239 121	22.11334	69.92853	2044990
440	193 600	20.97618	66.33250	2272727	490	240 100	22.13594	70.00000	2040816
441	194 481	21.00000	66.40783	2267574	491	241 081	22.15852	70.07139	2036660
442	195 364	21.02380	66.48308	2262443	492	242 064	22.18107	70.14271	2032520
443	196 249	21.04757	66.55825	2257336	493	243 049	22.20360	70.21396	2028398
444	197 136	21.07131	66.63332	2252252	494	244 036	22.22611	70.28513	2024291
445	198 025	21.09502	66.70832	2247191	495	245 025	22.24860	70.35624	2020202
446	198 916	21.11871	66.78323	2242152	496	246 016	22.27106	70.42727	2016129
447	199 809	21.14237	66.85806	2237136	497	247 009	22.29350	70.49823	2012072
448	200 704	21.16601	66.93280	2232143	498	248 004	22.31591	70.56912	2008032
449	201 601	21.18962	67.00746	2227171	499	249 001	22.33831	70.63993	2004008
450	202 500	21.21320	67.08204	2222222	500	250 000	22.36068	70.71068	2000000

Table B (Continued)

N	N²	√N	√10N	1/N .00	N	N²	√N	√10N	1/N .00
500	250 000	22.36068	70.71068	2000000	550	302 500	23.45208	74.16198	1818182
501	251 001	22.38303	70.78135	1996008	551	303 601	23.47339	74.22937	1814882
502	252 004	22.40536	70.85196	1992032	552	304 704	23.49468	74.29670	1811594
503	253 009	22.42766	70.92249	1988072	553	305 809	23.51595	74.36397	1808318
504	254 016	22.44994	70.99296	1984127	554	306 916	23.53720	74.43118	1805054
505	255 025	22.47221	71.06335	1980198	555	308 025	23.55844	74.49832	1801802
506	256 036	22.49444	71.13368	1976285	556	309 136	23.57965	74.56541	1798561
507	257 049	22.51666	71.20393	1972387	557	310 249	23.60085	74.63243	1795332
508	258 064	22.53886	71.27412	1968504	558	311 364	23.62202	74.69940	1792115
509	259 081	22.56103	71.34424	1964637	559	312 481	23.64318	74.76630	1788909
510	260 100	22.58318	71.41428	1960784	560	313 600	23.66432	74.83315	1785714
511	261 121	22.60531	71.48426	1956947	561	314 721	23.68544	74.89993	1782531
512	262 144	22.62742	71.55418	1953125	562	315 844	23.70654	74.96666	1779359
513	263 169	22.64950	71.62402	1949318	563	316 969	23.72762	75.03333	1776199
514	264 196	22.67157	71.69379	1945525	564	318 096	23.74868	75.09993	1773050
515	265 225	22.69361	71.76350	1941748	565	319 225	23.76973	75.16648	1769912
516	266 256	22.71563	71.83314	1937984	566	320 356	23.79075	75.23297	1766784
517	267 289	22.73763	71.90271	1934236	567	321 489	23.81176	75.29940	1763668
518	268 324	22.75961	71.97222	1930502	568	322 624	23.83275	75.36577	1760563
519	269 361	22.78157	72.04165	1926782	569	323 761	23.85372	75.43209	1757469
520	270 400	22.80351	72.11103	1923077	570	324 900	23.87467	75.49834	1754386
521	271 441	22.82542	72.18033	1919386	571	326 041	23.89561	75.56454	1751313
522	272 484	22.84732	72.24957	1915709	572	327 184	23.91652	75.63068	1748252
523	273 529	22.86919	72.31874	1912046	573	328 329	23.93742	75.69676	1745201
524	274 576	22.89105	72.38784	1908397	574	329 476	23.95830	75.76279	1742160
525	275 625	22.91288	72.45688	1904762	575	330 625	23.97916	75.82875	1739130
526	276 676	22.93469	72.52586	1901141	576	331 776	24.00000	75.89466	1736111
527	277 729	22.95648	72.59477	1897533	577	332 929	24.02082	75.96052	1733102
528	278 784	22.97825	72.66361	1893939	578	334 084	24.04163	76.02631	1730104
529	279 841	23.00000	72.73239	1890359	579	335 241	24.06242	76.09205	1727116
530	280 900	23.02173	72.80110	1886792	580	336 400	24.08319	76.15773	1724138
531	281 961	23.04344	72.86975	1883239	581	337 561	24.10394	76.22336	1721170
532	283 024	23.06513	72.93833	1879699	582	338 724	24.12468	76.28892	1718213
533	284 089	23.08679	73.00685	1876173	583	339 889	24.14539	76.35444	1715266
534	285 156	23.10844	73.07530	1872659	584	341 056	24.16609	76.41989	1712329
535	286 225	23.13007	73.14369	1869159	585	342 225	24.18677	76.48529	1709402
536	287 296	23.15167	73.21202	1865672	586	343 396	24.20744	76.55064	1706485
537	288 369	23.17326	73.28028	1862197	587	344 569	24.22808	76.61593	1703578
538	289 444	23.19483	73.34848	1858736	588	345 744	24.24871	76.68116	1700680
539	290 521	23.21637	73.41662	1855288	589	346 921	24.26932	76.74634	1697793
540	291 600	23.23790	73.48469	1851852	590	348 100	24.28992	76.81146	1694915
541	292 681	23.25941	73.55270	1848429	591	349 281	24.31049	76.87652	1692047
542	293 764	23.28089	73.62065	1845018	592	350 464	24.33105	76.94154	1689189
543	294 849	23.30236	73.68853	1841621	593	351 649	24.35159	77.00649	1686341
544	295 936	23.32381	73.75636	1838235	594	352 836	24.37212	77.07140	1683502
545	297 025	23.34524	73.82412	1834862	595	354 025	24.39262	77.13624	1680672
546	298 116	23.36664	73.89181	1831502	596	355 216	24.41311	77.20104	1677852
547	299 209	23.38803	73.95945	1828154	597	356 409	24.43358	77.26578	1675042
548	300 304	23.40940	74.02702	1824818	598	357 604	24.45404	77.33046	1672241
549	301 401	23.43075	74.09453	1821494	599	358 801	24.47448	77.39509	1669449
550	302 500	23.45208	74.16198	1818182	600	360 000	24.49490	77.45967	1666667

Table B (Continued)

N	N²	$\sqrt{N}$	$\sqrt{10N}$	1/N .00	N	N²	$\sqrt{N}$	$\sqrt{10N}$	1/N .00
600	360 000	24.49490	77.45967	1666667	650	422 500	25.49510	80.62258	1538462
601	361 201	24.51530	77.52419	1663894	651	423 801	25.51470	80.68457	1536098
602	362 404	24.53569	77.58866	1661130	652	425 104	25.53429	80.74652	1533742
603	363 609	24.55606	77.65307	1658375	653	426 409	25.55386	80.80842	1531394
604	364 816	24.57641	77.71744	1655629	654	427 716	25.57342	80.87027	1529052
605	366 025	24.59675	77.78175	1652893	655	429 025	25.59297	80.93207	1526718
606	367 236	24.61707	77.84600	1650165	656	430 336	25.61250	80.99383	1524390
607	368 449	24.63737	77.91020	1647446	657	431 649	25.63201	81.05554	1522070
608	369 664	24.65766	77.97435	1644737	658	432 964	25.65151	81.11720	1519757
609	370 881	24.67793	78.03845	1642036	659	434 281	25.67100	81.17881	1517451
610	372 100	24.69818	78.10250	1639344	660	435 600	25.69047	81.24038	1515152
611	373 321	24.71841	78.16649	1636661	661	436 921	25.70992	81.30191	1512859
612	374 544	24.73863	78.23043	1633987	662	438 244	25.72936	81.36338	1510574
613	375 769	24.75884	78.29432	1631321	663	439 569	25.74879	81.42481	1508296
614	376 996	24.77902	78.35815	1628664	664	440 896	25.76820	81.48620	1506024
615	378 225	24.79919	78.42194	1626016	665	442 225	25.78759	81.54753	1503759
616	379 456	24.81935	78.48567	1623377	666	443 556	25.80698	81.60882	1501502
617	380 689	24.83948	78.54935	1620746	667	444 889	25.82634	81.67007	1499250
618	381 924	24.85961	78.61298	1618123	668	446 224	25.84570	81.73127	1497006
619	383 161	24.87971	78.67655	1615509	669	447 561	25.86503	81.79242	1494768
620	384 400	24.89980	78.74008	1612903	670	448 900	25.88436	81.85353	1492537
621	385 641	24.91987	78.80355	1610306	671	450 241	25.90367	81.91459	1490313
622	386 884	24.93993	78.86698	1607717	672	451 584	25.92296	81.97561	1488095
623	388 129	24.95997	78.93035	1605136	673	452 929	25.94224	82.03658	1485884
624	389 376	24.97999	78.99367	1602564	674	454 276	25.96151	82.09750	1483680
625	390 625	25.00000	79.05694	1600000	675	455 625	25.98076	82.15838	1481481
626	391 876	25.01999	79.12016	1597444	676	456 976	26.00000	82.21922	1479290
627	393 129	25.03997	79.18333	1594896	677	458 329	26.01922	82.28001	1477105
628	394 384	25.05993	79.24645	1592357	678	459 684	26.03843	82.34076	1474926
629	395 641	25.07987	79.30952	1589825	679	461 041	26.05763	82.40146	1472754
630	396 900	25.09980	79.37254	1587302	680	462 400	26.07681	82.46211	1470588
631	398 161	25.11971	79.43551	1584786	681	463 761	26.09598	82.52272	1468429
632	399 424	25.13961	79.49843	1582278	682	465 124	26.11513	82.58329	1466276
633	400 689	25.15949	79.56130	1579779	683	466 489	26.13427	82.64381	1464129
634	401 956	25.17936	79.62412	1577287	684	467 856	26.15339	82.70429	1461988
635	403 225	25.19921	79.68689	1574803	685	469 225	26.17250	82.76473	1459854
636	404 496	25.21904	79.74961	1572327	686	470 596	26.19160	82.82512	1457726
637	405 769	25.23886	79.81228	1569859	687	471 969	26.21068	82.88546	1455604
638	407 044	25.25866	79.87490	1567398	688	473 344	26.22975	82.94577	1453488
639	408 321	25.27845	79.93748	1564945	689	474 721	26.24881	83.00602	1451379
640	409 600	25.29822	80.00000	1562500	690	476 100	26.26785	83.06624	1449275
641	410 881	25.31798	80.06248	1560062	691	477 481	26.28688	83.12641	1447178
642	412 164	25.33772	80.12490	1557632	692	478 864	26.30589	83.18654	1445087
643	413 449	25.35744	80.18728	1555210	693	480 249	26.32489	83.24662	1443001
644	414 736	25.37716	80.24961	1552795	694	481 636	26.34388	83.30666	1440922
645	416 025	25.39685	80.31189	1550388	695	483 025	26.36285	83.36666	1438849
646	417 316	25.41653	80.37413	1547988	696	484 416	26.38181	83.42661	1436782
647	418 609	25.43619	80.43631	1545595	697	485 809	26.40076	83.48653	1434720
648	419 904	25.45584	80.49845	1543210	698	487 204	26.41969	83.54639	1432665
649	421 201	25.47548	80.56054	1540832	699	488 601	26.43861	83.60622	1430615
650	422 500	25.49510	80.62258	1538462	700	490 000	26.45751	83.66600	1428571

Table B (Continued)

N	N²	√N	√10N	1/N .00	N	N²	√N	√10N	1/N .00
700	490 000	26.45751	83.66600	1428571	750	562 500	27.38613	86.60254	1333333
701	491 401	26.47640	83.72574	1426534	751	564 001	27.40438	86.66026	1331558
702	492 804	26.49528	83.78544	1424501	752	565 504	27.42262	86.71793	1329787
703	494 209	26.51415	83.84510	1422475	753	567 009	27.44085	86.77557	1328021
704	495 616	26.53300	83.90471	1420455	754	568 516	27.45906	86.83317	1326260
705	497 025	26.55184	83.96428	1418440	755	570 025	27.47726	86.89074	1324503
706	498 436	26.57066	84.02381	1416431	756	571 536	27.49545	86.94826	1322751
707	499 849	26.58947	84.08329	1414427	757	573 049	27.51363	87.00575	1321004
708	501 264	26.60827	84.14274	1412429	758	574 564	27.53180	87.06320	1319261
709	502 681	26.62705	84.20214	1410437	759	576 081	27.54995	87.12061	1317523
710	504 100	26.64583	84.26150	1408451	760	577 600	27.56810	87.17798	1315789
711	505 521	26.66458	84.32082	1406470	761	579 121	27.58623	87.23531	1314060
712	506 944	26.68333	84.38009	1404494	762	580 644	27.60435	87.29261	1312336
713	508 360	26.70206	84.43933	1402525	763	582 169	27.62245	87.34987	1310616
714	509 796	26.72078	84.49852	1400560	764	583 696	27.64055	87.40709	1308901
715	511 225	26.73948	84.55767	1398601	765	585 225	27.65863	87.46428	1307190
716	512 656	26.75818	84.61678	1396648	766	586 756	27.67671	87.52143	1305483
717	514 089	26.77686	84.67585	1394700	767	588 289	27.69476	87.57854	1303781
718	515 524	26.79552	84.73488	1392758	768	589 824	27.71281	87.63561	1302083
719	516 961	26.81418	84.79387	1390821	769	591 361	27.73085	87.69265	1300390
720	518 400	26.83282	84.85281	1388889	770	592 900	27.74887	87.74964	1298701
721	519 841	26.85144	84.91172	1386963	771	594 441	27.76689	87.80661	1297017
722	521 284	26.87006	84.97058	1385042	772	595 984	27.78489	87.86353	1295337
723	522 729	26.88866	85.02941	1383126	773	597 529	27.80288	87.92042	1293661
724	524 176	26.90725	85.08819	1381215	774	599 076	27.82086	87.97727	1291990
725	525 625	26.92582	85.14693	1379310	775	600 625	27.83882	88.03408	1290323
726	527 076	26.94439	85.20563	1377410	776	602 176	27.85678	88.09086	1288660
727	528 529	26.96294	85.26429	1375516	777	603 729	27.87472	88.14760	1287001
728	529 984	26.98148	85.32292	1373626	778	605 284	27.89265	88.20431	1285347
729	531 441	27.00000	85.38150	1371742	779	606 841	27.91057	88.26098	1283697
730	532 900	27.01851	85.44004	1369863	780	608 400	27.92848	88.31761	1282051
731	534 361	27.03701	85.49854	1367989	781	609 961	27.94638	88.37420	1280410
732	535 824	27.05550	85.55700	1366120	782	611 524	27.96426	88.43076	1278772
733	537 289	27.07397	85.61542	1364256	783	612 089	27.98211	88.48729	1277139
734	538 756	27.09243	85.67380	1362398	784	614 656	28.00000	88.54377	1275510
735	540 225	27.11088	85.73214	1360544	785	616 225	28.01785	88.60023	1273885
736	541 696	27.12932	85.79044	1358696	786	617 796	28.03569	88.65664	1272265
737	543 169	27.14774	85.84870	1356852	787	619 369	28.05352	88.71302	1270648
738	544 644	27.16616	85.90693	1355014	788	620 944	28.07134	88.76936	1269036
739	546 121	27.18455	85.96511	1353180	789	622 521	28.08914	88.82567	1267427
740	547 600	27.20294	86.02325	1351351	790	624 100	28.10694	88.88194	1265823
741	549 081	27.22132	86.08136	1349528	791	625 681	28.12472	88.93818	1264223
742	550 564	27.23968	86.13942	1347709	792	627 264	28.14249	88.99438	1262626
743	552 049	27.25803	86.19745	1345895	793	628 849	28.16026	89.05055	1261034
744	553 536	27.27636	86.25543	1344086	794	630 436	28.17801	89.10668	1259446
745	555 025	27.29469	86.31338	1342282	795	632 025	28.19574	89.16277	1257862
746	556 516	27.31300	86.37129	1340483	796	633 616	28.21347	89.21883	1256281
747	558 009	27.33130	86.42916	1338688	797	635 209	28.23119	89.27486	1254705
748	559 504	27.34959	86.48699	1336898	798	636 804	28.24889	89.33085	1253133
749	561 001	27.36786	86.54479	1335113	799	638 401	28.26659	89.38680	1251564
750	562 500	27.38613	86.60254	1333333	800	640 000	28.28427	89.44272	1250000

Table B (Continued)

N	N²	√N	√10N	1/N .00	N	N²	√N	√10N	1/N .00
800	640 000	28.28427	89.44272	1250000	850	722 500	29.15476	92.19544	1176471
801	641 601	28.30194	89.49860	1248439	851	724 201	29.17190	92.24966	1175088
802	643 204	28.31960	89.55445	1246883	852	725 904	29.18904	92.30385	1173709
803	644 809	28.33725	89.61027	1245330	853	727 609	29.20616	92.35800	1172333
804	646 416	28.35489	89.66605	1243781	854	729 316	29.22328	92.41212	1170960
805	648 025	28.37252	89.72179	1242236	855	731 025	29.24038	92.46621	1169591
806	649 636	28.39014	89.77750	1240695	856	732 736	29.25748	92.52027	1168224
807	651 249	28.40775	89.83318	1239157	857	734 449	29.27456	92.57429	1166861
808	652 864	28.42534	89.88882	1237624	858	736 164	29.29164	92.62829	1165501
809	654 481	28.44293	89.94443	1236094	859	737 881	29.30870	92.68225	1164144
810	656 100	28.46050	90.00000	1234568	860	739 600	29.32576	92.73618	1162791
811	657 721	28.47806	90.05554	1233046	861	741 321	29.34280	92.79009	1161440
812	659 344	28.49561	90.11104	1231527	862	743 044	29.35984	92.84396	1160093
813	660 969	28.51315	90.16651	1230012	863	744 769	29.37686	92.89779	1158749
814	662 596	28.53069	90.22195	1228501	864	746 496	29.39388	92.95160	1157407
815	664 225	28.54820	90.27735	1226994	865	748 225	29.41088	93.00538	1156069
816	665 856	28.56571	90.33272	1225490	866	749 956	29.42788	93.05912	1154734
817	667 489	28.58321	90.38805	1223990	867	751 689	29.44486	93.11283	1153403
818	669 124	28.60070	90.44335	1222494	868	753 424	29.46184	93.16652	1152074
819	670 761	28.61818	90.49862	1221001	869	755 161	29.47881	93.22017	1150748
820	672 400	28.63564	90.55385	1219512	870	756 900	29.49576	93.27379	1149425
821	674 041	28.65310	90.60905	1218027	871	758 641	29.51271	93.32738	1148106
822	675 684	28.67054	90.66422	1216545	872	760 384	29.52965	93.38094	1146789
823	677 329	28.68798	90.71935	1215067	873	762 129	29.54657	93.43447	1145475
824	678 976	28.70540	90.77445	1213592	874	763 876	29.56349	93.48797	1144165
825	680 625	28.72281	90.82951	1212121	875	765 625	29.58040	93.54143	1142857
826	682 276	28.74022	90.88454	1210654	876	767 376	29.59730	93.59487	1141553
827	683 929	28.75761	90.93954	1209190	877	769 129	29.61419	93.64828	1140251
828	685 584	28.77499	90.99451	1207729	878	770 884	29.63106	93.70165	1138952
829	687 241	28.79236	91.04944	1206273	879	772 641	29.64793	93.75500	1137656
830	688 900	28.80972	91.10434	1204819	880	774 400	29.66479	93.80832	1136364
831	690 561	28.82707	91.15920	1203369	881	776 161	29.68164	93.86160	1135074
832	692 224	28.84441	91.21403	1201923	882	777 924	29.69848	93.91486	1133787
833	693 889	28.86174	91.26883	1200480	883	779 689	29.71532	93.96808	1132503
834	695 556	28.87906	91.32360	1199041	884	781 456	29.73214	94.02127	1131222
835	697 225	28.89637	91.37833	1197605	885	783 225	29.74895	94.07444	1129944
836	698 896	28.91366	91.43304	1196172	886	784 996	29.76575	94.12757	1128668
837	700 569	28.93095	91.48770	1194743	887	786 769	29.78255	94.18068	1127396
838	702 244	28.94823	91.54234	1193317	888	788 544	29.79933	94.23375	1126126
839	703 921	28.96550	91.59694	1191895	889	790 321	29.81610	94.28680	1124859
840	705 600	28.98275	91.65151	1190476	890	792 100	29.83287	94.33981	1123596
841	707 281	29.00000	91.70605	1189061	891	793 881	29.84962	94.39280	1122334
842	708 964	29.01724	91.76056	1187648	892	795 664	29.86637	94.44575	1121076
843	710 649	29.03445	91.81503	1186240	893	797 449	29.88311	94.49868	1119821
844	712 336	29.05168	91.86947	1184834	894	799 236	29.89983	94.55157	1118568
845	714 025	29.06888	91.92388	1183432	895	801 025	29.91655	94.60444	1117318
846	715 716	29.08608	91.97826	1182033	896	802 816	29.93326	94.65728	1116071
847	717 409	29.10326	92.03260	1180638	897	804 609	29.94996	94.71008	1114827
848	719 104	29.12044	92.08692	1179245	898	806 404	29.96665	94.76286	1113586
849	720 801	29.13760	92.14120	1177856	899	808 201	29.98333	94.81561	1112347
850	722 500	29.15476	92.19544	1176471	900	810 000	30.00000	94.86833	1111111

Table B (Continued)

N	N²	√N	√10N	1/N .00	N	N²	√N	√10N	1/N .00
900	810 000	30.00000	94.86833	1111111	950	902 500	30.82207	97.46794	1052632
901	811 801	30.01666	94.92102	1109878	951	904 401	30.83829	97.51923	1051525
902	813 604	30.03331	94.97368	1108647	952	906 304	30.85450	97.57049	1050420
903	815 409	30.04996	95.02631	1107420	953	908 209	30.87070	97.62172	1049318
904	817 216	30.06659	95.07891	1106195	954	910.116	30.88689	97.67292	1048218
905	819 025	30.08322	95.13149	1104972	955	912 025	30.90307	97.72410	1047120
906	820 836	30.09983	95.18403	1103753	956	913 936	30.91925	97.77525	1046025
907	822 649	30.11644	95.23655	1102536	957	915 849	30.93542	97.82638	1044932
908	824 464	30.13304	95.28903	1101322	958	917 764	30.95158	97.87747	1043841
909	826 281	30.14963	95.34149	1100110	959	919 681	30.96773	97.92855	1042753
910	828 100	30.16621	95.39392	1098901	960	921 600	30.98387	97.97959	1041667
911	829 921	30.18278	95.44632	1097695	961	923 521	31.00000	98.03061	1040583
912	831 744	30.19934	95.49869	1096491	962	925 444	31.01612	98.08160	1039501
913	833 569	30.21589	95.55103	1095290	963	927 369	31.03224	98.13256	1038422
914	835 396	30.23243	95.60335	1094092	964	929 296	31.04835	98.18350	1037344
915	837 225	30.24897	95.65563	1092896	965	931 225	31.06445	98.23441	1036269
916	839 056	30.26549	95.70789	1091703	966	933 156	31.08054	98.28530	1035197
917	840 889	30.28201	95.76012	1090513	967	935 089	31.09662	98.33616	1034126
918	842 724	30.29851	95.81232	1089325	968	937 024	31.11270	98.38699	1033058
919	844 561	30.31501	95.86449	1088139	969	938 961	31.12876	98.43780	1031992
920	846 400	30.33150	95.91663	1086957	970	940 900	31.14482	98.48858	1030928
921	848 241	30.34798	95.96874	1085776	971	942 841	31.16087	98.53933	1029866
922	850 084	30.36445	96.02083	1084599	972	944 784	31.17691	98.59006	1028807
923	851 929	30.38092	96.07289	1083424	973	946 729	31.19295	98.64076	1027749
924	853 776	30.39737	96.12492	1082251	974	948 676	31.20897	98.69144	1026694
925	855 625	30.41381	96.17692	1081081	975	950 625	31.22499	98.74209	1025641
926	857 476	30.43025	96.22889	1079914	976	952 576	31.24100	98.79271	1024590
927	859 329	30.44667	96.28084	1078749	977	954 529	31.25700	98.84331	1023541
928	861 184	30.46309	96.33276	1077586	978	956 484	31.27299	98.89388	1022495
929	863 041	30.47950	96.38465	1076426	979	958 441	31.28898	98.94443	1021450
930	864 900	30.49590	96.43651	1075269	980	960 400	31.30495	98.99495	1020408
931	866 761	30.51229	96.48834	1074114	981	962 361	31.32092	99.04544	1019368
932	868 624	30.52868	96.54015	1072961	982	964 324	31.33688	99.09591	1018330
933	870 489	30.54505	96.59193	1071811	983	966 289	31.35283	99.14636	1017294
934	872 356	30.56141	96.64368	1070664	984	968 256	31.36877	99.19677	1016260
935	874 225	30.57777	96.69540	1069519	985	970 225	31.38471	99.24717	1015228
936	876 096	30.59412	96.74709	1068376	986	972 196	31.40064	99.29753	1014199
937	877 969	30.61046	96.79876	1067236	987	974 169	31.41656	99.34787	1013171
938	879 844	30.62679	96.85040	1066098	988	976 144	31.43247	99.39819	1012146
939	881 721	30.64311	96.90201	1064963	989	978 121	31.44837	99.44848	1011122
940	883 600	30.65942	96.95360	1063830	990	980 100	31.46427	99.49874	1010101
941	885 481	30.67572	97.00515	1062699	991	982 081	31.48015	99.54898	1009082
942	887 364	30.69202	97.05668	1061571	992	984 064	31.49603	99.59920	1008065
943	889 249	30.70831	97.10819	1060445	993	986 049	31.51190	99.64939	1007049
944	891 136	30.72458	97.15966	1059322	994	988 036	31.52777	99.69955	1006036
945	893 025	30.74085	97.21111	1058201	995	990 025	31.54362	99.74969	1005025
946	894 916	30.75711	97.26253	1057082	996	992 016	31.55947	99.79980	1004016
947	896 809	30.77337	97.31393	1055966	997	994 009	31.57531	99.84989	1003009
948	898 704	30.78961	97.36529	1054852	998	996 004	31.59114	99.89995	1002004
949	900 601	30.80584	97.41663	1053741	999	998 001	31.60696	99.94999	1001001
950	902 500	30.82207	97.46794	1052632	1000	1 000 000	31.62278	100.00000	1000000

Reproduced, by permission, from *Practical Business Statistics*, by Frederick E. Croxton and Dudley J. Cowden, published by Prentice-Hall, Inc., Englewood Cliffs, New Jersey, Second Edition, 1948, pp. 524–33.

C / Table of Factorials

N	$N!$
0	1
1	1
2	2
3	6
4	24
5	120
6	720
7	5040
8	40320
9	362880
10	3628800
11	39916800
12	479001600
13	6227020800
14	87178291200
15	1307674368000
16	20922789888000
17	355687428096000
18	6402373705728000
19	121645100408832000
20	2432902008176640000

D.1 / Binomial Distribution—Individual Terms

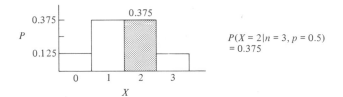

$P(X = 2|n = 3, p = 0.5)$
$= 0.375$

n	r	.01	.02	.04	.05	.06	.08	.10	.12	.14	.15	p .16	.18	.20	.22	.24	.25	.30	.35	.40	.45	.50	r
2	0	980	960	922	902	884	846	810	774	740	722	706	672	640	608	578	562	490	422	360	302	250	0
	1	020	039	077	095	113	147	180	211	241	255	269	295	320	343	365	375	420	455	480	495	500	1
	2	0+	0+	002	002	004	006	010	014	020	022	026	032	040	048	058	062	090	122	160	202	250	2
3	0	970	941	885	857	831	779	729	681	636	614	593	551	512	475	439	422	343	275	216	166	125	0
	1	029	058	111	135	159	203	243	279	311	325	339	363	384	402	416	422	441	444	432	408	375	1
	2	0+	001	005	007	010	018	027	038	051	057	065	080	096	113	131	141	189	239	288	334	375	2
	3	0+	0+	0+	0+	0+	001	001	002	003	003	004	006	008	011	014	016	027	043	064	091	125	3
4	0	961	922	849	815	781	716	656	600	547	522	498	452	410	370	334	316	240	179	130	092	063	0
	1	039	075	142	171	199	249	292	327	356	368	379	397	410	418	421	422	412	384	346	299	250	1
	2	001	002	009	014	019	033	049	067	087	098	108	131	154	177	200	211	265	311	346	368	375	2
	3	0+	0+	0+	0+	001	002	004	006	009	011	014	019	026	033	042	047	076	111	154	200	250	3
	4	0+	0+	0+	0+	0+	0+	0+	0+	0+	001	001	001	002	002	003	004	008	015	026	041	062	4
5	0	951	904	815	774	734	659	590	528	470	444	418	371	328	289	254	237	168	116	078	050	031	0
	1	048	092	170	204	234	287	328	360	383	392	398	407	410	407	400	396	360	312	259	206	156	1
	2	001	004	014	021	030	050	073	098	125	138	152	179	205	230	253	264	309	336	346	337	312	2
	3	0+	0+	001	001	002	004	008	013	020	024	029	039	051	065	080	088	132	181	230	276	312	3
	4	0+	0+	0+	0+	0+	0+	0+	001	002	002	003	004	006	009	013	015	028	049	077	113	156	4
	5	0+	0+	0+	0+	0+	0+	0+	0+	0+	0+	0+	0+	0+	001	001	001	002	005	010	018	031	5
6	0	941	886	783	735	690	606	531	464	405	377	351	304	262	225	193	178	118	075	047	028	016	0
	1	057	108	196	232	264	316	354	380	395	399	401	400	393	381	365	356	303	244	187	136	094	1
	2	001	006	020	031	042	069	098	130	161	176	191	220	246	269	288	297	324	328	311	278	234	2
	3	0+	0+	001	002	004	008	015	024	035	041	049	064	082	101	121	132	185	235	276	303	312	3
	4	0+	0+	0+	0+	0+	001	001	002	004	005	007	011	015	021	029	033	060	095	138	186	234	4
	5	0+	0+	0+	0+	0+	0+	0+	0+	0+	0+	001	001	002	002	004	004	010	020	037	061	094	5
	6	0+	0+	0+	0+	0+	0+	0+	0+	0+	0+	0+	0+	0+	0+	0+	0+	001	002	004	008	016	6
7	0	932	868	751	698	648	558	478	409	348	321	295	249	210	176	146	133	082	049	028	015	008	0
	1	066	124	219	257	290	340	372	390	396	396	393	393	367	347	324	311	247	185	131	087	055	1
	2	002	008	027	041	055	089	124	160	194	210	225	252	275	293	307	311	318	298	261	214	164	2
	3	0+	0+	002	004	006	013	023	036	053	062	071	092	115	138	161	173	227	268	290	292	273	3
	4	0+	0+	0+	0+	0+	001	003	005	009	011	014	020	029	039	051	058	097	144	194	239	273	4
	5	0+	0+	0+	0+	0+	0+	0+	0+	001	001	002	003	004	007	010	012	025	047	077	117	164	5
	6	0+	0+	0+	0+	0+	0+	0+	0+	0+	0+	0+	0+	0+	0+	001	001	004	008	017	032	055	6
	7	0+	0+	0+	0+	0+	0+	0+	0+	0+	0+	0+	0+	0+	0+	0+	0+	001	002	004	008		7
8	0	923	851	721	663	610	513	430	360	299	272	248	204	168	137	111	100	058	032	017	008	004	0
	1	075	139	240	279	311	357	383	392	390	385	378	335	336	309	281	267	198	137	090	055	031	1
	2	003	010	035	051	070	109	149	187	222	238	252	276	294	305	311	311	296	259	209	157	109	2
	3	0+	0+	003	005	009	019	033	051	072	084	096	121	147	172	196	208	254	279	279	257	219	3
	4	0+	0+	0+	0+	001	002	005	009	015	018	023	033	046	061	077	087	136	188	232	263	273	4
	5	0+	0+	0+	0+	0+	0+	0+	001	002	003	003	006	009	014	020	023	047	081	124	172	219	5
	6	0+	0+	0+	0+	0+	0+	0+	0+	0+	0+	0+	001	001	002	003	004	010	022	041	070	109	6
	7	0+	0+	0+	0+	0+	0+	0+	0+	0+	0+	0+	0+	0+	0+	0+	0+	001	003	008	016	031	7
	8	0+	0+	0+	0+	0+	0+	0+	0+	0+	0+	0+	0+	0+	0+	0+	0+	0+	001	002	004		8
9	0	914	834	693	630	573	472	387	316	257	232	208	168	134	107	085	075	040	021	010	005	002	0
	1	083	153	260	299	329	370	387	388	377	368	357	331	302	271	240	225	156	100	060	034	018	1
	2	003	013	043	063	084	129	172	212	245	260	272	291	302	306	304	300	267	216	161	111	070	2
	3	0+	001	004	008	013	026	045	067	093	107	121	149	176	201	224	234	267	272	251	212	164	3
	4	0+	0+	0+	001	001	003	007	014	023	028	035	049	066	085	106	117	172	219	251	260	246	4
	5	0+	0+	0+	0+	0+	0+	001	002	004	005	007	011	017	024	033	039	074	118	167	213	246	5
	6	0+	0+	0+	0+	0+	0+	0+	0+	0+	001	001	002	003	005	007	009	021	042	074	116	164	6
	7	0+	0+	0+	0+	0+	0+	0+	0+	0+	0+	0+	0+	0+	001	001	001	004	010	021	041	070	7
	8	0+	0+	0+	0+	0+	0+	0+	0+	0+	0+	0+	0+	0+	0+	0+	0+	001	004	009	018		8
	9	0+	0+	0+	0+	0+	0+	0+	0+	0+	0+	0+	0+	0+	0+	0+	0+	0+	0+	001	002		9

Table D.1 (Continued)

n	r										p												r
n	r	.01	.02	.04	.05	.06	.08	.10	.12	.14	.15	.16	.18	.20	.22	.24	.25	.30	.35	.40	.45	.50	r
10	0	904	817	665	599	539	434	349	279	221	197	175	137	107	083	064	056	028	013	006	003	001	0
	1	091	167	277	315	344	378	387	380	360	347	333	302	268	235	203	188	121	072	040	021	010	1
	2	004	015	052	075	099	148	194	233	264	276	286	298	302	298	288	282	233	176	121	076	044	2
	3	0+	001	006	010	017	034	057	085	115	130	145	174	201	224	243	250	267	252	215	166	117	3
	4	0+	0+	0+	001	002	005	011	020	033	040	048	067	088	111	134	146	200	238	251	238	205	4
	5	0+	0+	0+	0+	0+	001	001	003	006	008	011	018	026	037	051	058	103	154	201	234	246	5
	6	0+	0+	0+	0+	0+	0+	0+	0+	001	001	002	003	006	009	013	016	037	069	111	160	205	6
	7	0+	0+	0+	0+	0+	0+	0+	0+	0+	0+	0+	0+	001	001	002	003	009	021	042	075	117	7
	8	0+	0+	0+	0+	0+	0+	0+	0+	0+	0+	0+	0+	0+	0+	0+	0+	001	004	011	023	044	8
	9	0+	0+	0+	0+	0+	0+	0+	0+	0+	0+	0+	0+	0+	0+	0+	0+	0+	001	002	004	010	9
	10	0+	0+	0+	0+	0+	0+	0+	0+	0+	0+	0+	0+	0+	0+	0+	0+	0+	0+	0+	0+	001	10
11	0	895	801	638	569	506	400	314	245	190	167	147	113	086	065	049	042	020	009	004	001	0+	0
	1	099	180	293	329	355	382	384	368	341	325	308	272	236	202	170	155	093	052	027	013	005	1
	2	005	018	061	087	113	166	213	251	277	287	293	299	295	284	268	258	200	140	089	051	027	2
	3	0+	001	008	014	022	043	071	103	135	152	168	197	221	241	254	258	257	225	177	126	081	3
	4	0+	0+	001	001	003	008	016	028	044	054	064	086	111	136	160	172	220	243	236	206	161	4
	5	0+	0+	0+	0+	0+	001	002	005	010	013	017	027	039	054	071	080	132	183	221	236	226	5
	6	0+	0+	0+	0+	0+	0+	0+	001	002	002	003	006	010	015	022	027	057	099	147	193	226	6
	7	0+	0+	0+	0+	0+	0+	0+	0+	0+	0+	0+	001	002	003	005	006	017	038	070	113	161	7
	8	0+	0+	0+	0+	0+	0+	0+	0+	0+	0+	0+	0+	0+	0+	001	001	004	010	023	046	081	8
	9	0+	0+	0+	0+	0+	0+	0+	0+	0+	C+	0+	0+	0+	0+	0+	0+	001	002	005	013	027	9
	10	0+	0+	0+	0+	0+	0+	0+	0+	0+	0+	0+	0+	0+	0+	0+	0+	0+	0+	001	002	005	10
	11	0+	0+	0+	0+	0+	0+	0+	0+	0+	0+	0+	0+	0+	0+	0+	0+	0+	0+	0+	0+	0+	11
12	0	886	785	613	540	476	368	282	216	164	142	123	092	069	051	037	032	014	006	002	001	0+	0
	1	107	192	306	341	365	384	377	353	320	301	282	243	206	172	141	127	071	037	017	008	003	1
	2	006	022	070	099	128	183	230	265	286	292	296	294	283	266	244	232	168	109	064	034	016	2
	3	0+	001	010	017	027	053	085	120	155	172	188	215	236	250	257	258	240	195	142	092	054	3
	4	0+	0+	001	002	004	010	021	037	057	068	080	106	133	159	183	194	231	237	213	170	121	4
	5	0+	0+	0+	0+	0+	001	004	008	015	019	025	037	053	072	092	103	158	204	227	222	193	5
	6	0+	0+	0+	0+	0+	0+	0+	001	003	004	005	010	016	024	034	040	079	128	177	212	226	6
	7	0+	0+	0+	0+	0+	0+	0+	0+	0+	001	001	002	003	006	009	011	029	059	101	149	193	7
	8	0+	0+	0+	0+	0+	0+	0+	0+	0+	0+	0+	0+	001	001	002	002	008	020	042	076	121	8
	9	0+	0+	0+	0+	0+	0+	0+	0+	0+	0+	0+	0+	0+	0+	0+	0+	001	005	012	028	054	9
	10	0+	0+	0+	0+	0+	0+	0+	0+	0+	0+	0+	0+	0+	0+	0+	0+	0+	001	002	007	016	10
	11	0+	0+	0+	0+	0+	0+	0+	0+	0+	0+	0+	0+	0+	0+	0+	0+	0+	0+	0+	001	003	11
	12	0+	0+	0+	0+	0+	0+	0+	0+	0+	0+	0+	0+	0+	0+	0+	0+	0+	0+	0+	0+	0+	12
13	0	878	769	588	513	447	338	254	190	141	121	104	076	055	040	028	024	010	004	001	0+	0+	0
	1	115	204	319	351	371	382	367	336	298	277	257	216	179	145	116	103	054	026	011	004	002	1
	2	007	025	080	111	142	199	245	275	291	294	293	285	268	245	220	206	139	084	045	022	010	2
	3	0+	002	012	021	033	064	100	138	174	190	205	229	246	254	254	252	218	165	111	066	035	3
	4	0+	0+	001	003	005	014	028	047	071	084	098	126	154	179	201	210	234	222	184	135	087	4
	5	0+	0+	0+	0+	001	002	006	012	021	027	033	050	069	091	114	126	180	215	221	199	157	5
	6	0+	0+	0+	0+	0+	0+	001	002	004	006	008	015	023	034	048	056	103	155	197	217	209	6
	7	0+	0+	0+	0+	0+	0+	0+	0+	001	001	002	003	006	010	015	019	044	083	131	177	209	7
	8	0+	0+	0+	0+	0+	0+	0+	0+	0+	0+	0+	001	001	002	004	005	014	034	066	109	157	8
	9	0+	0+	0+	0+	0+	0+	0+	0+	0+	0+	0+	0+	0+	0+	001	001	003	010	024	050	087	9
	10	0+	0+	0+	0+	0+	0+	0+	0+	0+	0+	0+	0+	0+	0+	0+	0+	001	002	006	016	035	10
	11	0+	0+	0+	0+	0+	0+	0+	0+	0+	0+	0+	0+	0+	0+	0+	0+	0+	0+	001	004	010	11
	12	0+	0+	0+	0+	0+	0+	0+	0+	0+	0+	0+	0+	0+	0+	0+	0+	0+	0+	0+	0+	002	12
	13	0+	0+	0+	0+	0+	0+	0+	0+	0+	0+	0+	0+	0+	0+	0+	0+	0+	0+	0+	0+	0+	13
14	0	869	754	565	488	421	311	229	167	121	103	087	062	044	031	021	018	007	002	001	0+	0+	0
	1	123	215	329	359	376	379	356	319	276	254	232	191	154	122	095	083	041	018	007	003	001	1
	2	008	029	089	123	156	214	257	283	292	291	287	272	250	223	195	180	113	063	032	014	006	2
	3	0+	002	015	026	040	074	114	154	190	206	219	239	250	252	246	240	194	137	085	046	022	3
	4	0+	0+	002	004	007	018	035	058	085	100	115	144	172	195	214	220	229	202	155	104	061	4
	5	0+	0+	0+	0+	001	003	008	016	028	035	044	063	086	110	135	147	196	218	207	170	122	5
	6	0+	0+	0+	0+	0+	0+	001	003	007	009	012	021	032	047	064	073	126	176	207	209	183	6
	7	0+	0+	0+	0+	0+	0+	0+	001	001	002	003	005	009	015	023	028	062	108	157	195	209	7
	8	0+	0+	0+	0+	0+	0+	0+	0+	0+	0+	0+	001	002	004	006	008	023	051	092	140	183	8
	9	0+	0+	0+	0+	0+	0+	0+	0+	0+	0+	0+	0+	0+	001	001	002	007	018	041	076	122	9
	10	0+	0+	0+	0+	0+	0+	0+	0+	0+	0+	0+	0+	0+	0+	0+	0+	001	005	014	031	061	10
	11	0+	0+	0+	0+	0+	0+	0+	0+	0+	0+	0+	0+	0+	0+	0+	0+	0+	001	003	009	022	11
	12	0+	0+	0+	0+	0+	0+	0+	0+	0+	0+	0+	0+	0+	0+	0+	0+	0+	0+	001	002	006	12
	13	0+	0+	0+	0+	0+	0+	0+	0+	0+	0+	0+	0+	0+	0+	0+	0+	0+	0+	0+	0+	001	13
	14	0+	0+	0+	0+	0+	0+	0+	0+	0+	0+	0+	0+	0+	0+	0+	0+	0+	0+	0+	0+	0+	14

Table D.1 (Continued)

n	r	.01	.02	.04	.05	.06	.08	.10	.12	.14	.15	p .16	.18	.20	.22	.24	.25	.30	.35	.40	.45	.50	r
15	0	860	739	542	463	395	286	206	147	104	087	073	051	035	024	016	013	005	002	0+	0+	0+	0
	1	130	226	339	366	378	373	343	301	254	231	209	168	132	102	077	067	031	013	005	002	0+	1
	2	009	032	099	135	169	227	267	287	290	286	279	258	231	201	171	156	092	048	022	009	003	2
	3	0+	003	018	031	047	086	129	170	204	218	230	245	250	246	234	225	170	111	063	032	014	3
	4	0+	0+	002	005	009	022	043	069	100	116	131	162	188	208	221	225	219	179	127	078	042	4
	5	0+	0+	0+	001	001	004	010	021	036	045	055	078	103	129	154	165	206	212	186	140	092	5
	6	0+	0+	0+	0+	0+	001	002	005	010	013	017	029	043	061	081	092	147	191	207	191	153	6
	7	0+	0+	0+	0+	0+	0+	0+	001	002	003	004	008	014	022	033	039	081	132	177	201	196	7
	8	0+	0+	0+	0+	0+	0+	0+	0+	0+	001	001	002	003	006	010	013	035	071	118	165	196	8
	9	0+	0+	0+	0+	0+	0+	0+	0+	0+	0+	0+	0+	001	001	003	003	012	030	061	105	153	9
	10	0+	0+	0+	0+	0+	0+	0+	0+	0+	0+	0+	0+	0+	0+	0+	001	003	010	024	051	092	10
	11	0+	0+	0+	0+	0+	0+	0+	0+	0+	0+	0+	0+	0+	0+	0+	0+	001	002	007	019	042	11
	12	0+	0+	0+	0+	0+	0+	0+	0+	0+	0+	0+	0+	0+	0+	0+	0+	0+	0+	002	005	014	12
	13	0+	0+	0+	0+	0+	0+	0+	0+	0+	0+	0+	0+	0+	0+	0+	0+	0+	0+	0+	001	003	13
	14	0+	0+	0+	0+	0+	0+	0+	0+	0+	0+	0+	0+	0+	0+	0+	0+	0+	0+	0+	0+	0+	14
	15	0+	0+	0+	0+	0+	0+	0+	0+	0+	0+	0+	0+	0+	0+	0+	0+	0+	0+	0+	0+	0+	15
16	0	851	724	520	440	372	263	185	129	090	074	061	042	028	019	012	010	003	001	0+	0+	0+	0
	1	138	236	347	371	379	366	329	282	233	210	187	147	113	085	063	053	023	009	003	001	0+	1
	2	010	036	108	146	182	239	275	289	285	277	268	242	211	179	148	134	073	035	015	006	002	2
	3	0+	003	021	036	054	097	142	184	216	229	238	248	246	236	218	208	146	089	047	022	009	3
	4	0+	0+	003	006	011	027	051	081	114	131	147	177	200	216	224	225	204	155	101	057	028	4
	5	0+	0+	0+	001	002	006	014	027	045	056	067	093	120	146	170	180	210	201	162	112	067	5
	6	0+	0+	0+	0+	0+	001	003	007	013	018	023	037	055	076	098	110	165	198	198	168	122	6
	7	0+	0+	0+	0+	0+	0+	0+	001	003	005	006	012	020	030	044	052	101	152	189	197	175	7
	8	0+	0+	0+	0+	0+	0+	0+	0+	001	001	001	003	006	010	016	020	049	092	142	181	196	8
	9	0+	0+	0+	0+	0+	0+	0+	0+	0+	0+	0+	001	001	002	004	006	019	044	084	132	175	9
	10	0+	0+	0+	0+	0+	0+	0+	0+	0+	0+	0+	0+	0+	0+	001	001	006	017	039	075	122	10
	11	0+	0+	0+	0+	0+	0+	0+	0+	0+	0+	0+	0+	0+	0+	0+	0+	001	005	014	034	067	11
	12	0+	0+	0+	0+	0+	0+	0+	0+	0+	0+	0+	0+	0+	0+	0+	0+	0+	001	004	011	028	12
	13	0+	0+	0+	0+	0+	0+	0+	0+	0+	0+	0+	0+	0+	0+	0+	0+	0+	0+	001	003	009	13
	14	0+	0+	0+	0+	0+	0+	0+	0+	0+	0+	0+	0+	0+	0+	0+	0+	0+	0+	0+	001	002	14
	15	0+	0+	0+	0+	0+	0+	0+	0+	0+	0+	0+	0+	0+	0+	0+	0+	0+	0+	0+	0+	0+	15
	16	0+	0+	0+	0+	0+	0+	0+	0+	0+	0+	0+	0+	0+	0+	0+	0+	0+	0+	0+	0+	0+	16
17	0	843	709	500	418	349	242	167	114	077	063	052	034	023	015	009	008	002	001	0+	0+	0+	0
	1	145	246	354	374	379	358	315	264	213	189	167	128	096	070	051	043	017	006	002	001	0+	1
	2	012	040	118	158	194	249	280	288	278	267	255	225	191	158	128	114	058	026	010	004	001	2
	3	001	004	025	041	062	108	156	196	226	236	243	246	239	223	202	189	125	070	034	014	005	3
	4	0+	0+	004	008	014	033	060	094	129	146	162	189	209	221	223	221	187	132	080	041	018	4
	5	0+	0+	0+	001	002	007	017	033	054	067	080	108	136	162	183	191	208	185	138	087	047	5
	6	0+	0+	0+	0+	0+	001	004	009	018	024	031	047	068	091	116	128	178	199	184	143	094	6
	7	0+	0+	0+	0+	0+	0+	001	002	005	007	009	016	027	040	057	067	120	168	193	184	148	7
	8	0+	0+	0+	0+	0+	0+	0+	0+	001	002	002	006	008	014	023	028	064	113	161	188	185	8
	9	0+	0+	0+	0+	0+	0+	0+	0+	0+	0+	0+	001	002	004	007	009	028	061	107	154	185	9
	10	0+	0+	0+	0+	0+	0+	0+	0+	0+	0+	0+	0+	0+	001	002	002	009	026	057	101	148	10
	11	0+	0+	0+	0+	0+	0+	0+	0+	0+	0+	0+	0+	0+	0+	0+	001	003	009	024	052	094	11
	12	0+	0+	0+	0+	0+	0+	0+	0+	0+	0+	0+	0+	0+	0+	0+	0+	001	002	008	021	047	12
	13	0+	0+	0+	0+	0+	0+	0+	0+	0+	0+	0+	0+	0+	0+	0+	0+	0+	001	002	007	018	13
	14	0+	0+	0+	0+	0+	0+	0+	0+	0+	0+	0+	0+	0+	0+	0+	0+	0+	0+	001	002	005	14
	15	0+	0+	0+	0+	0+	0+	0+	0+	0+	0+	0+	0+	0+	0+	0+	0+	0+	0+	0+	0+	001	15
	16	0+	0+	0+	0+	0+	0+	0+	0+	0+	0+	0+	0+	0+	0+	0+	0+	0+	0+	0+	0+	0+	16
	17	0+	0+	0+	0+	0+	0+	0+	0+	0+	0+	0+	0+	0+	0+	0+	0+	0+	0+	0+	0+	0+	17
18	0	835	695	480	397	328	223	150	100	066	054	043	028	018	011	007	006	002	0+	0+	0+	0+	0
	1	152	255	360	376	377	349	300	246	194	170	149	111	081	058	041	034	013	004	001	0+	0+	1
	2	013	044	127	168	205	258	284	285	268	256	241	207	172	139	109	096	046	019	007	002	001	2
	3	001	005	028	047	070	120	168	207	233	241	244	243	230	209	184	170	105	055	025	009	003	3
	4	0+	0+	004	009	017	039	070	106	142	159	175	200	215	221	218	213	168	110	061	029	012	4
	5	0+	0+	001	001	003	009	022	040	065	079	093	123	151	175	193	199	202	166	115	067	033	5
	6	0+	0+	0+	0+	0+	002	005	012	023	030	038	058	082	107	132	144	187	194	166	118	071	6
	7	0+	0+	0+	0+	0+	0+	001	003	006	009	013	022	035	052	071	082	138	179	189	166	121	7
	8	0+	0+	0+	0+	0+	0+	0+	001	001	002	003	007	012	020	031	038	081	133	173	186	167	8
	9	0+	0+	0+	0+	0+	0+	0+	0+	0+	001	001	002	003	006	011	014	039	079	128	169	185	9
	10	0+	0+	0+	0+	0+	0+	0+	0+	0+	0+	0+	0+	001	002	003	004	015	038	077	125	167	10
	11	0+	0+	0+	0+	0+	0+	0+	0+	0+	0+	0+	0+	0+	0+	0+	001	005	015	037	074	121	11
	12	0+	0+	0+	0+	0+	0+	0+	0+	0+	0+	0+	0+	0+	0+	0+	0+	001	005	015	035	071	12
	13	0+	0+	0+	0+	0+	0+	0+	0+	0+	0+	0+	0+	0+	0+	0+	0+	0+	001	004	013	033	13
	14	0+	0+	0+	0+	0+	0+	0+	0+	0+	0+	0+	0+	0+	0+	0+	0+	0+	0+	001	004	012	14

Table D.1 (Continued)

n	r	.01	.02	.04	.05	.06	.08	.10	.12	.14	.15	.16	.18	.20	.22	.24	.25	.30	.35	.40	.45	.50	r
	15	0+	0+	0+	0+	0+	0+	0+	0+	0+	0+	0+	0+	0+	0+	0+	0+	0+	0+	0+	001	003	15
	16	0+	0+	0+	0+	0+	0+	0+	0+	0+	0+	0+	0+	0+	0+	0+	0+	0+	0+	0+	0+	001	16
	17	0+	0+	0+	0+	0+	0+	0+	0+	0+	0+	0+	0+	0+	0+	0+	0+	0+	0+	0+	0+	0+	17
	18	0+	0+	0+	0+	0+	0+	0+	0+	0+	0+	0+	0+	0+	0+	0+	0+	0+	0+	0+	0+	0+	18
19	0	826	681	460	377	309	205	135	088	057	046	036	023	014	009	005	004	001	0+	0+	0+	0+	0
	1	159	264	364	377	374	339	285	228	176	153	132	096	068	048	033	027	009	003	001	0+	0+	1
	2	014	049	137	179	215	265	285	280	258	243	226	190	154	121	093	080	036	014	005	001	0+	2
	3	001	006	032	053	078	131	180	217	238	243	244	236	218	194	166	152	087	042	017	006	002	3
	4	0+	0+	005	011	020	045	080	118	155	171	186	207	218	219	210	202	149	091	047	020	007	4
	5	0+	0+	001	002	004	012	027	048	076	091	106	137	164	185	199	202	192	147	093	050	022	5
	6	0+	0+	0+	0+	001	002	007	015	029	037	047	070	095	122	146	157	192	184	145	095	052	6
	7	0+	0+	0+	0+	0+	0+	001	004	009	012	017	029	044	064	086	097	153	184	180	144	096	7
	8	0+	0+	0+	0+	0+	0+	0+	001	002	003	005	009	017	027	041	049	098	149	180	177	144	8
	9	0+	0+	0+	0+	0+	0+	0+	0+	0+	001	001	003	005	009	016	020	051	098	146	177	176	9
	10	0+	0+	0+	0+	0+	0+	0+	0+	0+	0+	0+	001	001	003	005	007	022	053	098	145	176	10
	11	0+	0+	0+	0+	0+	0+	0+	0+	0+	0+	0+	0+	0+	001	001	002	008	023	053	097	144	11
	12	0+	0+	0+	0+	0+	0+	0+	0+	0+	0+	0+	0+	0+	0+	0+	0+	002	008	024	053	096	12
	13	0+	0+	0+	0+	0+	0+	0+	0+	0+	0+	0+	0+	0+	0+	0+	0+	001	002	008	023	052	13
	14	0+	0+	0+	0+	0+	0+	0+	0+	0+	0+	0+	0+	0+	0+	0+	0+	0+	001	002	008	022	14
	15	0+	0+	0+	0+	0+	0+	0+	0+	0+	0+	0+	0+	0+	0+	0+	0+	0+	0+	001	002	007	15
	16	0+	0+	0+	0+	0+	0+	0+	0+	0+	0+	0+	0+	0+	0+	0+	0+	0+	0+	0+	0+	002	16
	17	0+	0+	0+	0+	0+	0+	0+	0+	0+	0+	0+	0+	0+	0+	0+	0+	0+	0+	0+	0+	0+	17
	18	0+	0+	0+	0+	0+	0+	0+	0+	0+	0+	0+	0+	0+	0+	0+	0+	0+	0+	0+	0+	0+	18
	19	0+	0+	0+	0+	0+	0+	0+	0+	0+	0+	0+	0+	0+	0+	0+	0+	0+	0+	0+	0+	0+	19
20	0	818	668	442	358	290	189	122	078	049	039	031	019	012	007	004	003	001	0+	0+	0+	0+	0
	1	165	272	368	377	370	328	270	212	159	137	117	083	058	039	026	021	007	002	0+	0+	0+	1
	2	016	053	146	189	225	271	285	274	247	229	211	173	137	105	078	067	028	010	003	001	0+	2
	3	001	006	036	060	086	141	190	224	241	243	241	228	205	178	148	134	072	032	012	004	001	3
	4	0+	001	006	013	023	052	090	130	167	182	195	213	218	213	199	190	130	074	035	014	005	4
	5	0+	0+	001	002	005	015	032	057	087	103	119	149	175	192	201	202	179	127	075	036	015	5
	6	0+	0+	0+	0+	001	003	009	019	035	045	057	082	109	136	159	169	192	171	124	075	037	6
	7	0+	0+	0+	0+	0+	001	002	005	012	016	022	036	055	076	100	112	164	184	166	122	074	7
	8	0+	0+	0+	0+	0+	0+	0+	001	003	005	007	013	022	035	051	061	114	161	180	162	120	8
	9	0+	0+	0+	0+	0+	0+	0+	0+	001	001	002	004	007	013	022	027	065	116	160	177	160	9
	10	0+	0+	0+	0+	0+	0+	0+	0+	0+	0+	0+	001	002	004	008	010	031	069	117	159	176	10
	11	0+	0+	0+	0+	0+	0+	0+	0+	0+	0+	0+	0+	0+	001	002	003	012	034	071	119	160	11
	12	0+	0+	0+	0+	0+	0+	0+	0+	0+	0+	0+	0+	0+	0+	001	001	004	014	035	073	120	12
	13	0+	0+	0+	0+	0+	0+	0+	0+	0+	0+	0+	0+	0+	0+	0+	0+	001	004	015	037	074	13
	14	0+	0+	0+	0+	0+	0+	0+	0+	0+	0+	0+	0+	0+	0+	0+	0+	0+	001	005	015	037	14
	15	0+	0+	0+	0+	0+	0+	0+	0+	0+	0+	0+	0+	0+	0+	0+	0+	0+	0+	001	005	015	15
	16	0+	0+	0+	0+	0+	0+	0+	0+	0+	0+	0+	0+	0+	0+	0+	0+	0+	0+	0+	001	005	16
	17	0+	0+	0+	0+	0+	0+	0+	0+	0+	0+	0+	0+	0+	0+	0+	0+	0+	0+	0+	0+	001	17
	18	0+	0+	0+	0+	0+	0+	0+	0+	0+	0+	0+	0+	0+	0+	0+	0+	0+	0+	0+	0+	0+	18
	19	0+	0+	0+	0+	0+	0+	0+	0+	0+	0+	0+	0+	0+	0+	0+	0+	0+	0+	0+	0+	0+	19
	20	0+	0+	0+	0+	0+	0+	0+	0+	0+	0+	0+	0+	0+	0+	0+	0+	0+	0+	0+	0+	0+	20
21	0	810	654	424	341	273	174	109	068	042	033	026	015	010	005	003	002	001	0+	0+	0+	0+	0
	1	172	280	371	376	366	317	255	195	144	122	103	071	048	032	021	017	005	001	0+	0+	0+	1
	2	017	057	155	198	233	276	284	267	234	215	196	157	121	091	066	055	022	007	002	0+	0+	2
	3	001	007	041	066	094	152	200	230	242	241	236	218	192	162	132	117	058	024	009	003	001	3
	4	0+	001	008	016	027	059	100	141	177	191	202	215	216	205	187	176	113	059	026	009	003	4
	5	0+	0+	001	003	006	018	038	065	098	115	131	161	183	197	201	199	164	109	059	026	010	5
	6	0+	0+	0+	0+	001	004	011	024	043	054	067	094	122	148	169	177	188	156	105	057	026	6
	7	0+	0+	0+	0+	0+	001	003	007	015	020	027	044	065	089	114	126	172	180	149	101	055	7
	8	0+	0+	0+	0+	0+	0+	001	002	004	006	009	017	029	044	063	074	129	169	174	144	097	8
	9	0+	0+	0+	0+	0+	0+	0+	0+	001	002	002	005	010	018	029	036	080	132	168	170	140	9
	10	0+	0+	0+	0+	0+	0+	0+	0+	0+	0+	001	001	003	006	011	014	041	085	134	167	168	10
	11	0+	0+	0+	0+	0+	0+	0+	0+	0+	0+	0+	0+	001	002	003	005	018	046	089	137	168	11
	12	0+	0+	0+	0+	0+	0+	0+	0+	0+	0+	0+	0+	0+	0+	001	001	006	021	050	093	140	12
	13	0+	0+	0+	0+	0+	0+	0+	0+	0+	0+	0+	0+	0+	0+	0+	0+	002	008	023	053	097	13
	14	0+	0+	0+	0+	0+	0+	0+	0+	0+	0+	0+	0+	0+	0+	0+	0+	0+	002	009	025	055	14
	15	0+	0+	0+	0+	0+	0+	0+	0+	0+	0+	0+	0+	0+	0+	0+	0+	0+	001	003	009	026	15
	16	0+	0+	0+	0+	0+	0+	0+	0+	0+	0+	0+	0+	0+	0+	0+	0+	0+	0+	001	003	010	16
	17	0+	0+	0+	0+	0+	0+	0+	0+	0+	0+	0+	0+	0+	0+	0+	0+	0+	0+	0+	001	003	17
	18	0+	0+	0+	0+	0+	0+	0+	0+	0+	0+	0+	0+	0+	0+	0+	0+	0+	0+	0+	0+	001	18
	19	0+	0+	0+	0+	0+	0+	0+	0+	0+	0+	0+	0+	0+	0+	0+	0+	0+	0+	0+	0+	0+	19
	20	0+	0+	0+	0+	0+	0+	0+	0+	0+	0+	0+	0+	0+	0+	0+	0+	0+	0+	0+	0+	0+	20
	21	0+	0+	0+	0+	0+	0+	0+	0+	0+	0+	0+	0+	0+	0+	0+	0+	0+	0+	0+	0+	0+	21

Table D.1 (Continued)

n	r	.01	.02	.04	.05	.06	.08	.10	.12	.14	.15	P .16	.18	.20	.22	.24	.25	.30	.35	.40	.45	.50	r
22	0	802	641	407	324	256	160	098	060	036	028	022	013	007	004	002	002	0+	0+	0+	0+	0+	0
	1	178	288	373	375	360	306	241	180	130	109	090	061	041	026	017	013	004	001	0+	0+	0+	1
	2	019	062	163	207	241	279	281	258	222	201	181	141	107	078	055	046	017	005	001	0+	0+	2
	3	001	008	045	073	103	162	208	235	241	237	230	207	178	146	116	102	047	018	006	002	0+	3
	4	0+	001	009	018	031	067	110	152	186	199	208	216	211	196	174	161	096	047	019	006	002	4
	5	0+	0+	001	003	007	021	044	075	109	126	143	170	190	199	197	193	149	091	046	019	006	5
	6	0+	0+	0+	001	001	005	014	029	050	063	077	106	134	159	177	183	181	139	086	043	018	6
	7	0+	0+	0+	0+	0+	001	004	009	019	025	033	053	077	102	128	139	177	171	131	081	041	7
	8	0+	0+	0+	0+	0+	0+	001	002	006	008	012	022	036	054	075	087	142	173	164	125	076	8
	9	0+	0+	0+	0+	0+	0+	0+	0+	001	002	004	007	014	024	037	045	095	145	170	164	119	9
	10	0+	0+	0+	0+	0+	0+	0+	0+	0+	001	001	002	005	009	015	020	053	101	148	169	154	10
	11	0+	0+	0+	0+	0+	0+	0+	0+	0+	0+	0+	001	001	003	005	007	025	060	107	151	168	11
	12	0+	0+	0+	0+	0+	0+	0+	0+	0+	0+	0+	0+	0+	001	002	002	010	029	066	113	154	12
	13	0+	0+	0+	0+	0+	0+	0+	0+	0+	0+	0+	0+	0+	0+	001	001	003	012	034	071	119	13
	14	0+	0+	0+	0+	0+	0+	0+	0+	0+	0+	0+	0+	0+	0+	0+	0+	001	004	014	037	076	14
	15	0+	0+	0+	0+	0+	0+	0+	0+	0+	0+	0+	0+	0+	0+	0+	0+	001	001	016	041		15
	16	0+	0+	0+	0+	0+	0+	0+	0+	0+	0+	0+	0+	0+	0+	0+	0+	0+	001	006	018		16
	17	0+	0+	0+	0+	0+	0+	0+	0+	0+	0+	0+	0+	0+	0+	0+	0+	0+	0+	002	007		17
	18	0+	0+	0+	0+	0+	0+	0+	0+	0+	0+	0+	0+	0+	0+	0+	0+	0+	0+	0+	002		18
	19	0+	0+	0+	0+	0+	0+	0+	0+	0+	0+	0+	0+	0+	0+	0+	0+	0+	0+	0+	0+		19
	20	0+	0+	0+	0+	0+	0+	0+	0+	0+	0+	0+	0+	0+	0+	0+	0+	0+	0+	0+	0+		20
	21	0+	0+	0+	0+	0+	0+	0+	0+	0+	0+	0+	0+	0+	0+	0+	0+	0+	0+	0+	0+		21
	22	0+	0+	0+	0+	0+	0+	0+	0+	0+	0+	0+	0+	0+	0+	0+	0+	0+	0+	0+	0+		22
23	0	794	628	391	307	241	147	089	053	031	024	018	010	006	003	002	001	0+	0+	0+	0+	0+	0
	1	184	295	375	372	354	294	226	166	117	097	071	053	034	021	013	010	003	001	0+	0+	0+	1
	2	020	066	172	215	248	281	277	247	209	188	166	127	093	066	046	038	013	004	001	0+	0+	2
	3	001	009	050	079	111	171	215	237	238	232	222	195	163	131	101	088	038	014	004	001	0+	3
	4	0+	001	010	021	035	074	120	162	194	204	211	214	204	185	160	146	082	037	014	004	001	4
	5	0+	0+	002	004	009	025	051	084	120	137	153	179	194	198	192	185	133	076	035	013	004	5
	6	0+	0+	0+	001	002	006	017	034	059	073	087	118	145	168	182	185	171	122	070	032	012	6
	7	0+	0+	0+	0+	0+	001	005	011	023	031	040	063	088	115	139	150	178	160	113	064	029	7
	8	0+	0+	0+	0+	0+	0+	001	003	008	011	015	028	044	065	088	100	153	172	151	105	058	8
	9	0+	0+	0+	0+	0+	0+	0+	001	002	003	005	010	018	030	046	056	109	155	168	143	097	9
	10	0+	0+	0+	0+	0+	0+	0+	0+	0+	001	001	003	006	012	020	026	065	117	157	164	136	10
	11	0+	0+	0+	0+	0+	0+	0+	0+	0+	0+	0+	001	002	004	008	010	033	074	123	159	161	11
	12	0+	0+	0+	0+	0+	0+	0+	0+	0+	0+	0+	0+	0+	001	002	003	014	040	082	130	161	12
	13	0+	0+	0+	0+	0+	0+	0+	0+	0+	0+	0+	0+	0+	0+	001	001	005	018	046	090	136	13
	14	0+	0+	0+	0+	0+	0+	0+	0+	0+	0+	0+	0+	0+	0+	0+	0+	002	007	022	053	097	14
	15	0+	0+	0+	0+	0+	0+	0+	0+	0+	0+	0+	0+	0+	0+	0+	0+	002	009	026	058		15
	16	0+	0+	0+	0+	0+	0+	0+	0+	0+	0+	0+	0+	0+	0+	0+	0+	001	003	011	029		16
	17	0+	0+	0+	0+	0+	0+	0+	0+	0+	0+	0+	0+	0+	0+	0+	0+	0+	001	004	012		17
	18	0+	0+	0+	0+	0+	0+	0+	0+	0+	0+	0+	0+	0+	0+	0+	0+	0+	0+	001	004		18
	19	0+	0+	0+	0+	0+	0+	0+	0+	0+	0+	0+	0+	0+	0+	0+	0+	0+	0+	0+	001		19
	20	0+	0+	0+	0+	0+	0+	0+	0+	0+	0+	0+	0+	0+	0+	0+	0+	0+	0+	0+	0+		20
	21	0+	0+	0+	0+	0+	0+	0+	0+	0+	0+	0+	0+	0+	0+	0+	0+	0+	0+	0+	0+		21
	22	0+	0+	0+	0+	0+	0+	0+	0+	0+	0+	0+	0+	0+	0+	0+	0+	0+	0+	0+	0+		22
	23	0+	0+	0+	0+	0+	0+	0+	0+	0+	0+	0+	0+	0+	0+	0+	0+	0+	0+	0+	0+		23
24	0	786	616	375	292	227	135	080	047	027	020	015	009	005	003	001	001	0+	0+	0+	0+	0+	0
	1	190	302	375	369	347	282	213	152	105	086	070	045	028	017	010	008	002	0+	0+	0+	0+	1
	2	022	071	180	223	255	282	272	239	196	174	153	114	081	056	038	031	010	003	001	0+	0+	2
	3	002	011	055	086	119	180	221	238	234	225	213	183	149	117	088	075	031	010	003	001	0+	3
	4	0+	001	012	024	040	082	129	171	200	207	213	211	196	173	146	132	069	029	010	003	001	4
	5	0+	0+	002	005	010	029	057	093	130	147	162	185	196	195	184	176	118	062	027	009	003	5
	6	0+	0+	0+	001	002	008	020	040	067	082	098	129	155	174	184	185	160	106	056	024	008	6
	7	0+	0+	0+	0+	0+	002	006	014	028	037	048	073	100	126	149	159	176	147	096	050	021	7
	8	0+	0+	0+	0+	0+	0+	001	004	010	014	019	034	053	076	100	112	160	168	136	087	044	8
	9	0+	0+	0+	0+	0+	0+	0+	001	003	004	007	013	024	038	056	067	122	161	161	126	078	9
	10	0+	0+	0+	0+	0+	0+	0+	0+	001	001	002	004	009	016	027	033	079	130	161	155	117	10
	11	0+	0+	0+	0+	0+	0+	0+	0+	0+	0+	001	003	006	011	014	043	089	137	161	149		11
	12	0+	0+	0+	0+	0+	0+	0+	0+	0+	0+	0+	0+	001	002	004	005	020	052	099	143	161	12
	13	0+	0+	0+	0+	0+	0+	0+	0+	0+	0+	0+	0+	0+	0+	001	002	008	026	061	108	149	13
	14	0+	0+	0+	0+	0+	0+	0+	0+	0+	0+	0+	0+	0+	0+	0+	0+	003	011	032	069	117	14

Table D.1 (Continued)

n	r	.01	.02	.04	.05	.06	.08	.10	.12	.14	.15	P .16	.18	.20	.22	.24	.25	.30	.35	.40	.45	.50
24	15	0+	0+	0+	0+	0+	0+	0+	0+	0+	0+	0+	0+	0+	0+	0+	0+	001	004	014	038	078
	16	0+	0+	0+	0+	0+	0+	0+	0+	0+	0+	0+	0+	0+	0+	0+	0+	0+	001	005	017	044
	17	0+	0+	0+	0+	0+	0+	0+	0+	0+	0+	0+	0+	0+	0+	0+	0+	0+	0+	002	007	021
	18	0+	0+	0+	0+	0+	0+	0+	0+	0+	0+	0+	0+	0+	0+	0+	0+	0+	0+	0+	002	008
	19	0+	0+	0+	0+	0+	0+	0+	0+	0+	0+	0+	0+	0+	0+	0+	0+	0+	0+	0+	001	003
	20	0+	0+	0+	0+	0+	0+	0+	0+	0+	0+	0+	0+	0+	0+	0+	0+	0+	0+	0+	0+	001
	21	0+	0+	0+	0+	0+	0+	0+	0+	0+	0+	0+	0+	0+	0+	0+	0+	0+	0+	0+	0+	0+
	22	0+	0+	0+	0+	0+	0+	0+	0+	0+	0+	0+	0+	0+	0+	0+	0+	0+	0+	0+	0+	0+
	23	0+	0+	0+	0+	0+	0+	0+	0+	0+	0+	0+	0+	0+	0+	0+	0+	0+	0+	0+	0+	0+
	24	0+	0+	0+	0+	0+	0+	0+	0+	0+	0+	0+	0+	0+	0+	0+	0+	0+	0+	0+	0+	0+
25	0	778	603	360	277	213	124	072	041	023	017	013	007	004	002	001	001	0+	0+	0+	0+	0+
	1	196	308	375	365	340	270	199	140	094	076	061	038	024	014	008	006	001	0+	0+	0+	0+
	2	024	075	188	231	260	282	266	228	183	161	139	101	071	048	031	025	007	002	0+	0+	0+
	3	002	012	060	093	127	188	226	239	229	217	203	170	136	104	076	064	024	008	002	0+	0+
	4	0+	001	014	027	045	090	138	179	205	211	213	206	187	161	132	118	057	022	007	002	0+
	5	0+	0+	002	006	012	033	065	103	140	156	170	190	196	190	175	165	103	051	020	006	002
	6	0+	0+	0+	001	003	010	024	047	076	092	108	139	163	179	184	183	147	091	044	017	005
	7	0+	0+	0+	0+	0+	002	007	017	034	044	056	083	111	137	158	165	171	133	080	038	014
	8	0+	0+	0+	0+	0+	0+	002	005	012	017	024	041	062	087	112	124	165	161	120	070	032
	9	0+	0+	0+	0+	0+	0+	0+	001	004	006	009	017	029	046	067	078	134	163	151	108	061
	10	0+	0+	0+	0+	0+	0+	0+	0+	001	002	003	006	012	021	034	042	092	141	161	142	097
	11	0+	0+	0+	0+	0+	0+	0+	0+	0+	0+	001	002	004	008	015	019	054	103	147	158	133
	12	0+	0+	0+	0+	0+	0+	0+	0+	0+	0+	0+	0+	001	003	005	007	027	065	114	151	155
	13	0+	0+	0+	0+	0+	0+	0+	0+	0+	0+	0+	0+	0+	001	002	002	011	035	076	124	155
	14	0+	0+	0+	0+	0+	0+	0+	0+	0+	0+	0+	0+	0+	0+	0+	001	004	016	043	087	133
	15	0+	0+	0+	0+	0+	0+	0+	0+	0+	0+	0+	0+	0+	0+	0+	0+	001	006	021	052	097
	16	0+	0+	0+	0+	0+	0+	0+	0+	0+	0+	0+	0+	0+	0+	0+	0+	0+	002	009	027	061
	17	0+	0+	0+	0+	0+	0+	0+	0+	0+	0+	0+	0+	0+	0+	0+	0+	0+	001	003	012	032
	18	0+	0+	0+	0+	0+	0+	0+	0+	0+	0+	0+	0+	0+	0+	0+	0+	0+	0+	001	004	014
	19	0+	0+	0+	0+	0+	0+	0+	0+	0+	0+	0+	0+	0+	0+	0+	0+	0+	0+	0+	001	005
	20	0+	0+	0+	0+	0+	0+	0+	0+	0+	0+	0+	0+	0+	0+	0+	0+	0+	0+	0+	0+	002
	21	0+	0+	0+	0+	0+	0+	0+	0+	0+	0+	0+	0+	0+	0+	0+	0+	0+	0+	0+	0+	0+
	22	0+	0+	0+	0+	0+	0+	0+	0+	0+	0+	0+	0+	0+	0+	0+	0+	0+	0+	0+	0+	0+
	23	0+	0+	0+	0+	0+	0+	0+	0+	0+	0+	0+	0+	0+	0+	0+	0+	0+	0+	0+	0+	0+
	24	0+	0+	0+	0+	0+	0+	0+	0+	0+	0+	0+	0+	0+	0+	0+	0+	0+	0+	0+	0+	0+
	25	0+	0+	0+	0+	0+	0+	0+	0+	0+	0+	0+	0+	0+	0+	0+	0+	0+	0+	0+	0+	0+

Source: Abridged from "Tables of the Binomial Probability Distribution," in *Applied Mathematics Series No. 6*, National Bureau Standards (Washington D.C.: U.S. Government Printing Office, 1949).

D.2 / Binomial Distribution—Cumulative Terms

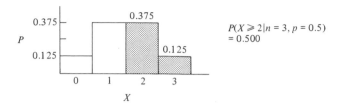

$$P(X \geqslant 2 | n = 3, p = 0.5)$$
$$= 0.500$$

		.01	.02	.04	.05	.06	.08	.10	.12	.14	.15	^{P}.16	.18	.20	.22	.24	.25	.30	.35	.40	.45	.50	
n	r																						r
2	0	1	1	1	1	1	1	1	1	1	1	1	1	1	1	1	1	1	1	1	1	1	0
	1	020	040	078	098	116	154	190	226	260	278	294	328	360	392	422	438	510	578	640	698	750	1
	2	0+	0+	002	002	004	006	010	014	020	022	026	032	040	048	058	062	090	122	160	202	250	2
3	0	1	1	1	1	1	1	1	1	1	1	1	1	1	1	1	1	1	1	1	1	1	0
	1	030	059	115	143	169	221	271	319	364	386	407	449	488	525	561	578	657	725	784	834	875	1
	2	0+	001	005	007	010	018	028	040	053	061	069	086	104	124	145	156	216	282	352	425	500	2
	3	0+	0+	0+	0+	0+	001	001	002	003	003	004	006	008	011	014	016	027	043	064	091	125	3
4	0	1	1	1	1	1	1	1	1	1	1	1	1	1	1	1	1	1	1	1	1	1	0
	1	039	078	151	185	219	284	344	400	453	478	502	548	590	630	666	684	760	821	870	908	938	1
	2	001	002	009	014	020	034	052	073	097	110	123	151	181	212	245	262	348	437	525	609	688	2
	3	0+	0+	0+	0+	001	002	004	006	010	012	014	020	027	036	045	051	084	126	179	241	312	3
	4	0+	0+	0+	0+	0+	0+	0+	0+	0+	001	001	001	002	002	003	004	008	015	026	041	062	4
5	0	1	1	1	1	1	1	1	1	1	1	1	1	1	1	1	1	1	1	1	1	1	0
	1	049	096	185	226	266	341	410	472	530	556	582	629	672	711	746	763	832	884	922	950	969	1
	2	001	004	015	023	032	054	081	112	147	165	183	222	263	304	346	367	472	572	663	744	812	2
	3	0+	0+	001	001	002	005	009	014	022	027	032	044	058	074	093	104	163	235	317	407	500	3
	4	0+	0+	0+	0+	0+	0+	0+	001	002	002	003	004	007	010	013	016	031	054	087	131	188	4
	5	0+	0+	0+	0+	0+	0+	0+	0+	0+	0+	0+	0+	0+	001	001	001	002	005	010	018	031	5
6	0	1	1	1	1	1	1	1	1	1	1	1	1	1	1	1	1	1	1	1	1	1	0
	1	059	114	217	265	310	394	469	536	595	623	649	696	738	775	807	822	882	925	953	972	984	1
	2	001	006	022	033	046	077	114	156	200	224	247	296	345	394	442	466	580	681	767	836	891	2
	3	0+	0+	001	002	004	009	016	026	039	047	056	076	099	125	154	169	256	353	456	558	656	3
	4	0+	0+	0+	0+	0+	001	001	003	005	006	007	012	017	024	033	038	070	117	179	255	344	4
	5	0+	0+	0+	0+	0+	0+	0+	0+	0+	0+	001	001	002	003	004	005	011	022	041	069	109	5
	6	0+	0+	0+	0+	0+	0+	0+	0+	0+	0+	0+	0+	0+	0+	0+	0+	001	002	004	008	016	6
7	0	1	1	1	1	1	1	1	1	1	1	1	1	1	1	1	1	1	1	1	1	1	0
	1	068	132	249	302	352	442	522	591	652	679	705	751	790	824	854	867	918	951	972	985	992	1
	2	002	008	029	044	062	103	150	201	256	283	311	368	423	478	530	555	671	766	841	898	938	2
	3	0+	0+	002	004	006	014	026	042	062	074	087	115	148	184	223	244	353	468	580	684	773	3
	4	0+	0+	0+	0+	0+	001	003	005	009	012	015	023	033	046	062	071	126	200	290	392	500	4
	5	0+	0+	0+	0+	0+	0+	0+	0+	001	001	002	003	005	007	011	013	029	056	096	153	227	5
	6	0+	0+	0+	0+	0+	0+	0+	0+	0+	0+	0+	0+	0+	001	001	001	004	009	019	036	062	6
	7	0+	0+	0+	0+	0+	0+	0+	0+	0+	0+	0+	0+	0+	0+	0+	0+	0+	001	002	004	008	7
8	0	1	1	1	1	1	1	1	1	1	1	1	1	1	1	1	1	1	1	1	1	1	0
	1	077	149	279	337	390	487	570	640	701	728	752	796	832	863	889	900	942	968	983	992	996	1
	2	003	010	038	057	079	130	187	248	311	343	374	437	497	554	608	633	745	831	894	937	965	2
	3	0+	0+	003	006	010	021	038	061	089	105	123	161	203	249	297	321	448	572	685	780	855	3
	4	0+	0+	0+	001	002	005	010	017	021	027	040	056	076	100	114	194	294	406	523	637	4
	5	0+	0+	0+	0+	0+	0+	001	002	003	004	007	010	016	023	027	058	106	174	260	363	5
	6	0+	0+	0+	0+	0+	0+	0+	0+	0+	0+	001	001	002	003	004	011	025	050	088	145	6
	7	0+	0+	0+	0+	0+	0+	0+	0+	0+	0+	0+	0+	0+	0+	0+	001	004	009	018	035	7
	8	0+	0+	0+	0+	0+	0+	0+	0+	0+	0+	0+	0+	0+	0+	0+	0+	0+	001	002	004	8
9	0	1	1	1	1	1	1	1	1	1	1	1	1	1	1	1	1	1	1	1	1	1	0
	1	086	166	307	370	427	528	613	684	743	768	792	832	866	893	915	925	960	979	990	995	998	1
	2	003	013	048	071	098	158	225	295	366	401	435	501	564	622	675	700	804	879	929	961	980	2
	3	0+	001	004	008	014	030	053	083	120	141	163	210	262	316	371	399	537	663	768	850	910	3
	4	0+	0+	0+	001	001	004	008	016	027	034	042	062	086	114	148	166	270	391	517	639	746	4
	5	0+	0+	0+	0+	0+	0+	001	002	004	006	007	012	020	029	042	049	099	172	267	379	500	5
	6	0+	0+	0+	0+	0+	0+	0+	0+	0+	001	001	002	003	005	008	010	025	054	099	166	254	6
	7	0+	0+	0+	0+	0+	0+	0+	0+	0+	0+	0+	0+	001	001	001	004	011	025	050	090	7
	8	0+	0+	0+	0+	0+	0+	0+	0+	0+	0+	0+	0+	0+	0+	0+	001	004	009	020	8
	9	0+	0+	0+	0+	0+	0+	0+	0+	0+	0+	0+	0+	0+	0+	0+	0+	0+	001	002	9

Table D.2 (Continued)

n	r	.01	.02	.04	.05	.06	.08	.10	.12	.14	.15	.16	.18	.20	.22	.24	.25	.30	.35	.40	.45	.50	r
10	0	1	1	1	1	1	1	1	1	1	1	1	1	1	1	1	1	1	1	1	1	1	0
	1	096	183	335	401	461	566	651	721	779	803	825	863	893	917	936	944	972	987	994	997	999	1
	2	004	016	058	086	118	188	264	342	418	456	492	561	624	682	733	756	851	914	954	977	989	2
	3	0+	001	006	012	019	040	070	109	155	180	206	263	322	383	444	474	617	738	833	900	945	3
	4	0+	0+	0+	001	002	006	013	024	040	050	061	088	121	159	201	224	350	486	618	734	828	4
	5	0+	0+	0+	0+	0+	001	002	004	007	010	013	021	033	048	067	078	150	249	367	496	623	5
	6	0+	0+	0+	0+	0+	0+	0+	0+	001	001	002	004	006	010	016	020	047	095	166	262	377	6
	7	0+	0+	0+	0+	0+	0+	0+	0+	0+	0+	0+	001	002	003	004	011	026	055	102	172	7
	8	0+	0+	0+	0+	0+	0+	0+	0+	0+	0+	0+	0+	0+	0+	0+	0+	002	005	012	027	055	8
	9	0+	0+	0+	0+	0+	0+	0+	0+	0+	0+	0+	0+	0+	0+	0+	0+	0+	001	002	005	011	9
	10	0+	0+	0+	0+	0+	0+	0+	0+	0+	0+	0+	0+	0+	0+	0+	0+	0+	0+	0+	0+	001	10
11	0	1	1	1	1	1	1	1	1	1	1	1	1	1	1	1	1	1	1	1	1	1	0
	1	105	199	362	431	494	600	686	755	810	833	853	887	914	935	951	958	980	991	996	999	1-	1
	2	005	020	069	102	138	218	303	387	469	508	545	615	678	733	781	803	887	939	970	986	994	2
	3	0+	001	008	015	025	052	090	137	191	221	252	316	383	449	513	545	687	800	881	935	967	3
	4	0+	0+	001	002	003	009	019	034	056	069	085	120	161	208	260	287	430	574	704	809	887	4
	5	0+	0+	0+	0+	0+	001	003	006	012	016	021	033	050	072	099	115	210	332	467	603	726	5
	6	0+	0+	0+	0+	0+	0+	001	001	002	003	004	007	012	019	028	034	078	149	247	367	500	6
	7	0+	0+	0+	0+	0+	0+	0+	0+	0+	0+	0+	001	002	004	006	008	022	050	099	174	274	7
	8	0+	0+	0+	0+	0+	0+	0+	0+	0+	0+	0+	0+	0+	0+	001	002	004	012	029	061	113	8
	9	0+	0+	0+	0+	0+	0+	0+	0+	0+	0+	0+	0+	0+	0+	0+	0+	001	002	006	015	033	9
	10	0+	0+	0+	0+	0+	0+	0+	0+	0+	0+	0+	0+	0+	0+	0+	0+	0+	0+	001	002	006	10
	11	0+	0+	0+	0+	0+	0+	0+	0+	0+	0+	0+	0+	0+	0+	0+	0+	0+	0+	0+	0+	0+	11
12	0	1	1	1	1	1	1	1	1	1	1	1	1	1	1	1	1	1	1	1	1	1	0
	1	114	215	387	460	524	632	718	784	836	858	877	908	931	949	963	968	986	994	998	999	1-	1
	2	006	023	081	118	160	249	341	431	517	557	595	664	725	778	822	842	915	958	980	992	997	2
	3	0+	002	011	020	032	065	111	167	230	264	299	370	442	511	578	609	747	849	917	958	981	3
	4	0+	0+	001	002	004	012	026	046	075	092	111	155	205	261	320	351	507	653	775	866	927	4
	5	0+	0+	0+	0+	0+	002	004	009	018	024	031	049	073	102	138	158	276	417	562	696	806	5
	6	0+	0+	0+	0+	0+	0+	001	001	003	005	006	012	019	030	045	054	118	213	335	473	613	6
	7	0+	0+	0+	0+	0+	0+	0+	0+	0+	001	001	002	004	007	011	014	039	085	158	261	387	7
	8	0+	0+	0+	0+	0+	0+	0+	0+	0+	0+	0+	0+	001	001	002	003	009	026	057	112	194	8
	9	0+	0+	0+	0+	0+	0+	0+	0+	0+	0+	0+	0+	0+	0+	001	001	002	006	015	036	073	9
	10	0+	0+	0+	0+	0+	0+	0+	0+	0+	0+	0+	0+	0+	0+	0+	0+	0+	001	003	008	019	10
	11	0+	0+	0+	0+	0+	0+	0+	0+	0+	0+	0+	0+	0+	0+	0+	0+	0+	0+	0+	001	003	11
	12	0+	0+	0+	0+	0+	0+	0+	0+	0+	0+	0+	0+	0+	0+	0+	0+	0+	0+	0+	0+	0+	12
13	0	1	1	1	1	1	1	1	1	1	1	1	1	1	1	1	1	1	1	1	1	1	0
	1	122	231	412	487	553	662	746	810	859	879	896	924	945	960	972	976	990	996	999	1-	1-	1
	2	007	027	093	135	181	279	379	474	561	602	640	708	766	815	856	873	936	970	987	995	998	2
	3	0+	002	014	025	039	080	134	198	270	308	346	423	498	570	636	667	798	887	942	973	989	3
	4	0+	0+	001	003	006	016	034	061	097	118	141	194	253	316	382	416	579	722	831	907	954	4
	5	0+	0+	0+	0+	001	003	006	014	026	034	044	068	099	137	182	206	346	499	647	772	867	5
	6	0+	0+	0+	0+	0+	0+	0+	001	002	006	014	020	030	046	068	080	165	284	426	573	709	6
	7	0+	0+	0+	0+	0+	0+	0+	0+	001	001	002	004	007	012	019	024	062	129	229	356	500	7
	8	0+	0+	0+	0+	0+	0+	0+	0+	0+	0+	0+	001	001	002	004	006	018	046	098	179	291	8
	9	0+	0+	0+	0+	0+	0+	0+	0+	0+	0+	0+	0+	0+	0+	001	001	004	013	032	070	133	9
	10	0+	0+	0+	0+	0+	0+	0+	0+	0+	0+	0+	0+	0+	0+	0+	0+	001	003	008	020	046	10
	11	0+	0+	0+	0+	0+	0+	0+	0+	0+	0+	0+	0+	0+	0+	0+	0+	0+	0+	001	004	011	11
	12	0+	0+	0+	0+	0+	0+	0+	0+	0+	0+	0+	0+	0+	0+	0+	0+	0+	0+	0+	001	002	12
	13	0+	0+	0+	0+	0+	0+	0+	0+	0+	0+	0+	0+	0+	0+	0+	0+	0+	0+	0+	0+	0+	13
14	0	1	1	1	1	1	1	1	1	1	1	1	1	1	1	1	1	1	1	1	1	1	0
	1	131	246	435	512	579	689	771	833	879	897	913	938	956	969	979	982	993	998	999	1-	1-	1
	2	008	031	106	153	204	310	415	514	603	643	681	747	802	847	884	899	953	979	992	997	999	2
	3	0+	002	017	030	048	096	158	232	311	352	393	474	552	624	689	719	839	916	960	983	994	3
	4	0+	0+	002	004	008	021	044	077	121	147	174	235	302	372	443	479	645	779	876	937	971	4
	5	0+	0+	0+	0+	001	004	009	020	036	047	059	091	130	176	230	258	416	577	721	833	910	5
	6	0+	0+	0+	0+	0+	0+	001	004	008	012	016	027	044	066	095	112	219	359	514	663	788	6
	7	0+	0+	0+	0+	0+	0+	0+	001	001	002	003	006	012	020	031	038	093	184	308	454	605	7
	8	0+	0+	0+	0+	0+	0+	0+	0+	0+	0+	001	001	002	005	008	010	031	075	150	259	395	8
	9	0+	0+	0+	0+	0+	0+	0+	0+	0+	0+	0+	0+	0+	001	002	002	008	024	058	119	212	9
	10	0+	0+	0+	0+	0+	0+	0+	0+	0+	0+	0+	0+	0+	0+	0+	0+	002	006	018	043	090	10
	11	0+	0+	0+	0+	0+	0+	0+	0+	0+	0+	0+	0+	0+	0+	0+	0+	0+	001	004	011	029	11
	12	0+	0+	0+	0+	0+	0+	0+	0+	0+	0+	0+	0+	0+	0+	0+	0+	0+	0+	001	002	006	12
	13	0+	0+	0+	0+	0+	0+	0+	0+	0+	0+	0+	0+	0+	0+	0+	0+	0+	0+	0+	0+	001	13
	14	0+	0+	0+	0+	0+	0+	0+	0+	0+	0+	0+	0+	0+	0+	0+	0+	0+	0+	0+	0+	0+	14

Table D.2 (Continued)

Cumulative binomial probabilities. The column headings give values of p.

n	r	.01	.02	.04	.05	.06	.08	.10	.12	.14	.15	.16	.18	.20	.22	.24	.25	.30	.35	.40	.45	.50	r
15	0	1	1	1	1	1	1	1	1	1	1	1	1	1	1	1	1	1	1	1	1	1	0
	1	140	261	458	537	605	714	794	853	896	913	927	949	965	976	984	987	995	998	1-	1-	1-	1
	2	010	035	119	171	226	340	451	552	642	681	718	781	833	874	906	920	965	986	995	998	1-	2
	3	0+	003	020	036	057	113	184	265	352	396	439	523	602	673	736	764	873	938	973	989	996	3
	4	0+	0+	002	005	010	027	056	096	148	177	209	278	352	427	502	539	703	827	909	958	982	4
	5	0+	0+	0+	001	001	005	013	026	048	062	078	117	164	219	281	314	485	648	783	880	941	5
	6	0+	0+	0+	0+	0+	001	002	006	012	017	023	039	061	090	127	148	278	436	597	739	849	6
	7	0+	0+	0+	0+	0+	0+	0+	001	002	004	005	010	018	030	046	057	131	245	390	548	696	7
	8	0+	0+	0+	0+	0+	0+	0+	0+	0+	001	001	002	004	008	013	017	050	113	213	346	500	8
	9	0+	0+	0+	0+	0+	0+	0+	0+	0+	0+	0+	0+	001	002	003	004	015	042	095	182	304	9
	10	0+	0+	0+	0+	0+	0+	0+	0+	0+	0+	0+	0+	0+	0+	001	001	004	012	034	077	151	10
	11	0+	0+	0+	0+	0+	0+	0+	0+	0+	0+	0+	0+	0+	0+	0+	0+	001	003	009	025	059	11
	12	0+	0+	0+	0+	0+	0+	0+	0+	0+	0+	0+	0+	0+	0+	0+	0+	0+	0+	002	006	018	12
	13	0+	0+	0+	0+	0+	0+	0+	0+	0+	0+	0+	0+	0+	0+	0+	0+	0+	0+	0+	001	004	13
	14	0+	0+	0+	0+	0+	0+	0+	0+	0+	0+	0+	0+	0+	0+	0+	0+	0+	0+	0+	0+	0+	14
	15	0+	0+	0+	0+	0+	0+	0+	0+	0+	0+	0+	0+	0+	0+	0+	0+	0+	0+	0+	0+	0+	15
16	0	1	1	1	1	1	1	1	1	1	1	1	1	1	1	1	1	1	1	1	1	1	0
	1	149	276	480	560	628	737	815	871	910	926	939	958	972	981	988	990	997	999	1-	1-	1-	1
	2	011	040	133	189	249	370	485	588	677	716	751	811	859	897	925	937	974	990	997	999	1-	2
	3	001	004	024	043	067	131	211	300	393	439	484	570	648	717	777	803	901	955	982	993	998	3
	4	0+	0+	003	007	013	034	068	116	176	210	246	322	402	481	558	595	754	866	935	972	989	4
	5	0+	0+	0+	001	002	007	017	035	062	079	099	146	202	265	334	370	550	711	833	915	962	5
	6	0+	0+	0+	0+	0+	001	003	008	017	024	032	053	082	119	164	190	340	510	671	802	895	6
	7	0+	0+	0+	0+	0+	0+	001	002	004	006	008	015	027	043	066	080	175	312	473	634	773	7
	8	0+	0+	0+	0+	0+	0+	0+	0+	001	001	002	004	007	013	021	027	074	159	284	437	598	8
	9	0+	0+	0+	0+	0+	0+	0+	0+	0+	0+	0+	001	001	003	006	007	026	067	142	256	402	9
	10	0+	0+	0+	0+	0+	0+	0+	0+	0+	0+	0+	0+	0+	001	001	002	007	023	058	124	227	10
	11	0+	0+	0+	0+	0+	0+	0+	0+	0+	0+	0+	0+	0+	0+	0+	0+	002	006	019	049	105	11
	12	0+	0+	0+	0+	0+	0+	0+	0+	0+	0+	0+	0+	0+	0+	0+	0+	0+	001	005	015	038	12
	13	0+	0+	0+	0+	0+	0+	0+	0+	0+	0+	0+	0+	0+	0+	0+	0+	0+	0+	001	003	011	13
	14	0+	0+	0+	0+	0+	0+	0+	0+	0+	0+	0+	0+	0+	0+	0+	0+	0+	0+	0+	001	002	14
	15	0+	0+	0+	0+	0+	0+	0+	0+	0+	0+	0+	0+	0+	0+	0+	0+	0+	0+	0+	0+	0+	15
	16	0+	0+	0+	0+	0+	0+	0+	0+	0+	0+	0+	0+	0+	0+	0+	0+	0+	0+	0+	0+	0+	16
17	0	1	1	1	1	1	1	1	1	1	1	1	1	1	1	1	1	1	1	1	1	1	0
	1	157	291	500	582	651	758	833	886	923	937	948	966	977	985	991	992	998	999	1-	1-	1-	1
	2	012	045	147	208	272	399	518	622	710	748	781	838	882	915	940	950	981	993	998	999	1-	2
	3	001	004	029	050	078	150	238	335	432	480	527	613	690	758	812	836	923	967	988	996	999	3
	4	0+	0+	004	009	016	042	083	138	207	244	284	367	451	533	611	647	798	897	954	982	994	4
	5	0+	0+	0+	001	003	009	022	045	078	099	122	178	242	313	388	426	611	765	874	940	975	5
	6	0+	0+	0+	0+	0+	001	005	011	023	032	042	069	106	151	205	235	403	580	736	853	928	6
	7	0+	0+	0+	0+	0+	0+	001	002	006	008	012	022	038	060	089	107	225	381	552	710	834	7
	8	0+	0+	0+	0+	0+	0+	0+	0+	001	002	003	006	011	019	032	040	105	213	359	526	685	8
	9	0+	0+	0+	0+	0+	0+	0+	0+	0+	0+	0+	001	003	005	009	012	040	099	199	337	500	9
	10	0+	0+	0+	0+	0+	0+	0+	0+	0+	0+	0+	0+	0+	001	002	003	013	038	092	183	315	10
	11	0+	0+	0+	0+	0+	0+	0+	0+	0+	0+	0+	0+	0+	0+	0+	001	003	012	035	083	166	11
	12	0+	0+	0+	0+	0+	0+	0+	0+	0+	0+	0+	0+	0+	0+	0+	0+	001	003	011	030	072	12
	13	0+	0+	0+	0+	0+	0+	0+	0+	0+	0+	0+	0+	0+	0+	0+	0+	0+	001	003	009	025	13
	14	0+	0+	0+	0+	0+	0+	0+	0+	0+	0+	0+	0+	0+	0+	0+	0+	0+	0+	0+	002	006	14
	15	0+	0+	0+	0+	0+	0+	0+	0+	0+	0+	0+	0+	0+	0+	0+	0+	0+	0+	0+	0+	001	15
	16	0+	0+	0+	0+	0+	0+	0+	0+	0+	0+	0+	0+	0+	0+	0+	0+	0+	0+	0+	0+	0+	16
	17	0+	0+	0+	0+	0+	0+	0+	0+	0+	0+	0+	0+	0+	0+	0+	0+	0+	0+	0+	0+	0+	17
18	0	1	1	1	1	1	1	1	1	1	1	1	1	1	1	1	1	1	1	1	1	1	0
	1	165	305	520	603	672	777	850	900	934	946	957	972	982	989	993	994	998	1-	1-	1-	1-	1
	2	014	050	161	226	294	428	550	654	740	776	808	861	901	931	952	961	986	995	999	1-	1-	2
	3	001	005	033	058	090	170	266	369	471	520	567	654	729	792	843	865	940	976	992	997	999	3
	4	0+	0+	005	011	020	051	098	162	238	280	323	411	499	582	659	694	835	922	967	988	996	4
	5	0+	0+	001	002	003	012	028	056	096	121	148	212	284	361	441	481	667	811	906	959	985	5
	6	0+	0+	0+	0+	0+	002	006	015	031	042	055	089	133	187	249	283	466	645	791	892	952	6
	7	0+	0+	0+	0+	0+	0+	001	003	008	012	017	031	051	080	117	139	278	451	626	774	881	7
	8	0+	0+	0+	0+	0+	0+	0+	001	002	003	004	009	016	028	046	057	141	272	437	609	760	8
	9	0+	0+	0+	0+	0+	0+	0+	0+	0+	001	001	002	004	008	015	019	060	139	263	422	593	9
	10	0+	0+	0+	0+	0+	0+	0+	0+	0+	0+	0+	0+	001	002	004	005	021	060	135	253	407	10
	11	0+	0+	0+	0+	0+	0+	0+	0+	0+	0+	0+	0+	0+	0+	001	001	006	021	058	128	240	11
	12	0+	0+	0+	0+	0+	0+	0+	0+	0+	0+	0+	0+	0+	0+	0+	0+	001	006	020	054	119	12
	13	0+	0+	0+	0+	0+	0+	0+	0+	0+	0+	0+	0+	0+	0+	0+	0+	0+	001	006	018	048	13
	14	0+	0+	0+	0+	0+	0+	0+	0+	0+	0+	0+	0+	0+	0+	0+	0+	0+	0+	001	005	015	14

Table D.2 (Continued)

n	r	.01	.02	.04	.05	.06	.08	.10	.12	.14	.15	P .16	.18	.20	.22	.24	.25	.30	.35	.40	.45	.50	r
	15	0+	0+	0+	0+	0+	0+	0+	0+	0+	0+	0+	0+	0+	0+	0+	0+	0+	0+	0+	001	004	15
	16	0+	0+	0+	0+	0+	0+	0+	0+	0+	0+	0+	0+	0+	0+	0+	0+	0+	0+	0+	0+	001	16
	17	0+	0+	0+	0+	0+	0+	0+	0+	0+	0+	0+	0+	0+	0+	0+	0+	0+	0+	0+	0+	0+	17
	18	0+	0+	0+	0+	0+	0+	0+	0+	0+	0+	0+	0+	0+	0+	0+	0+	0+	0+	0+	0+	0+	18
19	0	1	1	1	1	1	1	1	1	1	1	1	1	1	1	1	1	1	1	1	1	1	0
	1	174	319	540	623	691	795	865	912	943	954	964	977	986	991	995	996	999	1-	1-	1-	1-	1
	2	015	055	175	245	317	456	580	683	767	802	832	881	917	943	962	969	990	997	999	1-	1-	2
	3	001	006	038	067	102	191	295	403	509	559	606	691	763	822	869	889	954	983	995	998	1-	3
	4	0+	0+	006	013	024	060	115	187	271	316	362	455	545	628	703	737	867	941	977	992	998	4
	5	0+	0+	001	002	004	015	035	069	116	144	176	248	327	410	494	535	718	850	930	972	990	5
	6	0+	0+	0+	0+	001	003	009	020	040	054	070	111	163	225	295	332	526	703	837	922	968	6
	7	0+	0+	0+	0+	0+	0+	002	005	011	016	023	041	068	103	149	175	334	519	692	827	916	7
	8	0+	0+	0+	0+	0+	0+	0+	001	003	004	006	013	023	040	063	077	182	334	512	683	820	8
	9	0+	0+	0+	0+	0+	0+	0+	0+	001	001	001	003	007	013	022	029	084	185	333	506	676	9
	10	0+	0+	0+	0+	0+	0+	0+	0+	0+	0+	0+	001	002	003	007	009	033	087	186	329	500	10
	11	0+	0+	0+	0+	0+	0+	0+	0+	0+	0+	0+	0+	0+	001	002	002	011	035	088	184	324	11
	12	0+	0+	0+	0+	0+	0+	0+	0+	0+	0+	0+	0+	0+	0+	0+	0+	003	011	035	087	180	12
	13	0+	0+	0+	0+	0+	0+	0+	0+	0+	0+	0+	0+	0+	0+	0+	0+	001	003	012	034	084	13
	14	0+	0+	0+	0+	0+	0+	0+	0+	0+	0+	0+	0+	0+	0+	0+	0+	0+	001	003	011	032	14
	15	0+	0+	0+	0+	0+	0+	0+	0+	0+	0+	0+	0+	0+	0+	0+	0+	0+	0+	001	003	010	15
	16	0+	0+	0+	0+	0+	0+	0+	0+	0+	0+	0+	0+	0+	0+	0+	0+	0+	0+	0+	001	002	16
	17	0+	0+	0+	0+	0+	0+	0+	0+	0+	0+	0+	0+	0+	0+	0+	0+	0+	0+	0+	0+	0+	17
	18	0+	0+	0+	0+	0+	0+	0+	0+	0+	0+	0+	0+	0+	0+	0+	0+	0+	0+	0+	0+	0+	18
	19	0+	0+	0+	0+	0+	0+	0+	0+	0+	0+	0+	0+	0+	0+	0+	0+	0+	0+	0+	0+	0+	19
20	0	1	1	1	1	1	1	1	1	1	1	1	1	1	1	1	1	1	1	1	1	1	0
	1	182	332	558	642	710	811	878	922	951	961	969	981	988	993	996	997	999	1-	1-	1-	1-	1
	2	017	060	190	264	340	483	608	711	792	824	853	898	931	954	970	976	992	998	999	1-	1-	2
	3	001	007	044	075	115	212	323	437	545	595	642	725	794	849	891	909	965	988	996	999	1-	3
	4	0+	001	007	016	029	071	133	213	304	352	401	497	589	671	743	775	893	956	984	995	999	4
	5	0+	0+	001	003	006	018	043	083	137	170	206	265	370	458	544	585	762	882	949	981	994	5
	6	0+	0+	0+	0+	001	004	011	026	051	067	087	136	196	266	343	383	584	755	874	945	979	6
	7	0+	0+	0+	0+	0+	001	002	007	015	022	030	054	087	130	184	214	392	583	750	870	942	7
	8	0+	0+	0+	0+	0+	0+	0+	001	004	006	009	018	032	054	083	102	228	399	584	748	868	8
	9	0+	0+	0+	0+	0+	0+	0+	0+	001	001	002	005	010	019	032	041	113	238	404	586	748	9
	10	0+	0+	0+	0+	0+	0+	0+	0+	0+	0+	0+	001	003	005	010	014	048	122	245	409	588	10
	11	0+	0+	0+	0+	0+	0+	0+	0+	0+	0+	0+	0+	001	001	003	004	017	053	128	249	412	11
	12	0+	0+	0+	0+	0+	0+	0+	0+	0+	0+	0+	0+	0+	0+	001	001	005	020	057	131	252	12
	13	0+	0+	0+	0+	0+	0+	0+	0+	0+	0+	0+	0+	0+	0+	0+	0+	001	006	021	058	132	13
	14	0+	0+	0+	0+	0+	0+	0+	0+	0+	0+	0+	0+	0+	0+	0+	0+	0+	002	006	021	058	14
	15	0+	0+	0+	0+	0+	0+	0+	0+	0+	0+	0+	0+	0+	0+	0+	0+	0+	0+	002	006	021	15
	16	0+	0+	0+	0+	0+	0+	0+	0+	0+	0+	0+	0+	0+	0+	0+	0+	0+	0+	0+	002	006	16
	17	0+	0+	0+	0+	0+	0+	0+	0+	0+	0+	0+	0+	0+	0+	0+	0+	0+	0+	0+	0+	001	17
	18	0+	0+	0+	0+	0+	0+	0+	0+	0+	0+	0+	0+	0+	0+	0+	0+	0+	0+	0+	0+	0+	18
	19	0+	0+	0+	0+	0+	0+	0+	0+	0+	0+	0+	0+	0+	0+	0+	0+	0+	0+	0+	0+	0+	19
	20	0+	0+	0+	0+	0+	0+	0+	0+	0+	0+	0+	0+	0+	0+	0+	0+	0+	0+	0+	0+	0+	20
21	0	1	1	1	1	1	1	1	1	1	1	1	1	1	1	1	1	1	1	1	1	1	0
	1	190	346	576	659	727	826	891	932	958	967	974	985	991	995	997	998	999	1-	1-	1-	1-	1
	2	019	065	204	283	362	509	635	736	814	845	872	913	943	962	976	981	994	999	1-	1-	1-	2
	3	001	008	050	085	128	234	352	470	580	630	676	756	821	872	910	925	973	991	998	999	1-	3
	4	0+	001	009	019	034	082	152	240	338	389	440	538	630	710	779	808	914	967	989	997	999	4
	5	0+	0+	001	003	007	023	052	098	161	197	237	323	414	505	592	633	802	908	963	987	996	5
	6	0+	0+	0+	0+	001	005	014	033	063	083	106	162	231	308	391	433	637	799	904	961	987	6
	7	0+	0+	0+	0+	0+	001	003	009	020	029	039	068	109	160	222	256	449	643	800	904	961	7
	8	0+	0+	0+	0+	0+	0+	001	002	005	008	012	024	043	070	108	130	277	464	650	803	905	8
	9	0+	0+	0+	0+	0+	0+	0+	0+	001	002	003	007	014	026	044	056	148	294	476	659	808	9
	10	0+	0+	0+	0+	0+	0+	0+	0+	0+	0+	001	002	004	008	016	021	068	162	309	488	669	10
	11	0+	0+	0+	0+	0+	0+	0+	0+	0+	0+	0+	0+	001	002	005	006	026	077	174	321	500	11
	12	0+	0+	0+	0+	0+	0+	0+	0+	0+	0+	0+	0+	0+	001	001	002	009	031	085	184	332	12
	13	0+	0+	0+	0+	0+	0+	0+	0+	0+	0+	0+	0+	0+	0+	0+	0+	002	011	035	091	192	13
	14	0+	0+	0+	0+	0+	0+	0+	0+	0+	0+	0+	0+	0+	0+	0+	0+	001	003	012	038	095	14
	15	0+	0+	0+	0+	0+	0+	0+	0+	0+	0+	0+	0+	0+	0+	0+	0+	0+	001	004	013	039	15
	16	0+	0+	0+	0+	0+	0+	0+	0+	0+	0+	0+	0+	0+	0+	0+	0+	0+	0+	001	004	013	16
	17	0+	0+	0+	0+	0+	0+	0+	0+	0+	0+	0+	0+	0+	0+	0+	0+	0+	0+	0+	001	004	17
	18	0+	0+	0+	0+	0+	0+	0+	0+	0+	0+	0+	0+	0+	0+	0+	0+	0+	0+	0+	0+	001	18
	19	0+	0+	0+	0+	0+	0+	0+	0+	0+	0+	0+	0+	0+	0+	0+	0+	0+	0+	0+	0+	0+	19
	20	0+	0+	0+	0+	0+	0+	0+	0+	0+	0+	0+	0+	0+	0+	0+	0+	0+	0+	0+	0+	0+	20
	21	0+	0+	0+	0+	0+	0+	0+	0+	0+	0+	0+	0+	0+	0+	0+	0+	0+	0+	0+	0+	0+	21

Table D.2 (Continued)

The column headed **.16** falls under the centered label **p**.

n	r	.01	.02	.04	.05	.06	.08	.10	.12	.14	.15	.16	.18	.20	.22	.24	.25	.30	.35	.40	.45	.50	r
22	0	1	1	1	1	1	1	1	1	1	1	1	1	1	1	1	1	1	1	1	1	1	0
	1	198	359	593	676	744	840	902	940	964	972	978	987	993	996	998	998	1-	1-	1-	1-	1-	1
	2	020	071	219	302	384	535	661	760	834	863	888	926	952	970	981	985	996	999	1-	1-	1-	2
	3	001	009	056	095	142	256	380	502	612	662	707	785	846	892	926	939	979	994	998	1-	1-	3
	4	0+	001	011	022	040	094	172	267	372	425	477	578	668	746	810	838	932	975	992	998	1-	4
	5	0+	0+	002	004	009	027	062	115	186	226	270	362	457	550	637	677	835	928	973	992	998	5
	6	0+	0+	0+	001	002	006	018	041	077	100	127	191	267	351	439	483	687	837	928	973	992	6
	7	0+	0+	0+	0+	0+	001	004	012	026	037	050	085	133	193	263	301	506	698	842	929	974	7
	8	0+	0+	0+	0+	0+	0+	001	003	008	011	017	032	056	090	135	162	329	526	710	848	933	8
	9	0+	0+	0+	0+	0+	0+	0+	001	002	003	005	010	020	036	060	075	186	353	546	724	857	9
	10	0+	0+	0+	0+	0+	0+	0+	0+	0+	001	001	003	006	012	022	030	092	208	376	565	738	10
	11	0+	0+	0+	0+	0+	0+	0+	0+	0+	0+	0+	001	002	004	007	010	039	107	228	396	584	11
	12	0+	0+	0+	0+	0+	0+	0+	0+	0+	0+	0+	0+	0+	001	002	003	014	047	121	246	416	12
	13	0+	0+	0+	0+	0+	0+	0+	0+	0+	0+	0+	0+	0+	0+	0+	001	004	018	055	133	262	13
	14	0+	0+	0+	0+	0+	0+	0+	0+	0+	0+	0+	0+	0+	0+	0+	0+	001	006	021	062	143	14
	15	0+	0+	0+	0+	0+	0+	0+	0+	0+	0+	0+	0+	0+	0+	0+	0+	0+	0+	002	024	067	15
	16	0+	0+	0+	0+	0+	0+	0+	0+	0+	0+	0+	0+	0+	0+	0+	0+	0+	0+	002	008	026	16
	17	0+	0+	0+	0+	0+	0+	0+	0+	0+	0+	0+	0+	0+	0+	0+	0+	0+	0+	0+	002	008	17
	18	0+	0+	0+	0+	0+	0+	0+	0+	0+	0+	0+	0+	0+	0+	0+	0+	0+	0+	0+	0+	002	18
	19	0+	0+	0+	0+	0+	0+	0+	0+	0+	0+	0+	0+	0+	0+	0+	0+	0+	0+	0+	0+	0+	19
	20	0+	0+	0+	0+	0+	0+	0+	0+	0+	0+	0+	0+	0+	0+	0+	0+	0+	0+	0+	0+	0+	20
	21	0+	0+	0+	0+	0+	0+	0+	0+	0+	0+	0+	0+	0+	0+	0+	0+	0+	0+	0+	0+	0+	21
	22	0+	0+	0+	0+	0+	0+	0+	0+	0+	0+	0+	0+	0+	0+	0+	0+	0+	0+	0+	0+	0+	22
23	0	1	1	1	1	1	1	1	1	1	1	1	1	1	1	1	1	1	1	1	1	1	0
	1	206	372	609	693	759	853	911	947	969	976	982	990	994	997	998	999	1-	1-	1-	1-	1-	1
	2	022	077	234	321	405	559	685	781	852	880	902	937	960	975	985	988	997	999	1-	1-	1-	2
	3	002	011	062	105	157	278	408	533	643	692	736	810	867	909	939	951	984	996	999	1-	1-	3
	4	0+	001	012	026	046	107	193	295	405	460	514	615	703	778	838	863	946	982	995	999	1-	4
	5	0+	0+	002	005	011	033	073	133	212	256	303	401	499	593	678	717	864	945	981	995	999	5
	6	0+	0+	0+	001	002	008	023	050	092	119	150	222	305	395	487	532	731	869	946	981	995	6
	7	0+	0+	0+	0+	0+	002	006	015	033	046	062	104	160	227	305	346	560	747	876	949	983	7
	8	0+	0+	0+	0+	0+	0+	001	004	010	015	022	042	072	113	166	196	382	586	763	885	953	8
	9	0+	0+	0+	0+	0+	0+	0+	001	003	004	007	014	027	048	078	096	229	444	612	780	895	9
	10	0+	0+	0+	0+	0+	0+	0+	0+	001	001	002	004	009	017	031	041	120	259	444	636	798	10
	11	0+	0+	0+	0+	0+	0+	0+	0+	0+	0+	0+	001	001	003	005	011	055	142	287	472	661	11
	12	0+	0+	0+	0+	0+	0+	0+	0+	0+	0+	0+	0+	001	001	003	005	021	068	164	313	500	12
	13	0+	0+	0+	0+	0+	0+	0+	0+	0+	0+	0+	0+	0+	0+	001	001	007	028	081	184	339	13
	14	0+	0+	0+	0+	0+	0+	0+	0+	0+	0+	0+	0+	0+	0+	0+	0+	002	010	035	094	202	14
	15	0+	0+	0+	0+	0+	0+	0+	0+	0+	0+	0+	0+	0+	0+	0+	0+	001	003	013	041	105	15
	16	0+	0+	0+	0+	0+	0+	0+	0+	0+	0+	0+	0+	0+	0+	0+	0+	0+	001	004	015	047	16
	17	0+	0+	0+	0+	0+	0+	0+	0+	0+	0+	0+	0+	0+	0+	0+	0+	0+	0+	001	005	017	17
	18	0+	0+	0+	0+	0+	0+	0+	0+	0+	0+	0+	0+	0+	0+	0+	0+	0+	0+	0+	001	005	18
	19	0+	0+	0+	0+	0+	0+	0+	0+	0+	0+	0+	0+	0+	0+	0+	0+	0+	0+	0+	0+	001	19
	20	0+	0+	0+	0+	0+	0+	0+	0+	0+	0+	0+	0+	0+	0+	0+	0+	0+	0+	0+	0+	0+	20
	21	0+	0+	0+	0+	0+	0+	0+	0+	0+	0+	0+	0+	0+	0+	0+	0+	0+	0+	0+	0+	0+	21
	22	0+	0+	0+	0+	0+	0+	0+	0+	0+	0+	0+	0+	0+	0+	0+	0+	0+	0+	0+	0+	0+	22
	23	0+	0+	0+	0+	0+	0+	0+	0+	0+	0+	0+	0+	0+	0+	0+	0+	0+	0+	0+	0+	0+	23
24	0	1	1	1	1	1	1	1	1	1	1	1	1	1	1	1	1	1	1	1	1	1	0
	1	214	384	625	708	773	865	920	953	973	980	985	991	995	997	999	999	1-	1-	1-	1-	1-	1
	2	024	083	249	339	427	583	708	801	869	894	915	946	967	980	988	991	998	1-	1-	1-	1-	2
	3	002	012	069	116	172	301	436	563	673	720	763	833	885	924	950	960	988	997	999	1-	1-	3
	4	0+	001	014	030	053	121	214	324	439	495	550	650	736	807	862	885	958	987	996	999	1-	4
	5	0+	0+	002	006	013	039	085	153	239	287	337	439	540	634	717	753	889	958	987	996	999	5
	6	0+	0+	0+	001	002	010	028	060	109	139	174	254	344	439	533	578	771	896	960	987	997	6
	7	0+	0+	0+	0+	0+	002	007	019	041	057	076	126	189	264	349	393	611	789	904	964	989	7
	8	0+	0+	0+	0+	0+	0+	002	005	013	020	028	053	089	138	199	234	435	642	808	914	968	8
	9	0+	0+	0+	0+	0+	0+	0+	001	004	006	009	019	036	062	099	121	275	474	672	827	924	9
	10	0+	0+	0+	0+	0+	0+	0+	0+	001	002	002	006	013	024	042	055	153	313	511	701	846	10
	11	0+	0+	0+	0+	0+	0+	0+	0+	0+	0+	001	002	004	008	016	021	074	183	350	546	729	11
	12	0+	0+	0+	0+	0+	0+	0+	0+	0+	0+	0+	0+	001	002	005	007	031	094	213	385	581	12
	13	0+	0+	0+	0+	0+	0+	0+	0+	0+	0+	0+	0+	0+	001	001	002	012	042	114	242	419	13
	14	0+	0+	0+	0+	0+	0+	0+	0+	0+	0+	0+	0+	0+	0+	0+	001	004	016	053	134	271	14

Table D.2 (Continued)

P values indicated above the .16 column.

n	r	.01	.02	.04	.05	.06	.08	.10	.12	.14	.15	.16	.18	.20	.22	.24	.25	.30	.35	.40	.45	.50	r
24	15	0+	0+	0+	0+	0+	0+	0+	0+	0+	0+	0+	0+	0+	0+	0+	0+	001	005	022	065	154	15
	16	0+	0+	0+	0+	0+	0+	0+	0+	0+	0+	0+	0+	0+	0+	0+	0+	0+	002	008	027	076	16
	17	0+	0+	0+	0+	0+	0+	0+	0+	0+	0+	0+	0+	0+	0+	0+	0+	0+	0+	002	010	032	17
	18	0+	0+	0+	0+	0+	0+	0+	0+	0+	0+	0+	0+	0+	0+	0+	0+	0+	0+	001	003	011	18
	19	0+	0+	0+	0+	0+	0+	0+	0+	0+	0+	0+	0+	0+	0+	0+	0+	0+	0+	0+	001	003	19
	20	0+	0+	0+	0+	0+	0+	0+	0+	0+	0+	0+	0+	0+	0+	0+	0+	0+	0+	0+	0+	001	20
	21	0+	0+	0+	0+	0+	0+	0+	0+	0+	0+	0+	0+	0+	0+	0+	0+	0+	0+	0+	0+	0+	21
	22	0+	0+	0+	0+	0+	0+	0+	0+	0+	0+	0+	0+	0+	0+	0+	0+	0+	0+	0+	0+	0+	22
	23	0+	0+	0+	0+	0+	0+	0+	0+	0+	0+	0+	0+	0+	0+	0+	0+	0+	0+	0+	0+	0+	23
	24	0+	0+	0+	0+	0+	0+	0+	0+	0+	0+	0+	0+	0+	0+	0+	0+	0+	0+	0+	0+	0+	24
25	0	1	1	1	1	1	1	1	1	1	1	1	1	1	1	1	1	1	1	1	1	1	0
	1	222	397	640	723	787	876	928	959	977	983	987	993	996	998	999	999	1-	1-	1-	1-	1-	1
	2	026	089	264	358	447	605	729	820	883	907	926	955	973	984	991	993	998	1-	1-	1-	1-	2
	3	002	013	076	127	187	323	463	591	700	746	787	853	902	936	959	968	991	998	1-	1-	1-	3
	4	0+	001	017	034	060	135	236	352	471	529	584	683	766	832	883	904	967	990	998	1-	1-	4
	5	0+	0+	003	007	015	045	098	173	267	318	371	477	579	672	752	786	910	968	991	998	1-	5
	6	0+	0+	0+	001	003	012	033	071	127	162	200	288	383	482	577	622	807	917	971	991	998	6
	7	0+	0+	0+	0+	001	003	009	024	051	070	092	149	220	303	393	439	659	827	926	974	993	7
	8	0+	0+	0+	0+	0+	001	002	007	017	025	036	066	109	166	235	273	488	694	846	936	978	8
	9	0+	0+	0+	0+	0+	0+	0+	002	005	008	012	025	047	079	123	149	323	533	726	866	946	9
	10	0+	0+	0+	0+	0+	0+	0+	0+	001	002	003	008	017	033	056	071	189	370	575	758	885	10
	11	0+	0+	0+	0+	0+	0+	0+	0+	0+	0+	001	002	006	012	022	030	098	229	414	616	788	11
	12	0+	0+	0+	0+	0+	0+	0+	0+	0+	0+	0+	001	002	004	008	011	044	125	268	457	655	12
	13	0+	0+	0+	0+	0+	0+	0+	0+	0+	0+	0+	0+	0+	001	002	003	017	060	154	306	500	13
	14	0+	0+	0+	0+	0+	0+	0+	0+	0+	0+	0+	0+	0+	0+	001	001	006	025	078	183	345	14
	15	0+	0+	0+	0+	0+	0+	0+	0+	0+	0+	0+	0+	0+	0+	0+	0+	002	009	034	096	212	15
	16	0+	0+	0+	0+	0+	0+	0+	0+	0+	0+	0+	0+	0+	0+	0+	0+	0+	003	013	044	115	16
	17	0+	0+	0+	0+	0+	0+	0+	0+	0+	0+	0+	0+	0+	0+	0+	0+	0+	001	004	017	054	17
	18	0+	0+	0+	0+	0+	0+	0+	0+	0+	0+	0+	0+	0+	0+	0+	0+	0+	0+	001	006	022	18
	19	0+	0+	0+	0+	0+	0+	0+	0+	0+	0+	0+	0+	0+	0+	0+	0+	0+	0+	0+	002	007	19
	20	0+	0+	0+	0+	0+	0+	0+	0+	0+	0+	0+	0+	0+	0+	0+	0+	0+	0+	0+	0+	002	20
	21	0+	0+	0+	0+	0+	0+	0+	0+	0+	0+	0+	0+	0+	0+	0+	0+	0+	0+	0+	0+	0+	21
	22	0+	0+	0+	0+	0+	0+	0+	0+	0+	0+	0+	0+	0+	0+	0+	0+	0+	0+	0+	0+	0+	22
	23	0+	0+	0+	0+	0+	0+	0+	0+	0+	0+	0+	0+	0+	0+	0+	0+	0+	0+	0+	0+	0+	23
	24	0+	0+	0+	0+	0+	0+	0+	0+	0+	0+	0+	0+	0+	0+	0+	0+	0+	0+	0+	0+	0+	24
	25	0+	0+	0+	0+	0+	0+	0+	0+	0+	0+	0+	0+	0+	0+	0+	0+	0+	0+	0+	0+	0+	25

Source: Abridged from "Tables of the Binomial Probability Distribution," in *Applied Mathematics Series No. 6*, National Bureau of Standards (Washington D.C.: U.S. Government Printing Office, 1949).

Reproduced with permission from William A. Spurr and Charles P. Bonini, *Statistical Analysis for Business Decisions* (Homewood, Ill.: Richard D. Irwin, Inc., 1973 ©), pp. 690–695.

E.1 / Poisson Distribution—Individual Terms

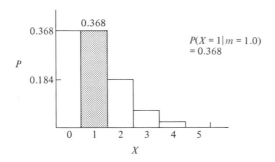

$$P(X = 1 \mid m = 1.0) = 0.368$$

A blank space is left for values less than .0005.

x	.001	.002	.003	.004	.005	.006	.007	.008	.009	.01	.02	.03	.04	.05	.06	.07	.08	.09	.10	.15	x
0	999	998	997	996	995	994	993	992	991	990	980	970	961	951	942	932	923	914	905	861	0
1	001	002	003	004	005	006	007	008	009	010	020	030	038	048	057	065	074	082	090	129	1
2													001	001	002	002	003	004	005	010	2

x	.20	.25	.30	.40	.50	.60	.70	.80	.90	1.0	1.1	1.2	1.3	1.4	1.5	1.6	1.7	1.8	1.9	2.0	x
0	819	779	741	670	607	549	497	449	407	368	333	301	273	247	223	202	183	165	150	135	0
1	164	195	222	268	303	329	348	359	366	368	366	361	354	345	335	323	311	298	284	271	1
2	016	024	033	054	076	099	122	144	165	184	201	217	230	242	251	258	264	268	270	271	2
3	001	002	003	007	013	020	028	038	049	061	074	087	100	113	126	138	150	161	171	180	3
4			001	002	003	005	008	011	015	020	026	032	039	047	055	063	072	081	090	4
5						001	001	002	003	004	006	008	011	014	018	022	026	031	036	5
6								001	001	001	002	003	004	005	006	008	010	012	6
7											001	001	001	001	002	003	003	7
8															001	001	8

x	2.1	2.2	2.3	2.4	2.5	2.6	2.7	2.8	2.9	3.0	3.1	3.2	3.3	3.4	3.5	3.6	3.7	3.8	3.9	4.0	x
0	122	111	100	091	082	074	067	061	055	050	045	041	037	033	030	027	025	022	020	018	0
1	257	244	231	218	205	193	181	170	160	149	140	130	122	113	106	098	091	085	079	073	1
2	270	268	265	261	257	251	245	238	231	224	216	209	201	193	185	177	169	162	154	147	2
3	189	197	203	209	214	218	220	222	224	224	224	223	221	219	216	212	209	205	200	195	3
4	099	108	117	125	134	141	149	156	162	168	173	178	182	186	189	191	193	194	195	195	4
5	042	048	054	060	067	074	080	087	094	101	107	114	120	126	132	138	143	148	152	156	5
6	015	017	021	024	028	032	036	041	045	050	056	061	066	072	077	083	088	094	099	104	6
7	004	005	007	008	010	012	014	016	019	022	025	028	031	035	039	042	047	051	055	060	7
8	001	002	002	002	003	004	005	006	007	008	010	011	013	015	017	019	022	024	027	030	8
9				001	001	001	001	002	002	003	003	004	005	006	007	008	009	010	012	013	9
10									001	001	001	001	002	002	002	003	003	004	005	005	10
11														001	001	001	001	001	002	002	11
12																			001	001	12

Table E.1 (Continued)

m is indicated above column 5.1.

x	4.1	4.2	4.3	4.4	4.5	4.6	4.7	4.8	4.9	5.0	5.1	5.2	5.3	5.4	5.5	5.6	5.7	5.8	5.9	6.0	x
0	017	015	014	012	011	010	009	008	007	007	006	006	005	005	004	004	003	003	003	002	0
1	068	063	058	054	050	046	043	040	036	034	031	029	026	024	022	021	019	018	016	015	1
2	139	132	125	119	112	106	100	095	089	084	079	075	070	066	062	058	054	051	048	045	2
3	190	185	180	174	169	163	157	152	146	140	135	129	124	119	113	108	103	098	094	089	3
4	195	194	193	192	190	188	185	182	179	175	172	168	164	160	156	152	147	143	138	134	4
5	160	163	166	169	171	173	174	175	175	175	175	175	174	173	171	170	168	166	163	161	5
6	109	114	119	124	128	132	136	140	143	146	149	151	154	156	157	158	159	160	160	161	6
7	064	069	073	078	082	087	091	096	100	104	109	113	116	120	123	127	130	133	135	138	7
8	033	036	039	043	046	050	054	058	061	065	069	073	077	081	085	089	092	096	100	103	8
9	015	017	019	021	023	026	028	031	033	036	039	042	045	049	052	055	059	062	065	069	9
10	006	007	008	009	010	012	013	015	016	018	020	022	024	026	029	031	033	036	039	041	10
11	002	003	003	004	004	005	006	006	007	008	009	010	012	013	014	016	017	019	021	023	11
12	001	001	001	001	002	002	002	003	003	003	004	005	005	006	007	007	008	009	010	011	12
13					001	001	001	001	001	001	002	002	002	002	003	003	004	004	005	005	13
14											001	001	001	001	001	001	001	002	002	002	14
15																	001	001	001	001	15

m is indicated above column 7.1.

x	6.1	6.2	6.3	6.4	6.5	6.6	6.7	6.8	6.9	7.0	7.1	7.2	7.3	7.4	7.5	8.0	8.5	9.0	9.5	10.0	x
0	002	002	002	002	002	001	001	001	001	001	001	001	001	001	001						0
1	014	013	012	011	010	009	008	008	007	006	006	005	005	005	004	003	002	001	001		1
2	042	039	036	034	032	030	028	026	024	022	021	019	018	017	016	011	007	005	003	002	2
3	085	081	077	073	069	065	062	058	055	052	049	046	044	041	039	029	021	015	011	008	3
4	129	125	121	116	112	108	103	099	095	091	087	084	080	076	073	057	044	034	025	019	4
5	158	155	152	149	145	142	138	135	131	128	124	120	117	113	109	092	075	061	048	038	5
6	160	160	159	159	157	156	155	153	151	149	147	144	142	139	137	122	107	091	076	063	6
7	140	142	144	145	146	147	148	149	149	149	149	149	148	147	146	140	129	117	104	090	7
8	107	110	113	116	119	121	124	126	128	130	132	134	135	136	137	140	138	132	123	113	8
9	072	076	079	082	086	089	092	095	098	101	104	107	110	112	114	124	130	132	130	125	9
10	044	047	050	053	056	059	062	065	068	071	074	077	080	083	086	099	110	119	124	125	10
11	024	026	029	031	033	035	038	040	043	045	048	050	053	056	059	072	085	097	107	114	11
12	012	014	015	016	018	019	021	023	025	026	028	030	032	034	037	048	060	073	084	095	12
13	006	007	007	008	009	010	011	012	013	014	015	017	018	020	021	030	040	050	062	073	13
14	003	003	003	004	004	005	005	006	006	007	008	009	009	010	011	017	024	032	042	052	14
15	001	001	001	002	002	002	002	003	003	003	004	004	005	005	006	009	014	019	027	035	15
16			001	001	001	001	001	001	001	001	002	002	002	002	003	005	007	011	016	022	16
17									001	001	001	001	001	001	001	002	004	006	009	013	17
18																001	002	003	005	007	18
19																	001	001	002	004	19
20																		001	001	002	20
21																				001	21

Source: Abridged from more detailed tables, such as available in _Poisson's Exponential Binomial Limit_, E. C. Molina (New York: Van Nostrand, 1942).

Reproduced with permission from William A. Spurr and Charles P. Bonini, _Statistical Analysis for Business Decisions_ (Homewood, Ill.: Richard D. Irwin, 1973 ©), pp. 696–697.

E.2 / Poisson Distribution—Cumulative Terms

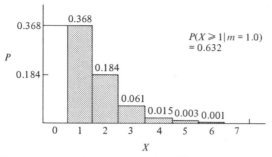

$$P(X \geqslant 1 \mid m = 1.0) = 0.632$$

This table presents the Poisson probabilities of X or more occurrences per unit of measurement for selected values of m, the mean number of occurrences per unit of measurement.

The numeral 1 indicates a value less than 1 but greater than 0.9995. A blank space is left for values less than 0.0005.

x	.001	.002	.003	.004	.005	.006	.007	.008	.009	.01	.02	.03	.04	.05	.06	.07	.08	.09	.10	.15	x
0	1	1	1	1	1	1	1	1	1	1	1	1	1	1	1	1	1	1	1	1	0
1	001	002	003	004	005	006	007	008	009	010	020	030	039	049	058	068	077	086	095	139	1
2													001	001	002	002	003	004	005	010	2
3																				001	3

x	.20	.25	.30	.40	.50	.60	.70	.80	.90	1.0	1.1	1.2	1.3	1.4	1.5	1.6	1.7	1.8	1.9	2.0	x
0	1	1	1	1	1	1	1	1	1	1	1	1	1	1	1	1	1	1	1	1	0
1	181	221	259	330	393	451	503	551	593	632	667	699	727	753	777	798	817	835	850	865	1
2	018	026	037	062	090	122	156	191	228	264	301	337	373	408	442	475	507	537	566	594	2
3	001	002	004	008	014	023	034	047	063	080	100	121	143	167	191	217	243	269	296	323	3
4				001	002	003	006	009	013	019	026	034	043	054	066	079	093	109	125	143	4
5							001	001	002	004	005	008	011	014	019	024	030	036	044	053	5
6										001	001	002	002	003	004	006	008	010	013	017	6
7														001	001	001	002	003	003	005	7
8																		001	001	001	8

x	2.1	2.2	2.3	2.4	2.5	2.6	2.7	2.8	2.9	3.0	3.1	3.2	3.3	3.4	3.5	3.6	3.7	3.8	3.9	4.0	x
0	1	1	1	1	1	1	1	1	1	1	1	1	1	1	1	1	1	1	1	1	0
1	878	889	900	909	918	926	933	939	945	950	955	959	963	967	970	973	975	978	980	982	1
2	620	645	669	692	713	733	751	769	785	801	815	829	841	853	864	874	884	893	901	908	2
3	350	377	404	430	456	482	506	531	554	577	599	620	641	660	679	697	715	731	747	762	3
4	161	181	201	221	242	264	286	308	330	353	375	397	420	442	463	485	506	527	547	567	4
5	062	072	084	096	109	123	137	152	168	185	202	219	237	256	275	294	313	332	352	371	5
6	020	025	030	036	042	049	057	065	074	084	094	105	117	129	142	156	170	184	199	215	6
7	006	007	009	012	014	017	021	024	029	034	039	045	051	058	065	073	082	091	101	111	7
8	001	002	003	003	004	005	007	008	010	012	014	017	020	023	027	031	035	040	045	051	8
9			001	001	001	001	002	002	003	004	005	006	007	008	010	012	014	016	019	021	9
10							001	001	001	001	001	002	002	003	003	004	005	006	007	008	10
11													001	001	001	001	002	002	002	003	11
12																		001	001	001	12

Table E.2 (Continued)

m

x	4.1	4.2	4.3	4.4	4.5	4.6	4.7	4.8	4.9	5.0	5.1	5.2	5.3	5.4	5.5	5.6	5.7	5.8	5.9	6.0	x
0	1	1	1	1	1	1	1	1	1	1	1	1	1	1	1	1	1	1	1	1	0
1	983	985	986	988	989	990	991	992	993	993	994	994	995	995	996	997	997	997	997	998	1
2	915	922	928	934	939	944	948	952	956	960	963	966	969	971	973	976	978	979	981	983	2
3	776	790	803	815	826	837	848	857	867	875	884	891	898	905	912	918	923	928	933	938	3
4	586	605	623	641	658	674	690	706	721	735	749	762	775	787	798	809	820	830	840	849	4
5	391	410	430	449	468	487	505	524	542	560	577	594	610	627	642	658	673	687	701	715	5
6	231	247	263	280	297	314	332	349	366	384	402	419	437	454	471	488	505	522	538	554	6
7	121	133	144	156	169	182	195	209	223	238	253	268	283	298	314	330	346	362	378	394	7
8	057	064	071	079	087	095	104	113	123	133	144	155	167	178	191	203	216	229	242	256	8
9	024	028	032	036	040	045	050	056	062	068	075	082	089	097	106	114	123	133	143	153	9
10	010	011	013	015	017	020	022	025	028	032	036	040	044	049	054	059	065	071	077	084	10
11	003	004	005	006	007	008	009	010	012	014	016	018	020	023	025	028	031	035	039	042	11
12	001	001	002	002	002	003	003	004	005	005	006	007	008	010	011	012	014	016	018	020	12
13			001	001	001	001	001	001	002	002	002	003	003	004	004	005	006	007	008	009	13
14								001	001	001	001	001	001	001	002	002	002	003	003	004	14
15															001	001	001	001	001	001	15
16																				001	16

m

x	6.1	6.2	6.3	6.4	6.5	6.6	6.7	6.8	6.9	7.0	7.1	7.2	7.3	7.4	7.5	8.0	8.5	9.0	9.5	10.0	x
0	1	1	1	1	1	1	1	1	1	1	1	1	1	1	1	1	1	1	1	1	0
1	998	998	998	998	998	999	999	999	999	999	999	999	999	999	999	1-	1-	1-	1-	1-	1
2	984	985	987	988	989	990	991	991	992	993	993	994	994	995	995	997	998	999	999	1-	2
3	942	946	950	954	957	960	963	966	968	970	973	975	976	978	980	986	991	994	996	997	3
4	857	866	874	881	888	895	901	907	913	918	923	928	933	937	941	958	970	979	985	990	4
5	728	741	753	765	776	787	798	808	818	827	836	844	853	860	868	900	926	945	960	971	5
6	570	586	601	616	631	645	659	673	686	699	712	724	736	747	759	809	850	884	911	933	6
7	410	426	442	458	473	489	505	520	535	550	565	580	594	608	622	687	744	793	835	870	7
8	270	284	298	313	327	342	357	372	386	401	416	431	446	461	475	547	614	676	731	780	8
9	163	174	185	197	208	220	233	245	258	271	284	297	311	324	338	407	477	544	608	667	9
10	091	098	106	114	123	131	140	150	159	170	180	190	201	212	224	283	347	413	478	542	10
11	047	051	056	061	067	073	079	085	092	099	106	113	121	129	138	184	237	294	355	417	11
12	022	025	028	031	034	037	041	045	049	053	058	063	068	074	079	112	151	197	248	303	12
13	010	011	013	014	016	018	020	022	024	027	030	033	036	039	043	064	091	124	164	208	13
14	004	005	005	006	007	008	009	010	011	013	014	016	018	020	022	034	051	074	102	136	14
15	002	002	002	003	003	003	004	004	005	006	006	007	008	009	010	017	027	041	060	083	15
16	001	001	001	001	001	001	002	002	002	002	003	003	004	004	005	008	014	022	033	049	16
17							001	001	001	001	001	001	001	002	002	004	007	011	018	027	17
18											001	001	001	001	002	003	005	009	014	18	
19																001	001	002	004	007	19
20																	001	001	002	003	20
21																			001	002	21
22																				001	22

Source: Abridged from more detailed tables, such as available in *Poisson's Exponential Binomial Limit*, E. C. Molina (New York: Van Nostrand, 1942).

F / Values of e^{-x}

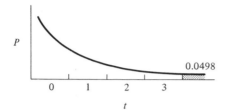

If $\lambda = 1$ and $t = 3$, then
$$e^{-1 \times 3} = e^{-3} = 0.0498$$

x	e^{-x}	x	e^{-x}
0.0	1.0000	3.5	0.0302
0.1	0.9048	3.6	0.0273
0.2	0.8187	3.7	0.0247
0.3	0.7408	3.8	0.0224
0.4	0.6703	3.9	0.0202
0.5	0.6065	4.0	0.0183
0.6	0.5488	4.1	0.0166
0.7	0.4966	4.2	0.0150
0.8	0.4493	4.3	0.0136
0.9	0.4066	4.4	0.0123
1.0	0.3679	4.5	0.0111
1.1	0.3329	4.6	0.0101
1.2	0.3012	4.7	0.0091
1.3	0.2725	4.8	0.0082
1.4	0.2466	4.9	0.0074
1.5	0.2231	5.0	0.0067
1.6	0.2019	5.1	0.0061
1.7	0.1827	5.2	0.0055
1.8	0.1653	5.3	0.0050
1.9	0.1496	5.4	0.0045
2.0	0.1353	5.5	0.0041
2.1	0.1225	5.6	0.0037
2.2	0.1108	5.7	0.0033
2.3	0.1003	5.8	0.0030
2.4	0.0907	5.9	0.0027
2.5	0.0821	6.0	0.0025
2.6	0.0743	6.1	0.0022
2.7	0.0672	6.2	0.0020
2.8	0.0608	6.3	0.0018
2.9	0.0550	6.4	0.0017
3.0	0.0498	6.5	0.0015
3.1	0.0450	6.6	0.0014
3.2	0.0408	6.7	0.0012
3.3	0.0369	6.8	0.0011
3.4	0.0334	6.9	0.0010
		7.0	0.0009

From Park J. Ewart, James S. Ford, and Chi-Yuan Lin, *Probability for Statistical Decision Making*, © 1974, p. 330. Reprinted by permission of Prentice-Hall, Inc., Englewood Cliffs, N.J.

G / Areas under the Normal Curve

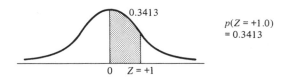

$$p(Z = +1.0) = 0.3413$$

$\frac{x}{s}$ or $\frac{z}{\sigma}$	.00	.01	.02	.03	.04	.05	.06	.07	.08	.09
0.0	.0000	.0040	.0080	.0120	.0160	.0199	.0239	.0279	.0319	.0359
0.1	.0398	.0438	.0478	.0517	.0557	.0596	.0636	.0675	.0714	.0753
0.2	.0793	.0832	.0871	.0910	.0948	.0987	.1026	.1064	.1103	.1141
0.3	.1179	.1217	.1255	.1293	.1331	.1368	.1406	.1443	.1480	.1517
0.4	.1554	.1591	.1628	.1664	.1700	.1736	.1772	.1808	.1844	.1879
0.5	.1915	.1950	.1985	.2019	.2054	.2088	.2123	.2157	.2190	.2224
0.6	.2257	.2291	.2324	.2357	.2389	.2422	.2454	.2486	.2518	.2549
0.7	.2580	.2612	.2642	.2673	.2704	.2734	.2764	.2794	.2823	.2852
0.8	.2881	.2910	.2939	.2967	.2995	.3023	.3051	.3078	.3106	.3133
0.9	.3159	.3186	.3212	.3238	.3264	.3289	.3315	.3340	.3365	.3389
1.0	.3413	.3438	.3461	.3485	.3508	.3531	.3554	.3577	.3599	.3621
1.1	.3643	.3665	.3686	.3708	.3729	.3749	.3770	.3790	.3810	.3830
1.2	.3849	.3869	.3888	.3907	.3925	.3944	.3962	.3980	.3997	.4015
1.3	.4032	.4049	.4066	.4082	.4099	.4115	.4131	.4147	.4162	.4177
1.4	.4192	.4207	.4222	.4236	.4251	.4265	.4279	.4292	.4306	.4319
1.5	.4332	.4345	.4357	.4370	.4382	.4394	.4406	.4418	.4429	.4441
1.6	.4452	.4463	.4474	.4484	.4495	.4505	.4515	.4525	.4535	.4545
1.7	.4554	.4564	.4573	.4582	.4591	.4599	.4608	.4616	.4625	.4633
1.8	.4641	.4649	.4656	.4664	.4671	.4678	.4686	.4693	.4699	.4706
1.9	.4713	.4719	.4726	.4732	.4738	.4744	.4750	.4756	.4761	.4767
2.0	.4772	.4778	.4783	.4788	.4793	.4798	.4803	.4808	.4812	.4817
2.1	.4821	.4826	.4830	.4834	.4838	.4842	.4846	.4850	.4854	.4857
2.2	.4861	.4864	.4868	.4871	.4875	.4878	.4881	.4884	.4887	.4890
2.3	.4893	.4896	.4898	.4901	.4904	.4906	.4909	.4911	.4913	.4916
2.4	.4918	.4920	.4922	.4925	.4927	.4929	.4931	.4932	.4934	.4936
2.5	.4938	.4940	.4941	.4943	.4945	.4946	.4948	.4949	.4951	.4952
2.6	.4953	.4955	.4956	.4957	.4959	.4960	.4961	.4962	.4963	.4964
2.7	.4965	.4966	.4967	.4968	.4969	.4970	.4971	.4972	.4973	.4974
2.8	.4974	.4975	.4976	.4977	.4977	.4978	.4979	.4979	.4980	.4981
2.9	.4981	.4982	.4982	.4983	.4984	.4984	.4985	.4985	.4986	.4986
3.0	.49865	.4987	.4987	.4988	.4988	.4989	.4989	.4989	.4990	.4990
3.1	.49903	.4991	.4991	.4991	.4992	.4992	.4992	.4992	.4993	.4993
3.2	.4993129									
3.3	.4995166									
3.4	.4996631									
3.5	.4997674									
3.6	.4998409									
3.7	.4998922									
3.8	.4999277									
3.9	.4999519									
4.0	.4999683									
4.5	.4999966									
5.0	.4999997133									

Reprinted, by permission, from *Applied General Statistics*, by Frederick E. Croxton, Dudley J. Cowden, and Sidney Klein, published by Prentice-Hall, Inc., Englewood Cliffs, N.J., third edition, 1967, p. 666.

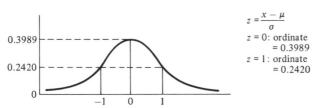

H / Ordinates of the Normal Curve

$$z = \frac{x - \mu}{\sigma}$$

$z = 0$: ordinate
  $= 0.3989$
$z = 1$: ordinate
  $= 0.2420$

z	.00	.01	.02	.03	.04	.05	.06	.07	.08	.09
.0	.3989	.3989	.3989	.3988	.3986	.3984	.3982	.3980	.3977	.3973
.1	.3970	.3965	.3961	.3956	.3951	.3945	.3939	.3932	.3925	.3918
.2	.3910	.3902	.3894	.3885	.3876	.3867	.3857	.3847	.3836	.3825
.3	.3814	.3802	.3790	.3778	.3765	.3752	.3739	.3725	.3712	.3697
.4	.3683	.3658	.3653	.3637	.3621	.3605	.3589	.3572	.3555	.3538
.5	.3521	.3503	.3485	.3467	.3448	.3429	.3410	.3391	.3372	.3352
.6	.3332	.3312	.3292	.3271	.3251	.3230	.3209	.3187	.3166	.3144
.7	.3123	.3101	.3079	.3056	.3034	.3011	.2989	.2966	.2943	.2920
.8	.2897	.2874	.2850	.2827	.2803	.2780	.2756	.2732	.2709	.2685
.9	.2661	.2637	.2613	.2589	.2565	.2541	.2516	.2492	.2468	.2444
1.0	.2420	.2396	.2371	.2347	.2323	.2299	.2275	.2251	.2227	.2203
1.1	.2179	.2155	.2131	.2107	.2083	.2059	.2036	.2012	.1989	.1965
1.2	.1942	.1919	.1895	.1872	.1849	.1826	.1804	.1781	.1758	.1736
1.3	.1714	.1691	.1669	.1647	.1626	.1604	.1582	.1561	.1539	.1518
1.4	.1497	.1476	.1456	.1435	.1415	.1394	.1374	.1354	.1334	.1315
1.5	.1295	.1276	.1257	.1238	.1219	.1200	.1182	.1163	.1145	.1127
1.6	.1109	.1092	.1074	.1057	.1040	.1023	.1006	.0989	.0973	.0957
1.7	.0940	.0925	.0909	.0893	.0878	.0863	.0848	.0833	.0818	.0804
1.8	.0790	.0775	.0761	.0748	.0734	.0721	.0707	.0694	.0681	.0669
1.9	.0656	.0644	.0632	.0620	.0608	.0596	.0584	.0573	.0562	.0551
2.0	.0540	.0529	.0519	.0508	.0498	.0488	.0478	.0468	.0459	.0449
2.1	.0440	.0431	.0422	.0413	.0404	.0396	.0387	.0379	.0371	.0363
2.2	.0355	.0347	.0339	.0332	.0325	.0317	.0310	.0303	.0297	.0290
2.3	.0283	.0277	.0270	.0264	.0258	.0252	.0246	.0241	.0235	.0229
2.4	.0224	.0219	.0213	.0208	.0203	.0198	.0194	.0189	.0184	.0180
2.5	.0175	.0171	.0167	.0163	.0158	.0154	.0151	.0147	.0143	.0139
2.6	.0136	.0132	.0129	.0126	.0122	.0119	.0116	.0113	.0110	.0107
2.7	.0104	.0101	.0099	.0096	.0093	.0091	.0088	.0086	.0084	.0081
2.8	.0079	.0077	.0075	.0073	.0071	.0069	.0067	.0065	.0063	.0061
2.9	.0060	.0058	.0056	.0055	.0053	.0051	.0050	.0048	.0047	.0046
3.0	.0044	.0043	.0042	.0040	.0039	.0038	.0037	.0036	.0035	.0034
3.1	.0033	.0032	.0031	.0030	.0029	.0028	.0027	.0026	.0025	.0025
3.2	.0024	.0023	.0022	.0022	.0021	.0020	.0020	.0019	.0018	.0018
3.3	.0017	.0017	.0016	.0016	.0015	.0015	.0014	.0014	.0013	.0013
3.4	.0012	.0012	.0012	.0011	.0011	.0010	.0010	.0010	.0009	.0009
3.5	.0009	.0008	.0008	.0008	.0008	.0007	.0007	.0007	.0007	.0006
3.6	.0006	.0006	.0006	.0005	.0005	.0005	.0005	.0005	.0005	.0004
3.7	.0004	.0004	.0004	.0004	.0004	.0004	.0003	.0003	.0003	.0003
3.8	.0003	.0003	.0003	.0003	.0003	.0002	.0002	.0002	.0002	.0002
3.9	.0002	.0002	.0002	.0002	.0002	.0002	.0002	.0002	.0001	.0001

From Taro Yamane, *Statistics: An Introductory Analysis,* third edition, © 1973, p. 1079. Reprinted by permission of Harper & Row, New York, N.Y.

I / Unit-Loss Normal Function $L_n(D)$

The value of D represents the deviation of the break-even point $\bar{X}_b$ from the mean μ expressed as a ratio to the standard deviation, or

$$D = \left| \frac{\bar{X}_b - \mu}{\sigma} \right|$$

Example: For $D = .5$, $Ln(D) = .1978$ For $D = 3.05$, $Ln(D) = .0003199$

u	.00	.01	.02	.03	.04	.05	.06	.07	.08	.09
.0	.3989	.3940	.3890	.3841	.3793	.3744	.3697	.3649	.3602	.3556
.1	.3509	.3464	.3418	.3373	.3328	.3284	.3240	.3197	.3154	.3111
.2	.3069	.3027	.2986	.2944	.2904	.2863	.2824	.2784	.2745	.2706
.3	.2668	.2630	.2592	.2555	.2518	.2481	.2445	.2409	.2374	.2339
.4	.2304	.2270	.2236	.2203	.2169	.2137	.2104	.2072	.2040	.2009
.5	.1978	.1947	.1917	.1887	.1857	.1828	.1799	.1771	.1742	.1714
.6	.1687	.1659	.1633	.1606	.1580	.1554	.1528	.1503	.1478	.1453
.7	.1429	.1405	.1381	.1358	.1334	.1312	.1289	.1267	.1245	.1223
.8	.1202	.1181	.1160	.1140	.1120	.1100	.1080	.1061	.1042	.1023
.9	.1004	.09860	.09680	.09503	.09328	.09156	.08986	.08819	.08654	.08491
1.0	.08332	.08174	.08019	.07866	.07716	.07568	.07422	.07279	.07138	.06999
1.1	.06862	.06727	.06595	.06465	.06336	.06210	.06086	.05964	.05844	.05726
1.2	.05610	.05496	.05384	.05274	.05165	.05059	.04954	.04851	.04750	.04650
1.3	.04553	.04457	.04363	.04270	.04179	.04090	.04002	.03916	.03831	.03748
1.4	.03667	.03587	.03508	.03431	.03356	.03281	.03208	.03137	.03067	.02998
1.5	.02931	.02865	.02800	.02736	.02674	.02612	.02552	.02494	.02436	.02380
1.6	.02324	.02270	.02217	.02165	.02114	.02064	.02015	.01967	.01920	.01874
1.7	.01829	.01785	.01742	.01699	.01658	.01617	.01578	.01539	.01501	.01464
1.8	.01428	.01392	.01357	.01323	.01290	.01257	.01226	.01195	.01164	.01134
1.9	.01105	.01077	.01049	.01022	$.0^{2}9957$	$.0^{2}9698$	$.0^{2}9445$	$.0^{2}9198$	$.0^{2}8957$	$.0^{2}8721$
2.0	$.0^{2}8491$	$.0^{2}8266$	$.0^{2}8046$	$.0^{2}7832$	$.0^{2}7623$	$.0^{2}7418$	$.0^{2}7219$	$.0^{2}7024$	$.0^{2}6835$	$.0^{2}6649$
2.1	$.0^{2}6468$	$.0^{2}6292$	$.0^{2}6120$	$.0^{2}5952$	$.0^{2}5788$	$.0^{2}5628$	$.0^{2}5472$	$.0^{2}5320$	$.0^{2}5172$	$.0^{2}5028$
2.2	$.0^{2}4887$	$.0^{2}4750$	$.0^{2}4616$	$.0^{2}4486$	$.0^{2}4358$	$.0^{2}4235$	$.0^{2}4114$	$.0^{2}3996$	$.0^{2}3882$	$.0^{2}3770$
2.3	$.0^{2}3662$	$.0^{2}3556$	$.0^{2}3453$	$.0^{2}3352$	$.0^{2}3255$	$.0^{2}3159$	$.0^{2}3067$	$.0^{2}2977$	$.0^{2}2889$	$.0^{2}2804$
2.4	$.0^{2}2720$	$.0^{2}2640$	$.0^{2}2561$	$.0^{2}2484$	$.0^{2}2410$	$.0^{2}2337$	$.0^{2}2267$	$.0^{2}2199$	$.0^{2}2132$	$.0^{2}2067$

D	.00	.01	.02	.03	.04	.05	.06	.07	.08	.09
2.5	$.0^{2}2004$	$.0^{2}1943$	$.0^{2}1883$	$.0^{2}1826$	$.0^{2}1769$	$.0^{2}1715$	$.0^{2}1662$	$.0^{2}1610$	$.0^{2}1560$	$.0^{2}1511$
2.6	$.0^{2}1464$	$.0^{2}1418$	$.0^{2}1373$	$.0^{2}1330$	$.0^{2}1288$	$.0^{2}1247$	$.0^{2}1207$	$.0^{2}1169$	$.0^{2}1132$	$.0^{2}1095$
2.7	$.0^{2}1060$	$.0^{2}1026$	$.0^{3}9928$	$.0^{3}9607$	$.0^{3}9295$	$.0^{3}8992$	$.0^{3}8699$	$.0^{3}8414$	$.0^{3}8138$	$.0^{3}7870$
2.8	$.0^{3}7611$	$.0^{3}7359$	$.0^{3}7115$	$.0^{3}6879$	$.0^{3}6650$	$.0^{3}6428$	$.0^{3}6213$	$.0^{3}6004$	$.0^{3}5802$	$.0^{3}5606$
2.9	$.0^{3}5417$	$.0^{3}5233$	$.0^{3}5055$	$.0^{3}4883$	$.0^{3}4716$	$.0^{3}4555$	$.0^{3}4398$	$.0^{3}4247$	$.0^{3}4101$	$.0^{3}3959$
3.0	$.0^{3}3822$	$.0^{3}3689$	$.0^{3}3560$	$.0^{3}3436$	$.0^{3}3316$	$.0^{3}3199$	$.0^{3}3087$	$.0^{3}2978$	$.0^{3}2873$	$.0^{3}2771$
3.1	$.0^{3}2673$	$.0^{3}2577$	$.0^{3}2485$	$.0^{3}2396$	$.0^{3}2311$	$.0^{3}2227$	$.0^{3}2147$	$.0^{3}2070$	$.0^{3}1995$	$.0^{3}1922$
3.2	$.0^{3}1852$	$.0^{3}1785$	$.0^{3}1720$	$.0^{3}1657$	$.0^{3}1596$	$.0^{3}1537$	$.0^{3}1480$	$.0^{3}1426$	$.0^{3}1373$	$.0^{3}1322$
3.3	$.0^{3}1273$	$.0^{3}1225$	$.0^{3}1179$	$.0^{3}1135$	$.0^{3}1093$	$.0^{3}1051$	$.0^{3}1012$	$.0^{4}9734$	$.0^{4}9365$	$.0^{4}9009$
3.4	$.0^{4}8666$	$.0^{4}8335$	$.0^{4}8016$	$.0^{4}7709$	$.0^{4}7413$	$.0^{4}7127$	$.0^{4}6852$	$.0^{4}6587$	$.0^{4}6331$	$.0^{4}6085$
3.5	$.0^{4}5848$	$.0^{4}5620$	$.0^{4}5400$	$.0^{4}5188$	$.0^{4}4984$	$.0^{4}4788$	$.0^{4}4599$	$.0^{4}4417$	$.0^{4}4242$	$.0^{4}4073$
3.6	$.0^{4}3911$	$.0^{4}3755$	$.0^{4}3605$	$.0^{4}3460$	$.0^{4}3321$	$.0^{4}3188$	$.0^{4}3059$	$.0^{4}2935$	$.0^{4}2816$	$.0^{4}2702$
3.7	$.0^{4}2592$	$.0^{4}2486$	$.0^{4}2385$	$.0^{4}2287$	$.0^{4}2193$	$.0^{4}2103$	$.0^{4}2016$	$.0^{4}1933$	$.0^{4}1853$	$.0^{4}1776$
3.8	$.0^{4}1702$	$.0^{4}1632$	$.0^{4}1563$	$.0^{4}1498$	$.0^{4}1435$	$.0^{4}1375$	$.0^{4}1317$	$.0^{4}1262$	$.0^{4}1208$	$.0^{4}1157$
3.9	$.0^{4}1108$	$.0^{4}1061$	$.0^{4}1016$	$.0^{5}9723$	$.0^{5}9307$	$.0^{5}8908$	$.0^{5}8525$	$.0^{5}8158$	$.0^{5}7806$	$.0^{5}7469$
4.0	$.0^{5}7145$	$.0^{5}6835$	$.0^{5}6538$	$.0^{5}6253$	$.0^{5}5980$	$.0^{5}5718$	$.0^{5}5468$	$.0^{5}5227$	$.0^{5}4997$	$.0^{5}4777$
4.1	$.0^{5}4566$	$.0^{5}4364$	$.0^{5}4170$	$.0^{5}3985$	$.0^{5}3807$	$.0^{5}3637$	$.0^{5}3475$	$.0^{5}3319$	$.0^{5}3170$	$.0^{5}3027$
4.2	$.0^{5}2891$	$.0^{5}2760$	$.0^{5}2635$	$.0^{5}2516$	$.0^{5}2402$	$.0^{5}2292$	$.0^{5}2188$	$.0^{5}2088$	$.0^{5}1992$	$.0^{5}1901$
4.3	$.0^{5}1814$	$.0^{5}1730$	$.0^{5}1650$	$.0^{5}1574$	$.0^{5}1501$	$.0^{5}1431$	$.0^{5}1365$	$.0^{5}1301$	$.0^{5}1241$	$.0^{5}1183$
4.4	$.0^{5}1127$	$.0^{5}1074$	$.0^{5}1024$	$.0^{6}9756$	$.0^{6}9296$	$.0^{6}8857$	$.0^{6}8437$	$.0^{6}8037$	$.0^{6}7655$	$.0^{6}7290$
4.5	$.0^{6}6942$	$.0^{6}6610$	$.0^{6}6294$	$.0^{6}5992$	$.0^{6}5704$	$.0^{6}5429$	$.0^{6}5167$	$.0^{6}4917$	$.0^{6}4679$	$.0^{6}4452$
4.6	$.0^{6}4236$	$.0^{6}4029$	$.0^{6}3833$	$.0^{6}3645$	$.0^{6}3467$	$.0^{6}3297$	$.0^{6}3135$	$.0^{6}2981$	$.0^{6}2834$	$.0^{6}2694$
4.7	$.0^{6}2560$	$.0^{6}2433$	$.0^{6}2313$	$.0^{6}2197$	$.0^{6}2088$	$.0^{6}1984$	$.0^{6}1884$	$.0^{6}1790$	$.0^{6}1700$	$.0^{6}1615$
4.8	$.0^{6}1533$	$.0^{6}1456$	$.0^{6}1382$	$.0^{6}1312$	$.0^{6}1246$	$.0^{6}1182$	$.0^{6}1122$	$.0^{6}1065$	$.0^{6}1011$	$.0^{7}9588$
4.9	$.0^{7}9096$	$.0^{7}8629$	$.0^{7}8185$	$.0^{7}7763$	$.0^{7}7362$	$.0^{7}6982$	$.0^{7}6620$	$.0^{7}6276$	$.0^{7}5950$	$.0^{7}5640$

From Robert Schlaifer, *Probability and Statistics for Business Decision*. New York McGraw-Hill Book Co., Inc., © 1959, pp. 706–707. These tables are reproduced here by specific permission of the copyright holder, the President and Fellow of Harvard College.

J / Critical Values of *t*

	Level of significance for one-tailed test					
	.10	.05	.025	.01	.005	.0005
df	Level of significance for two-tailed test					
	.20	.10	.05	.02	.01	.001
1	3.078	6.314	12.706	31.821	63.657	636.619
2	1.886	2.920	4.303	6.965	9.925	31.598
3	1.638	2.353	3.182	4.541	5.841	12.941
4	1.533	2.132	2.776	3.747	4.604	8.610
5	1.476	2.015	2.571	3.365	4.032	6.859
6	1.440	1.943	2.447	3.143	3.707	5.959
7	1.415	1.895	2.365	2.998	3.499	5.405
8	1.397	1.860	2.306	2.896	3.355	5.041
9	1.383	1.833	2.262	2.821	3.250	4.781
10	1.372	1.812	2.228	2.764	3.169	4.587
11	1.363	1.796	2.201	2.718	3.106	4.437
12	1.356	1.782	2.179	2.681	3.055	4.318
13	1.350	1.771	2.160	2.650	3.012	4.221
14	1.345	1.761	2.145	2.624	2.977	4.140
15	1.341	1.753	2.131	2.602	2.947	4.073
16	1.337	1.746	2.120	2.583	2.921	4.015
17	1.333	1.740	2.110	2.567	2.898	3.965
18	1.330	1.734	2.101	2.552	2.878	3.922
19	1.328	1.729	2.093	2.539	2.861	3.883
20	1.325	1.725	2.086	2.528	2.845	3.850
21	1.323	1.721	2.080	2.518	2.831	3.819
22	1.321	1.717	2.074	2.508	2.819	3.792
23	1.319	1.714	2.069	2.500	2.807	3.767
24	1.318	1.711	2.064	2.492	2.797	3.745
25	1.316	1.708	2.060	2.485	2.787	3.725
26	1.315	1.706	2.056	2.479	2.779	3.707
27	1.314	1.703	2.052	2.473	2.771	3.690
28	1.313	1.701	2.048	2.467	2.763	3.674
29	1.311	1.699	2.045	2.462	2.756	3.659
30	1.310	1.697	2.042	2.457	2.750	3.646
40	1.303	1.684	2.021	2.423	2.704	3.551
60	1.296	1.671	2.000	2.390	2.660	3.460
120	1.289	1.658	1.980	2.358	2.617	3.373
∞	1.282	1.645	1.960	2.326	2.576	3.291

Table J is abridged from Table III of Fisher and Yates, *Statistical Tables for Biological, Agricultural, and Medical Research*, published by Longman Group Ltd., 6th ed., 1974, previously published by Oliver and Boyd Ltd., Edinburgh, by permission of the authors and publishers.

K / Random Numbers

6063	2353	8531	8892	4109	5782	2283	1385	0699	5927
6305	1326	4551	2815	8937	2908	0698	5509	4303	9911
0143	0187	8127	2026	8313	8341	2479	4722	6602	2236
1031	0754	7989	4948	1804	3025	0997	9562	3674	7876
2022	3227	2147	5613	2857	8859	4941	7274	9412	0620
9149	0806	9751	8870	9677	9676	1854	8094	7658	7012
5863	0513	1402	3866	8696	9142	6063	2252	7818	2477
8724	0806	9644	8284	7010	0868	9076	4915	5751	9214
6783	4207	2958	5295	3175	3396	8117	5918	1037	4319
0862	1620	4690	0036	9654	4078	1918	8721	8454	7671
9394	2466	6427	5395	9393	0520	7074	0634	5578	4023
3220	3058	7787	7706	4094	5603	3303	8300	6185	8705
1491	3503	0584	7221	6176	0116	0309	1975	0910	3535
4368	5705	8579	5790	7244	6547	8495	7973	1805	7251
2325	4026	2919	8327	0267	2616	6572	8620	8245	6257
0591	1775	5134	8709	7373	3332	0507	5525	7640	2840
3471	1461	1149	6798	6070	9930	1862	3672	6718	3849
2600	9885	6219	3668	1005	5418	5832	0416	4220	4692
9572	7874	6034	4514	2628	1693	0628	2200	9006	3795
0822	2790	9386	5783	2689	2565	1565	0349	3410	5216
4329	3028	2549	2529	9434	3083	6800	8569	9290	8298
9289	5212	2355	9367	1297	1638	9282	3720	7178	2695
3932	9960	3399	1700	8253	1375	4594	6024	1223	5383
2282	0648	7561	7528	5870	7907	0713	8608	9682	8576
9933	3416	5957	2574	5553	5534	4707	3206	0963	2459
9015	6416	6603	2967	7591	5013	2878	8424	5452	4659
1539	0719	2637	9969	8450	4489	3528	3364	1459	9708
6849	5595	7969	2582	5627	1920	9772	8560	0892	6500
2523	7769	3536	9611	1079	1694	1254	4195	5799	5928
0701	7355	0587	8878	3446	1137	7690	0647	1407	6362
2163	8543	4594	6022	0496	8648	2999	1262	6702	0811
0327	5727	1070	5996	8660	9024	2135	9799	8414	9136
2169	3160	8707	6361	6339	4054	3251	7397	3480	5805
8393	8147	5360	4150	2990	3380	1789	7436	4781	0337
9726	9151	2064	0609	5878	9095	9737	2897	6510	8891
0515	2296	2636	9756	5313	7754	0916	6066	3905	1298
0649	8398	5614	0140	3155	2211	4988	3674	7663	0620
0026	9426	8005	8579	5774	7962	5092	5856	1626	0980
3422	0092	1626	1298	2475	1997	9796	7076	1541	1731
8191	1983	9164	1885	5468	8216	4327	8109	5880	9804
7408	0486	7654	4829	2711	6592	4785	5901	7147	9314
8261	9440	8118	6338	8157	9052	9093	8449	4066	4894
9274	8838	8342	3114	0455	6212	8862	6701	0099	0501
2699	0383	1400	3484	1492	4683	5369	3851	5870	0903
8740	0349	3502	3971	9960	6325	6727	4715	2945	9938
0247	2372	0424	0578	0036	1619	4479	7108	8520	1487
5136	9444	8343	1152	3615	1420	8923	7307	3978	5724
4844	8931	0964	2878	8212	9328	2656	1965	4805	0634
0205	8457	4333	2555	5353	9201	1606	2715	4014	1877
2517	5061	7642	3891	7713	7066	5435	1200	7455	5562

Table K (Continued)

2271	2572	8665	3272	9033	8256	2822	3646	7599	0270
3025	0788	5311	7792	1837	4739	4552	3234	5572	9885
3382	6151	1011	3778	9951	7709	8060	2258	8536	2290
7870	5799	6032	9043	4526	8100	1957	9539	5370	0046
1697	0002	2340	6959	1915	1626	1297	1533	6572	3835
3395	3381	1862	3250	8614	5683	6757	5628	2551	6971
6081	6526	3028	2338	5702	8819	3679	4829	9909	4712
3470	9879	2935	1141	6398	6387	5634	9589	3212	7963
0432	8641	5020	6612	1038	1547	0948	4278	0020	6509
4995	5596	8286	8377	8567	8237	3520	8244	5694	3326
8246	6718	3851	5870	1216	2107	1387	1621	5509	5772
7825	8727	2849	3501	3551	1001	0123	7873	5926	6078
6258	2450	2962	1183	3666	4156	4454	8239	4551	2920
3235	5783	2701	2378	7460	3398	1223	4688	3674	7872
2525	9008	5997	0885	1053	2340	7066	5328	6412	5054
5852	9739	1457	8999	2789	9068	9829	1336	3148	7875
0440	3769	7864	4029	4494	9829	1339	4910	1303	9161
0820	4641	2375	2542	4093	5364	1145	2848	2792	0431
7114	2842	8554	6881	6377	9427	8216	1193	8042	8449
6558	9301	9096	0577	8520	5923	4717	0188	8545	8745
0345	9937	5569	0279	8951	6183	7787	7808	5149	2185
7430	2074	9427	8422	4082	5629	2971	9456	0649	7981
8030	7345	3389	4739	5911	1022	9189	2565	1982	8577
6272	6718	3849	4715	3156	2823	4174	8733	5600	7702
4894	9847	5611	4763	8755	3388	5114	3274	6681	3657
2676	5984	6806	2692	4012	0934	2436	0869	9557	2490
9305	2074	9378	7670	8284	7431	7361	2912	2251	7395
5138	2461	7213	1905	7775	9881	8782	6272	0632	4418
2452	4200	8674	9202	0812	3986	1143	7343	2264	9072
8882	3033	8746	7390	8609	1144	2531	6944	8869	1570
1087	9336	8020	9166	4472	8293	2904	7949	3165	7400
5666	2841	8134	9588	2915	4116	2802	6917	3993	8764
9790	2228	9702	1690	7170	7511	1937	0723	4505	7155
3250	8860	3294	2684	6572	3415	5750	8726	2647	6596
5450	3922	0950	0890	6434	2306	2781	1066	3681	2404
5765	0765	7311	5270	5910	7009	0240	7435	4568	6484
8408	1939	0599	5347	2160	7376	4696	6969	0787	3838
8460	7658	6906	9177	1492	4680	3719	3456	8681	6736
4198	7244	3849	4819	1008	6781	3388	5253	7041	6712
9872	4441	6712	9614	2736	5533	9062	2534	0855	7946
6485	0487	0004	5563	1481	1546	8245	6116	6920	0990
2064	0512	9509	0341	8131	7778	8609	9417	1216	4189
9927	8987	5321	3125	9992	9449	5951	5872	2057	5731
4918	9690	6121	8770	6053	6931	7252	5409	1869	4229
8099	5821	3899	2685	6781	3178	0096	2986	8878	8991
1901	4974	1262	6810	4673	8772	6616	2632	7891	9970
8273	6675	4925	3924	2274	3860	1662	7480	8674	4503
2878	8213	3170	5126	0434	9481	7029	8688	4027	3340
6088	1182	3242	0835	1765	8819	3462	9820	5759	4189
5773	6600	5306	0354	8295	0148	6608	9064	3421	8570

Table K (Continued)

5500	2276	6307	2346	1285	7000	5306	0414	3383	2137
3251	8902	8843	2112	8567	8131	8116	5270	5994	7445
4675	1435	2192	0874	2897	0262	5092	5541	4014	2086
3543	6130	4247	4859	2660	7852	9096	0578	0097	4746
3521	8772	6612	0721	3899	2999	1263	7017	8057	4983
5573	9396	3464	1702	9204	3389	5678	2589	0288	4633
7478	7569	7551	3380	2152	5411	2647	7242	2800	6183
3339	2854	9691	9562	3252	9848	6030	8472	2266	1270
5505	8474	3167	8552	5409	1556	4247	4652	2953	5394
6381	2086	5457	7703	2758	2963	8167	6712	9820	5654
6975	5239	0762	5846	2431	0543	4956	8787	9651	2605
7185	4019	7332	2820	4853	8636	9505	6575	0365	6648
4510	1658	5615	2194	1901	4975	1895	4383	0415	3771
7752	0105	4769	2994	7445	0781	4960	4253	9451	6518
4834	4043	6591	3646	8918	4603	1970	9145	7615	3905
8866	6036	9755	4508	9061	2080	3406	9856	1298	6281
6622	4612	2030	7299	8414	8822	5176	9443	6054	6462
9094	8973	3335	2183	5192	1630	0959	8143	9182	8012
5618	6445	2983	0375	2540	2735	4901	5515	4787	7058
2705	2693	1944	8074	2015	3261	5529	7193	5401	9531
1797	4334	3293	2632	3770	1675	9363	7795	3331	8995
9448	5174	5869	0448	8613	4400	6938	5161	8691	2838
3461	1304	9682	8577	4449	1896	8328	1698	7138	1141
7092	5007	5596	8522	2580	4495	4728	8948	4434	2438
5533	4294	0939	4050	1225	6414	5895	0148	7053	5935
7852	8988	5951	4919	7404	2426	4450	2358	3082	4561
8313	8456	9892	0981	6736	8021	6226	5573	1664	9489
1158	2241	9861	7588	2669	5480	9160	4267	1690	7278
9338	7226	0025	8844	8181	5565	2418	9394	0837	3106
7711	1336	3251	8902	8425	5766	3262	5848	3545	7073
2656	1863	3884	6516	6922	1808	1896	8853	0964	3089
7980	9370	2850	3818	7281	8352	9637	0618	2430	6525
1409	7865	5908	4296	1888	2792	4014	1667	1295	0814
7657	6630	5000	1493	5459	5869	0315	8134	9587	2184
2863	5450	1329	8787	8795	4604	2615	0075	1433	7707
3988	2042	2906	8995	0818	9288	1650	0803	8319	2533
4551	2815	8941	4893	8612	4844	0042	3890	7069	8512
5772	4732	2829	3931	9540	6256	5420	2179	9448	5489
9150	1435	3817	8975	4276	9569	0175	6663	0045	5549
5764	7914	8280	1337	3779	8197	9105	5985	1054	2866
5895	0044	5021	3846	7599	0398	5212	9509	0134	4656
6857	1174	8085	6503	5355	3027	1708	3626	7059	0167
2538	2669	3746	3270	1214	9983	8434	1344	1160	3292
9983	1387	1410	8891	2523	8705	9190	2986	7654	5142
5061	9529	2922	2199	8310	6954	8090	5371	0672	6281
9999	4226	2815	8817	5606	5190	0495	7867	9968	5951
9078	5936	2393	7875	6871	3163	9203	2863	5693	9973
4823	2291	8925	6306	1717	0320	2549	3107	5488	0303
1232	1384	5698	9313	3501	3238	7227	0220	6118	7655
7694	6484	0279	8528	7214	1750	0577	8418	0698	5403

Table K (Continued)

```
0366   6390   2107   3875   4488   2911   1727   8108   3484   6370
3686   8812   8754   2758   3079   2994   3642   1580   1475   0366
4195   4602   1481   7324   8570   6913   6228   1934   6165   0554
8180   5460   0134   4469   8619   7723   8084   3293   1895   4886
1498   7883   5280   0692   7202   1273   3334   1554   3303   8569

9428   8633   9606   7679   4182   4035   6849   5593   6712   9822
9630   5879   9342   9618   8513   4399   9734   7744   4600   0224
1086   8918   7713   5909   2620   6612   0616   1298   2476   2386
2478   3551   1247   8004   0301   6672   6176   0682   2493   6381
2808   1133   5853   8737   9804   2404   7400   5904   8803   0377

8934   2047   4963   4531   6391   9064   3526   2482   9328   5556
1156   1191   1182   3032   8640   4681   3932   6975   4926   4870
5677   7494   0987   8870   4837   5267   4119   4163   1953   3553
3719   3586   5775   7309   5111   0919   7721   7032   1164   2105
6556   8472   1848   1056   3670   7509   0854   7210   9336   8127

1246   3476   4027   3654   2444   9040   5331   2363   4738   9822
6591   3387   4109   7956   5837   6914   6435   2624   8610   4005
8197   9026   4868   6372   2695   7143   2783   1925   3383   9060
5035   4569   7158   8531   8891   0975   6329   1329   8746   0989
1563   9650   2139   7696   7511   1725   7292   0664   8440   8593

6034   4512   1505   3857   0290   3270   8389   9612   1892   8707
2435   0238   6478   5727   0862   1621   5228   5038   2000   0433
9418   4486   5992   7172   8353   6516   6605   6387   8126   1603
3116   1295   0563   6475   4382   9902   6621   9209   8060   1787
5426   5517   5603   3722   4965   5892   8135   5214   9877   6429

2494   6696   5881   1198   2055   4624   4592   4788   7477   7149
1362   2650   8867   6503   5250   7622   5989   5909   2623   7875
5622   8415   9553   7882   1402   4723   7101   1917   8305   0440
6687   5386   9837   9111   8123   3859   1134   6321   4756   1325
0045   5546   2340   7068   6692   3802   8740   0563   8253   1589

3441   4562   1126   6427   7674   6564   1996   9167   4995   6200
9354   3914   6037   7309   5111   3080   3616   2152   2426   4450
8655   6422   1264   7859   3622   8979   7253   4257   5523   4808
0143   0292   0220   2205   4773   4964   5055   5460   0240   7505
5860   4714   6437   3670   5881   1131   7609   9690   3736   7266

8400   6939   5684   7116   3472   4006   1069   5272   5209   8271
3262   4214   5901   1064   7064   4286   1038   1178   3658   4628
2220   1426   2920   8956   8142   4642   3008   9816   5548   7753
9734   7954   9700   1489   3213   8400   7043   7552   4019   9938
3178   1061   8942   8397   4898   3793   6603   2864   6014   5225

4189   6015   5328   8242   0427   1270   1992   4789   8075   7632
4774   5282   1202   5496   8949   8940   9032   6872   3581   6631
9541   6606   6881   4916   5257   3207   9530   4546   9880   0479
4560   8877   8779   1690   6959   1916   2049   7214   0761   5111
2719   2098   7631   2574   5660   8600   2922   1570   6442   8082

7081   8366   4236   6582   9193   4328   8842   1588   1391   7714
2300   5410   2186   6846   4440   6180   6021   5258   3080   3723
4090   3091   2193   1295   0563   6579   6249   9151   1959   8949
2656   1861   2833   0067   2726   3697   5862   6058   8434   1240
9465   8924   6068   1461   0656   2718   1468   5401   9638   0931
```

From Donald B. Owen, *Handbook of Statistical Tables*, © 1962. Reprinted by permission of Addison-Wesley, Reading, Mass.

GLOSSARY OF SYMBOLS

Symbols	Description
A	An event or an event set
$A_1, A_2, \ldots A_n$	Alternative decision problems
a	(1) A fixed value in linear decision model (2) Lower limit for range of uniform random variable (3) Any given value in exponential problem function; $P(t = a)$
A_1 OR A_2 A_1 union A_2 $A_1 \cup A_2$ A_1 cup A_2	Event whose set is represented by the union of the event sets for A_1 and A_2
A_1 AND A_2 $A_1 \cap A_2$ A_1 cap A_2	Event whose set is represented by the intersection of the event sets for A_1 and A_2
$\sim A, \bar{A}, A^*, A^0$	Complementary events
$\sim A$	Course of action other than A
α(alpha)	(1) Size of Type I error. Type I error occurs when the true hypothesis is rejected (2) Critical value in testing hypothesis
b	(1) Net contribution (price—variable cost) in linear decision model (2) Upper limit for range of uniform random distribution (3) Variable cost in sampling
β(beta)	Size of Type II error. Type II error occurs when false hypothesis is accepted
D	(1) Domain in set theory (2) Unit normal loss deviate $= \left\lvert \dfrac{\mu_k - \mu}{\sigma} \right\rvert$

Symbols	Descriptions
d	Per unit opportunity loss or slope of loss function in linear decision model $= \lvert b_1 - b_2 \rvert$
ε (epsilon)	"Element of"
	Empty set
E	Error range $= \pm(S_{\bar{x}})(Z_\alpha)$
e	Base of the natural log $= 2.718$
$E(X)$	(1) Expected value of random variable X, indicating long-run average value $= \sum\limits_{i=1}^{n} P(X_i) \cdot X_i$ (2) Expected value of uniform probability distribution $= \dfrac{a + b}{2}$
$E(t)$	Mean of exponential probability distribution $= 1/\lambda$
EOL	Expected opportunity loss. Expected loss due to uncertainty
EVPI_0	Prior expected value of perfect information
EVPI_1	Posterior expected value of perfect information
EVSI	Expected value of sampling information
ENGS	Expected net gain from sample $=$ EVSI $-$ Cost of sample
F	Event of "failure"
$f(x)$	Function f at x
H	"Head" in tossing a coin
H_0	Null hypothesis
H_1	Alternative hypothesis
i	Subscripts in summation
I_0	Degree of information on state of nature existing prior to sampling $= \dfrac{1}{S_0{}^2}$
$I_{\bar{x}}$	Degree of information on state of nature which is resulted from the sampling information $= \dfrac{1}{S_{\bar{x}}{}^2}$
I_1	Degree of information that is improved by posterior analysis $= I_0 + I_{\bar{x}} = 1/S_1{}^2$

Symbols	Descriptions
Ip_0	Prior information adjustment factor $= \dfrac{I_0}{I_1}$
Ip_1	Sampling information adjustment factor $= I_{\bar{x}}/I_1$
k	Subscript in break-even value
L	Opportunity loss in continuous decision function
$L_n(D)$	Unit normal loss function in linear decision model, or probability of D
λ (Lambda)	Average number of success in Poisson process $= np$
$1/\lambda$	Average times between arrivals
μ (mu)	(1) Arithmetic mean of population (2) Mean of binomial probability distribution $= np$ (3) Random variable in linear decision model
μ_1	Posterior mean $= \dfrac{1}{I_1}(I_0 M_0 + I_{\bar{x}}\bar{X})$
μ_t	True mean
μ_h	Hypothetical mean
μ_k	Break-even value of random variable μ $\mu_k = \dfrac{a_2 - a_1}{b_1 - b_2}$
m	Mean of Poisson probability distribution $= np$
MC	Marginal cost of sample
MG	Marginal gain from sample
n	(1) Sample size (2) Necessary sample size $= \left[\dfrac{(S)(Z_\alpha)}{E}\right]^2$
n_x	(1) The maximum number of samples within EVSI limit (2) The maximum number of samples within $EVPI_0$ limit
N	Finite population size
o	Subscripts in EVPI indicating an event prior to sampling
$P(X = x)$	Probability of event x
$P(X \geq x)$	Cumulative probability that X assumes any value greater than or equal to x
$P(A_1 \cup A_2)$	Probability of union of A_1 and A_2. It occurs when A_1, A_2 or both A_1 and A_2 occurs

Symbols	Descriptions
$P(A\|B)$	Conditional probability of event A given that event B already occurred
$P(A \cap B)$	Joint probability that events A and B both occur
$P(A_j\|B)$	Bayesian probability of events A_j given B. $$P(A_j\|B) = \frac{P(A_j) \times P(B\|A_j)}{\sum\limits_{i=1}^{n} (P(A_i) \times P(B\|A_i))}$$ It is also known as posterior probability, revised probability or inverse probability.
q	Random variable
$q(X)$	$1 - p(X)$, where $p(X)$ is probability of event X
R	Range in set theory
S	(1) Event of success (2) Set (3) Sample space (4) Standard deviation of Poisson probability $= \sqrt{v} = \sqrt{m} = \sqrt{np}$ (5) Standard deviation of sample mean $$= \sqrt{\frac{\sum fX_i^2 - (\sum fX_i)^2/n}{n-1}}$$
$S_{\bar{x}}$	Standard error of sample mean $= \dfrac{S}{\sqrt{n}}$
S_1	Standard deviation of posterior mean $= \sqrt{\dfrac{(S_0)^2(S_x)^2}{S_{\bar{x}}^2 + S_0^2}}$
S_*	Amount of uncertainty that was reduced by the samples $= \sqrt{S_0^2 - S_1^2}$
σ (Sigma)	(1) Standard deviation of population mean (μ): $\sigma = \sqrt{V(X)} = \sqrt{E(X^2) - [E(X)]^2}$ (2) Standard deviation of mean of binomial probability distribution $= \sqrt{npq}$
$\sum$	Summation sign
t	(1) Length of time or space between successive events in Poisson process (2) Student t-distribution $= \dfrac{(\bar{X} - \mu)\sqrt{n}}{S}$

Symbols	Descriptions
U	Utile
$E(U)$	Expected utility index
$U(L)$	Utility of lottery L
v	Aspiration level or satisfying point which is used to locate the point of inflection in utility curve
$V(t)$	Variance of random variable t under exponential probability density function $= 1/\lambda^2$
$V(X)$	(1) Variance of random variable X: $$= E(X - \mu)^2 = E(X^2) - [E(X)]^2 = E(X - \mu)^2$$ (2) Variance of random variable X under binomial probability distribution $= npq$ (3) Variance of random variable X under Poisson probability distribution $= m = np$ (4) Variance of random variable X under uniform probability distribution $= \dfrac{(b - a)^2}{12}$
$\mid$ (vertical line)	(1) Used in set theory and interpret as "such that" (2) Used in conditional probability and interpret as "given"
w	Ratio of variance of sample mean (S^2) to prior mean $(S_0{}^2)$ or $$w = \frac{S^2}{S_0{}^2}$$
W	$$= \frac{1}{\sqrt{2\pi}} e^{\left[\left(-\frac{1}{2}\right)\left(\frac{X - \mu}{\sigma}\right)\right]^2}$$
X_i	(1) Random variables (2) Alternative acts in decision problem
$\bar{X}$	Sample mean $= \dfrac{\sum f X_i}{n}$
y	$= a + b\mu$. Linear decision model x where a and b are constant and μ is random variable
Y	Ordinate (expected frequency) of normal distribution at a given value of X
Z	Standard normal deviation $= \dfrac{X - \mu}{\sigma}$

345

INDEX